UAV-Assisted Intelligent Vehicular Networks

UAV-Assisted Intelligent Vehicular Networks

Guest Editors

Dawei Wang
Ruonan Zhang

Basel • Beijing • Wuhan • Barcelona • Belgrade • Novi Sad • Cluj • Manchester

Guest Editors

Dawei Wang
School of Electronics
and Information
Northwestern Polytechnical
University
Xi'an
China

Ruonan Zhang
School of Electronics
and Information
Northwestern Polytechnical
University
Xi'an
China

Editorial Office
MDPI AG
Grosspeteranlage 5
4052 Basel, Switzerland

This is a reprint of the Special Issue, published open access by the journal *Drones* (ISSN 2504-446X), freely accessible at: https://www.mdpi.com/journal/drones/special_issues/UAV_vehicular_networks.

For citation purposes, cite each article independently as indicated on the article page online and as indicated below:

Lastname, A.A.; Lastname, B.B. Article Title. *Journal Name* **Year**, *Volume Number*, Page Range.

ISBN 978-3-7258-3583-6 (Hbk)
ISBN 978-3-7258-3584-3 (PDF)
https://doi.org/10.3390/books978-3-7258-3584-3

Contents

About the Editors

Dawei Wang

Dawei Wang received his B.S. degree from University of Jinan, China, in 2011 and PhD degree from Xi'an Jiaotong University, China in 2018. From 2016 to 2017, he was a Visiting Student at the School of Engineering, The University of British Columbia. He is currently an Associate Professor at the School of Electronics and Information, Northwestern Polytechnical University, Xi'an, China. He has served as a Technical Program Committee (TPC) member for many international conferences, such as IEEE GLOBECOM, IEEE ICC, etc. His research interests include physical-layer security, integrated sensing and communication, NOMA and UAV communications, and resource allocation.

Ruonan Zhang

Ruonan Zhang received his B.S. and M.Sc. degrees in electrical and electronics engineering from Xi'an Jiaotong University, Xi'an, China, in 2000 and 2003, respectively, and PhD degree in electrical and electronics engineering from the University of Victoria, Victoria, BC, Canada, in 2010. He was an IC Design Engineer for Motorola Inc. and Freescale Semiconductor Inc., Tianjin, China, from 2003 to 2006. Since 2010, he has been working at the Department of Communication Engineering, Northwestern Polytechnical University, Xi'an, where he is currently a Professor. His research interests include wireless channel measurement and modeling, architecture and protocol design of wireless networks, and satellite communications. Dr. Zhang was a recipient of the New Century Excellent Talent Grant from the Ministry of Education of China. He also served as the Local Arrangement Co-Chair for the IEEE/CIC International Conference on Communications in China (ICCC) in 2013, the Industry Track and Workshop Chair for the IEEE International Conference on High Performance Switching and Routing (HPSR) in 2019, and as an Associate Editor for the *Journal of Communications and Networks*.

Preface

The rapid evolution of unmanned aerial vehicle (UAV) networks, vehicle-to-everything (V2X) communication systems, and the integration of emerging technologies such as edge computing, blockchain, and artificial intelligence has given rise to a wide range of groundbreaking research areas. This Special Issue compiles several papers that explore innovative solutions in these domains, with a particular focus on optimizing communication networks, ensuring security, and improving resource management. The subjects covered in this collection address various pressing challenges, including resource allocation in UAV networks, relay selection in MEC-aided ultra-dense networks, secure transmissions for 6G vehicular IoT services, and the integration of intelligent algorithms in UAV-to-ground systems.

This Special Issue aims to provide valuable insights into cutting-edge methods and algorithms, including deep learning, optimization techniques, and security protocols, to advance the design and performance of UAV-assisted systems. It reflects the interdisciplinary nature of the field, bringing together contributions that address critical aspects of modern communication and computational networks. The goal is to foster the development of more efficient, robust, and secure communication frameworks for UAV-assisted applications, contributing to the ongoing revolution in mobile and wireless technologies.

We would like to express our sincere gratitude to all the contributing authors, whose exceptional research and innovative ideas have enriched this Special Issue. Additionally, we acknowledge the support and guidance from our colleagues, peer reviewers, and all those who have assisted in the preparation of these works. Without their invaluable contributions, this Special Issue would not have been possible.

We hope that this Special Issue serves as a reference for researchers, practitioners, and academics alike, advancing the state of the art in UAV and IoT networks, and inspiring future work in this exciting and dynamic area.

Dawei Wang and Ruonan Zhang
Guest Editors

 *drones*

Article

Intelligent Resource Allocation Using an Artificial Ecosystem Optimizer with Deep Learning on UAV Networks

Ahsan Rafiq [1], Reem Alkanhel [2,*], Mohammed Saleh Ali Muthanna [3], Evgeny Mokrov [4], Ahmed Aziz [5,6] and Ammar Muthanna [4]

[1] School of Automation, Chongqing University of Posts and Telecommunications, Chongqing 400065, China; l201710003@stu.cqupt.edu.cn
[2] Department of Information Technology, College of Computer and Information Sciences, Princess Nourah bint Abdulrahman University, Riyadh 11671, Saudi Arabia
[3] Institute of Computer Technologies and Information Security, Southern Federal University, 347922 Taganrog, Russia; muthanna@sfedu.ru
[4] Department of Telecommunication Systems, Peoples' Friendship University of Russia (RUDN University), 6 Miklukho-Maklaya, 117198 Moscow, Russia; mokrov-ev@rudn.ru (E.M.); muthanna.asa@sut.ru (A.M.)
[5] Department of Computer Science, Faculty of Computer and Artificial Intelligence, Benha University, Banha 13511, Egypt; ahmed.aziz@fci.bu.edu.eg
[6] Department of International Business Management, Tashkent State University of Economics, Tashkent 100066, Uzbekistan
* Correspondence: rialkanhal@pnu.edu.sa

Abstract: An Unmanned Aerial Vehicle (UAV)-based cellular network over a millimeter wave (mmWave) frequency band addresses the necessities of flexible coverage and high data rate in the next-generation network. But, the use of a wide range of antennas and higher propagation loss in mmWave networks results in high power utilization and UAVs are limited by low-capacity onboard batteries. To cut down the energy cost of UAV-aided mmWave networks, Energy Harvesting (EH) is a promising solution. But, it is a challenge to sustain strong connectivity in UAV-based terrestrial cellular networks due to the random nature of renewable energy. With this motivation, this article introduces an intelligent resource allocation using an artificial ecosystem optimizer with a deep learning (IRA-AEODL) technique on UAV networks. The presented IRA-AEODL technique aims to effectually allot the resources in wireless UAV networks. In this case, the IRA-AEODL technique focuses on the maximization of system utility over all users, combined user association, energy scheduling, and trajectory design. To optimally allocate the UAV policies, the stacked sparse autoencoder (SSAE) model is used in the UAV networks. For the hyperparameter tuning process, the AEO algorithm is used for enhancing the performance of the SSAE model. The experimental results of the IRA-AEODL technique are examined under different aspects and the outcomes stated the improved performance of the IRA-AEODL approach over recent state of art approaches.

Keywords: UAV networks; resource allocation; deep learning; artificial ecosystem optimizer; wireless networks

Citation: Rafiq, A.; Alkanhel, R.; Muthanna, M.S.A.; Mokrov, E.; Aziz, A.; Muthanna, A. Intelligent Resource Allocation Using an Artificial Ecosystem Optimizer with Deep Learning on UAV Networks. *Drones* **2023**, *7*, 619. https://doi.org/10.3390/drones7100619

Academic Editors: Dawei Wang and Ruonan Zhang

Received: 29 August 2023
Revised: 26 September 2023
Accepted: 27 September 2023
Published: 3 October 2023

1. Introduction

Unmanned aerial vehicle (UAV)-assisted communication presents a line-of-sight (LoS) wireless connection with controllable and flexible utilization [1]. In this regard, UAVs were mainly utilized to enrich the capacity and network coverage for ground users. As well, in wireless powered networks (WPN), UAVs are used as mobile charging stations to deliver radio frequency (RF)-energy supply to lower power user gadgets [2]. As a UAV generally utilizes limited-capacity batteries to carry out tasks, like flying, hovering, and offering services, it was vital to make the trade-offs between their coverage area and energy utilization along with service time [3]. Specifically, UAV-based aerial platforms that provide wireless services have allured the wide industry and research efforts concerning control, deployment problems, and navigation. To enhance the coverage and energy efficiency for UAV-aided communication networks, resource allocation, namely subchannels, transmit power, and serving users, is essential [4].

Furthermore, consider a multiple-UAV-based wireless communication network (multi-UAV network) where a joint model to optimize trajectory and resource allocation was analyzed as a means to guarantee fairness by optimizing the minimal output throughput among users [5]. In this study, the author to strike tradeoffs between the sum rate and delay of sensing errands for multi-UAV based uplink single cell network devised a hybrid trajectory design and subchannel assignment method [6]. Human interference is constrained for the control design of UAVs because of the maneuverability and versatility of UAVs. Hence, to boost the outcome of UAV-enabled communication networks, machine learning (ML)-based intelligent control of UAVs is a priority [7]. Neural networks (NNs)-based trajectory design is taken into account where concerned from the viewpoint of UAVs' manufactured structures. Likewise, based on reinforcement learning (RL), a UAV routing design method was developed.

To build data distributions, the Gaussian mixture model was used where a weight expectation-related predictive on-demand deployment algorithm of UAV was devised for reducing the transmit power. As previously mentioned, ML is an auspicious power tool to offer potential and autonomous solutions smartly to boost the UAV-assisted communication network. But, several pieces of research focused on the trajectory and deployment models of UAVs in communication networks [8]. However, resource allocation methods like sub-channels and transmit power are taken into account as well as the previous research concentrated on time-independent scenarios. Furthermore, for time-dependent cases, the capacities of ML-based resource allocation techniques were inspected [9]. But, many ML techniques concentrated on multi-or-single UAV scenarios by assuming the accessibility of whole network data for all UAVs.

This article introduces an intelligent resource allocation using an artificial ecosystem optimizer with a deep learning (IRA-AEODL) technique on UAV networks. The presented IRA-AEODL technique aims to effectually allot the resources in the wireless UAV network. In such cases, the IRA-AEODL technique focuses on the maximization of system utility over all users, combined user association, energy scheduling, and trajectory design. To optimally allocate the UAV policies, the stacked sparse autoencoder (SSAE) model is used in the UAV networks. For the hyper-parameter tuning process, the AEO algorithm is used to enhance the performance of the SSAE model. The experimental results of the IRA-AEODL technique are examined under different aspects.

The highlights of this article include the use of unmanned aerial vehicles (UAVs) as a solution for flexible coverage and high data rates in next-generation networks, the challenge of energy consumption and limited battery capacity in UAVs, and the introduction of an intelligent resource allocation technique using an artificial ecosystem optimizer with deep learning (IRA-AEODL) on UAV networks. The research motivation behind this article is to find a solution to the energy cost issue in UAV-aided mmWave networks by utilizing energy harvesting and an intelligent resource allocation technique.

2. Related Works

In [10], the authors examine the Resource Allocation (RA) issue in UAV-assisted EH-powered D2D Cellular Networks (UAV-EH-DCNs). The main goal is to enhance power effectiveness and, at the same time, ensure the gratification of Ground Users (GUs). Also, the LSTM network is implemented to ease the rapidity of conjunction by taking out the prior data of GUs' gratification in regulating the present RA policy. Chang et al. [11] suggest an ML-founded policy RA protocol that encompasses RL and DL to devise the maximum strategy of the comprehensive UAV. Then, the authors also introduce a Multi-Agent (MA) DRL system for dispersed employment without being aware of a previous idea of the dynamic behavior of networks. Li et al. [12] suggest a novel DRL-founded Flight Resource Allocation Framework (FRA) to lessen the comprehensive information packet loss in a sequential activity space. Also, a state classification layer, leveraging LSTM, is established in forecasting network dynamics, outcoming from time-varying airborne channels and power arrivals at the devices on the ground.

In [13], the authors concentrate on a downlink cellular network, where several UAVs play as aerial base stations for the users on the ground over Frequency Division Multiple Access (FDMA). Targeting maximizing both fairness and comprehensive throughput, the authors prototype RA and route design as a Decentralized Partially Observable Markov Decision Process (Dec-POMDP) and suggest MARL as a resolution. In [14], a MADRL-founded approach is introduced to accomplish the optimum long-term network utility while gratifying the customer's device value of service needs. However, considering that the efficacy of every UAV was determined founded on the atmosphere of the network and several other UAV activities, the JTDPA issue is prototyped as the stochastic game.

In [15], the authors present their IRA-AEODL framework, which combines the Intra-Routing Algorithm (IRA) and Aerial Edge-mounted On-Demand Learning (AEODL). The IRA allows UAVs in the network to organize them for routing, while AEODL leverages machine learning to enhance dynamic route optimization. Afterward, the authors evaluate the performance of their proposed IRA-AEODL network, comparing it against existing UAV network solutions. They perform numerical simulations to evaluate the end-to-end delay, network throughput, and packet delivery ratio. They also analyze the mobile edge computing capabilities of their proposed network.

In [16], the authors examine the anti-jamming issue with integrated channel and energy distribution for UAV networks. Specifically, the authors concentrate on discarding both shared intrusion amongst exterior malevolent jamming and UAVs to optimize the scheme Quality of Experience (QoE) related to energy utilization. Then, the authors suggest a joint MA Layered Q Learning (MALQL) founded anti-jamming transmission protocol in minimizing the huge dimensionality of the activity space and examine the asymptotic convergence of the suggested protocol. In [16], the novelty of this research lies in its ability to address the total energy reduction issue in a non-convex way, while also incorporating several advanced protocols, such as a central MARL protocol and an MA Federated RL protocol, into an MEC scheme with multiple UAVs. By doing so, the authors propose a new and innovative approach that can potentially reduce energy consumption and improve the overall energy efficiency. The author [17] presents a stochastic geometry-based analysis of an integrated aerial-ground network, enabled by multi-UAVs. The novelty of this paper is that the exact distribution of the network throughput is derived and explored under various system parameters. However, the analysis is restricted to Rayleigh fading and a single interfering UAV.

Overall, the literature survey highlights a research gap in the area of resource allocation in UAV-assisted networks. While there have been previous studies focusing on using algorithms such as LSTM, RL, and DRL for efficient resource allocation, there is still a need for further investigation in this area. Furthermore, there is also a need for exploring the use of multi-agent reinforcement learning (MARL) in resource allocation as it has shown promising results in other areas of machine learning. There is also a gap in the evaluation of these proposed resource allocation techniques as most existing studies use simulation-based results rather than real-world implementation and testing. Therefore, further research in this field can contribute to the development of more efficient and adaptive resource allocation policies for UAV-assisted networks.

3. The Proposed Model

In this article, we proposed a novel IRA-AEODL technique for efficient resource allocation in UAV networks. A key advantage of the proposed IRA-AEODL technique compared to existing solutions is its ability to maximize system utility over a set of users by combining user association, energy scheduling, and trajectory design. Figure 1 visually demonstrates the overall architecture of the IRA-AEODL approach. Furthermore, a 3D Cartesian coordinate system is used to ensure optimal coverage for each user. The user set and UAV swarm are represented as U and M, respectively, with |M| = M and |U| = U. The trajectory of each UAV is modeled through time slots, t, with t∈{1,2,T}. Additionally, the constellation of UAVs is assumed to fly at a fixed height H. Finally, the base station or satellite is responsible for the learning procedure required to ensure optimization within the IRA-AEODL approach. The main motivators for using this technique are (i) the availability and ease of access to unlabeled data; (ii) the potential for notable enhancements in the model's performance by including a significant amount of unlabeled data in training; and (iii) the practical constraints of human resources in terms of labeling data. To assess the efficacy of this approach, we conducted a practical analysis on a genuine dataset which showcased the considerable boost in the overall classification accuracy of the SSAE model through the inclusion of a substantial quantity of unlabeled data in the pre-training stage.

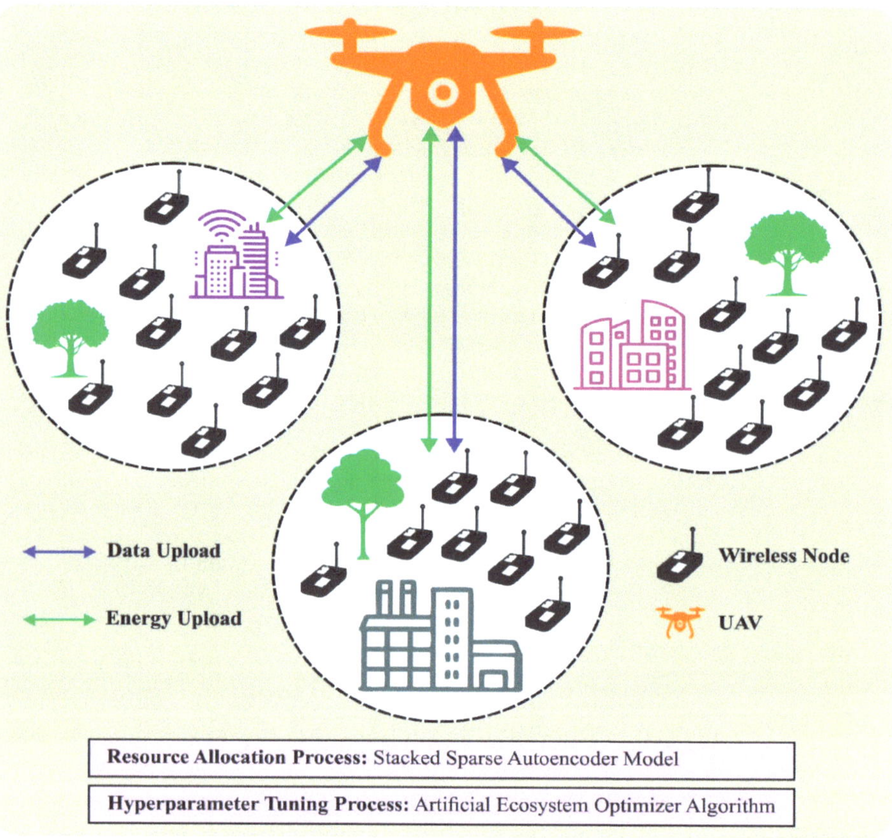

Figure 1. Overall procedure of the IRA-AEODL system.

3.1. System Model

Assume $M > 1$ UAVs share a similar frequency spectrum and a group of $U > 1 GUs$. The GU set and UAV swam are represented as $\mathcal{U}$ and M, respectively [18]. We have $|M| = M$ and $|\mathcal{U}| = U$. Each UAV provides service to the user in successive time slots. We represent the time slot as t; $t \in \{1, 2, T\}$. The total period was represented as $\mathcal{T}$. In the presented model, take a 3D Cartesian coordinate system but the predetermined position of every GU u represented by vertical and horizontal coordinates, for example, $\varphi_u = [x_u, y_u]^T \in \mathbb{R}^{2\times1}$, $u \in \mathcal{U}$. Each UAV is considered to fly at a fixed distance $d_h = H$ above ground and the coordinate of UAVs m at t time was represented as

$\psi_m(t) = [x_m(t), y_m(t)]^T \in \mathbb{R}^{2 \times 1}$. Assume a base controller is performing the learning procedure that could be *BS* or satellite. Furthermore, the UAV is capable of communicating within the swam.

Assume each UAV will fly back to the base hence the trajectory needs to fulfil the subsequent constraints

$$\psi_m(1) = \psi_m(T). \tag{1}$$

Moreover, the trajectory of the UAV is also subjected to specific constraints of distance and speed, which are the following:

$$\|\psi_m(t+1) - \psi_m(t)\| \leq V_{\max}, \tag{2}$$

$$\|\psi_m(t) - \psi_j(t)\| \geq S_{\min}, \tag{3}$$

where $V_{\max}$ denotes the maximal speed of UAV and S_{in} represents the minimal inter-UAV distance to prevent specific collision or interference. Consequently, the distance between UAV m and user u in the t time slot is shown below:

$$d_{m,u}(t) = \sqrt{H^2 + \|\psi_m(t) - \psi_u\|^2}. \tag{4}$$

3.1.1. Path Loss Model

The UAV is capable of establishing an *LoS* link with *GU*. Since the changes in real-time environments (urban, rural, suburban, and so on) are generally unpredictable, the randomness related to *LoS* and Non-LoS (NLoS) in a specific time must be considered while developing the UAV. Consequently, consider the GU connection with UAV through the *LoS* connection with specific probability which we represent as *LoS* probability. The *LoS* probability depends on the environment and the location of *GU* and UAV.

$$\rho_{m,u}^{los}(t) = \frac{1}{1 + \xi_1 \exp[-\xi_2(\theta_{m,u}(t) - \xi_1)]}, \tag{5}$$

In Equation (5), ξ_1 and ξ_2 denote the constant; the value depends on the environment and carrier frequency. $\theta_{m,u}(t)$ indicates the elevation angle as follows:

$$\theta_{m,u}(t) = \frac{180}{\pi sir1\left(\dfrac{H}{d_{m,u}(t)}\right)}. \tag{6}$$

The *LoS* and NLoS path loss methods between the user u and UAV m are shown below:

$$\hat{L}_{m,u}(t) = \begin{cases} \eta_1 \left(\dfrac{4\pi f_c d_{mu}(t)}{c}\right)^{\alpha}, & LoS\ link, \\ \eta_2 \left(\dfrac{4\pi f_c d_{mu}(t)}{c}\right)^{\alpha}, & NLoS\ link, \end{cases} \tag{7}$$

In Equation (7), η_1 and η_2 denote the excess coefficient in *LoS* and NLoS links, respectively. f_c indicates the carrier frequency, c represents the light speed, and α shows the path loss exponent. Assuming the UAV and GU locations, it is challenging to define whether it is *LoS* or NLoS path loss method that must be utilized in the UAVs technique.

$$L_{m,u}(t) = \rho_{m,u}^{los}(t)\eta_1\left(\frac{4\pi f_c d_{m,u}(t)}{c}\right)^{\alpha} + (1 - \rho_{m,u}^{los}(t))\eta_2\left(\frac{4\pi f_c d_{m,u}(t)}{c}\right)^{\alpha}. \tag{8}$$

3.1.2. Transmission Model

A binary parameter $\beta_{m,u}(t)$ is determined as the user association indicator to express the user relationship between GU and UAV, which is

$$\beta_{m,u}(t) = \begin{cases} 1, & \textit{if GU u associates with UAV m,} \\ 0, & \textit{otherwise.} \end{cases} \tag{9}$$

Consider one GU passing through one UAV in a provided time slot, viz., $\sum_{m=1}^{M} \beta_{m,u}(t) \leq 1$. Furthermore, the transmit power of UAV m for u was represented by $p_{m,u}(t)$ and the channel gain between UAV m and user u is represented by $h_{m,u}(t)$.

$$R_u(t) = \sum_{m=1}^{M} \beta_{m,u}(t) log_2(1 + \gamma_{m,u}(t)), \tag{10}$$

As a result, different UAVs could cause interference with GU u, $\gamma_{m,u}(t)$, modeled as SINR of the relationship between m and u, as follows:

$$\gamma_{m,u}(t) = \frac{p_{m,u}(t,)h_{m,u}(,t)L_{m,u'}^{-1}(t)}{\sum_{j=1, j\neq m}^{M} p_{ju}(t)h_{ju}(t)L_{ju}^{-1}(t) + \sigma^2}, \tag{11}$$

In Equation (11), σ^2 denotes the noise variance. It should be such that the transmit power, channel state, and trajectory of the UAV are continuous. Next, after quantizing and partitioning the value into distinct levels within the range, in every t time slot, the value of this variable is understood as a discrete counterpart.

3.2. SSAE-Based Resource Allocation Scheme

To optimally allocate the UAV policies, the SSAE model is used in the UAV networks. The building block of SSAE in the AE is an archetypal NN that learns to map the input X to output Y [18]. The entire AE is split into decoder and encoder parts: the encoded part (W_X, B_X), which maps the input X to the code I_c, and the decoder part (W_Y, B_Y), which maps the code to the reconstruction data Y. The architecture of SSAE was demonstrated in Figure 2, the decoding part is with weighted W_Y and bias B_Y and the encoding part is with W_X weight and B_X bias. Thus

$$I_C = g_{LS}(W_X X + B_X), \tag{12}$$

$$Y = g_{LS}(W_Y I_C + B_Y), \tag{13}$$

where the output Y represents the estimate of input X and g_{LS} indicates the log sigmoid function:

$$g_{LS}(z) = \frac{1}{1 + exp(-z)}. \tag{14}$$

The SAE is different from the AE model. The sparsity could assist AE to attain the best performance. To minimalize the error between the output Y and the input vector X, the raw loss function of AE is assumed as follows:

$$J_{raw}(W_x, W_y, B_x, B_Y) = \frac{1}{N_S}\|Y - X^2\|, \tag{15}$$

In Equation (15), N_S denotes the number of training instances. From Equations (12) and (13), the output Y is formulated as follows

$$Y = g_{AE}(X|W_X, W_Y, B_X, B_Y), \tag{16}$$

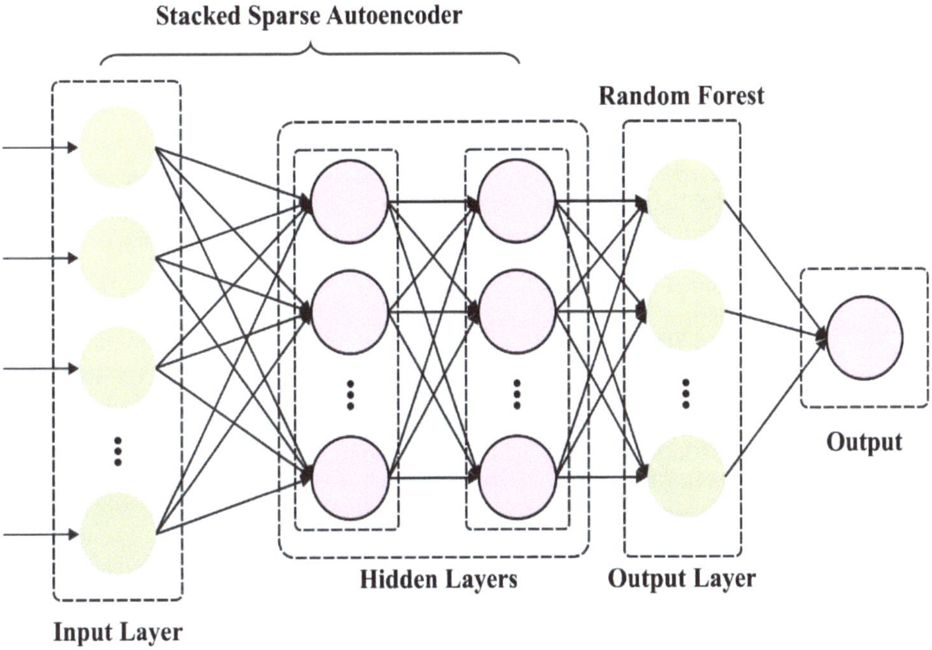

Figure 2. Structure of SSAE.

In Equation (16), g_{AE} denotes the abstract of the AE function. Thus, Equation (15) is formulated as follows:

$$J_{raw}(W_X, W_Y, B_X, B_Y) = \frac{1}{N_S}\|.g_{AE}(X|W_X, W_Y, B_X, B_Y) - X\|^2. \tag{17}$$

To learn a trivial mapping or prevent over-complete mapping, we determine one regularized term Γ_s of the sparsity constraint and one L_2 regularization term Γ_w of the weight (W_X, W_y) and it is expressed below:

$$J(W_X, W_Y, B_X, B_Y) = \frac{1}{N_S}\|g_{AE}(X|W_X, W_Y, B_X, B_Y) - X\|^2 + a_s \times \Gamma_s + a_w \times \Gamma_{wy} \tag{18}$$

In Equation (18), a_s and a_w refer to the sparsity and weight regulation factors. The sparsity regularization term can be represented as follows:

$$\Gamma_s = \sum_{j=1}^{|I|} g_{KL}(\rho_J\hat{\rho}) = \sum_{j=1}^{|I|} \rho \, log \, \frac{\rho}{\hat{\rho}_j} + (1-\rho)log_J^{\frac{1-\rho}{1-\hat{\rho}}} \tag{19}$$

In Equation (19), $\hat{\rho}_j$ denotes the *j-th* neuron's average activation value over each N_s trained sample, $|I|$ denotes the number of components of internal code output I_C, ρ indicates the desirable value, termed the sparsity proportion factor, and g_{KL} represents the Kullback–Leibler divergence function. The weight regularization term can be represented as follows:

$$\Gamma_w = \frac{1}{2} \times \|W_X W_Y\|_2^2. \tag{20}$$

SAE is utilized as a key component and the last SSAE classifiers by subsequent three processes are constructed by the following actions: (i) append the softmax layer at the end of the AI method; (ii) involve preprocessing, input, vectorization, and 2D-FrFE layers; and (iii) stack the available SAE. In the classifier stage, four SAE blocks with many neurons of

(N_1, N_2, N_3, N_4) are applied. As a result of the trial-and-error method, we apply four SAE blocks. Lastly, the softmax layer with the neuron of N_c is appended, where N_c denotes the number of fruit classes.

3.3. Hyperparameter Tuning using the AEO Algorithm

For the hyperparameter tuning process, the AEO algorithm is used for enhancing the performance of the SSAE model. The AEO is an innovative nature-inspired metaheuristic algorithm that hinges on the energy transmission model among living creatures that assist to maintain species stability [19]. The three operators that are utilized to obtain solutions are decomposition, production, and consumption. The energy flow in an ecosystem consists of decomposers, producers, and consumers.

3.3.1. Production

In AEO, the producer represents the worse individual in the population. Thus, it needs to be upgraded concerning the optimal individual by the lower and upper boundaries such that it helps others to find other areas. Through the production operator, a new individual is produced, among randomly generated (x_{rand}) and the best (x) individuals by substituting the prior one. The mathematical representation of the production operator is shown below:

$$x_1(t+1) = (1-\alpha)x_n(t) + \alpha x_{rand}(t) \tag{21}$$

$$\alpha = \left(1 - \frac{t}{T}\right)r_1 \tag{22}$$

$$x_{rand} = \bar{r}(Ub - Lb) + Lb \tag{23}$$

Here, n represents the population size, T signifies the iteration number, Ub and Lb denote upper and lower boundaries, and r_1 signifies a random integer that lies between $[0,1]$. $\bar{r}$ and α denote a random vector within $[0, 1]$ and a linear weight coefficient. The α coefficient provided in Equation (21) assists to drift the individual linearly from the random location to the optimal individual through iteration.

3.3.2. Consumption

The consumers perform this operation and then the production operator finishes the production. Each consumer may eat an arbitrarily selective consumer taking low energy or a producer for obtaining energy. A Lévy flight is a random walk termed as a consumption factor (C) and was determined as follows for enhancing the exploration ability:

$$C = \frac{1}{2}\frac{v_1}{|v_2|} \tag{24}$$

$$v_1 \sim N(O, 1), v_2 \sim N(O, 1) \tag{25}$$

$N(0, 1)$ represents the normal distribution for the mean and SD equivalent to *zero* and one, respectively. Distinct approaches can be implemented with various kinds of users. A consumer eats only the producer in case of being arbitrarily selective as a herbivore (x_2 and x_5 are herbivore consumers, therefore, consume only producer x_1). This strategy was depicted in Equation (26).

$$x_i(t+1) = x_i(t) + C \cdot (x_i(t) - x_1(t)), i \in [2, \cdots, n] \tag{26}$$

A consumer only eats another consumer with a high energy level once it can be selective as a carnivore arbitrarily (a consumer in individuals of x_2–x_5 are consumed by consumer x_6 as the last is a carnivore and takes a lower energy level than individuals of x_2–x_6). A carnivore performance was demonstrated as follows:

$$x_i(t+1) = x_i(t) + C \cdot (x_i(t) - x_i(t)), +i \in [3, \cdots, n] \tag{27}$$

$$j = randi([2i - 1]) \tag{28}$$

Uniquely from the last two performances, a consumer with a high level of energy or producer is arbitrarily eaten by the user when it can be selected as an omnivore arbitrarily (either the producer x_1 or arbitrarily selected users in x_2–x_6 is eaten by x_7 since it can be an omnivore and is the low energy level of x_2–x_6).

$$x_i(t + 1) = x_i(t) + C \cdot (r_2 \cdot (x_i(t) - x_1(t))) + (1 - r_2)(x_i(t) - x_j(t)), \; i \in [3, \cdots, n] \tag{29}$$

$$j = randi([2i - 1]) \tag{30}$$

whereas, r_2 implies the random number from the range of zero to one. A searching individual's place was upgraded in terms of both arbitrarily selective and worse individuals from the population utilizing the consumption operator. Hence, it permits the technique for executing a global search.

3.3.3. Decomposition

SDecomposition is a vital procedure for taking a suitably working ecosystem. The decomposer breaks down every dead individual continuously from the population for providing needed nutrients for the producer's development. The decomposition feature of D together with weighted coefficients of h and e can be intended for the mathematical model. Individuals' parameters support upgrading the location of x_i (*ith* individual) by the location of x_n (the decomposer position). Besides, every individual's next position has been permitted for spreading nearby the decomposer (optimum individual). The mathematical formula is provided as follows:

$$x_i(t + 1) = x_n(t) + D \cdot (e \cdot x_n(t) - h \cdot x_i(t)), i \in 1, \cdots, n \tag{31}$$

$$D = 3u, u \sim N(0, 1) \tag{32}$$

$$e = r_3 \cdot randi([1 \; 2]) - 1 \tag{33}$$

$$h = 2 \cdot r_3 - 1. \tag{34}$$

4. Results and Discussion

In this section, the experimental validation of the IRA-AEODL technique is examined under various aspects. Table 1 and Figure 3 report a comparative average throughput (ATHRO) study of the IRA-AEODL technique with recent models [20]. The outcomes indicate the increasing ATHRO values of the IRA AEODL technique under all K values. For K = 2, the IRA-AEODL technique obtains a higher ATHO value of 1.62 bps while the MP, RP, MAB, DQL, and MADDPG [21] models accomplish reduced ATHO values of 0.71 bps, 0.72 bps, 1.43 bps, 1.50 bps, and 1.57 bps, respectively. Similarly, with K = 6, the IRA-AEODL technique reaches improving ATHO of 1.72 bps while the MP, RP, MAB, DQL, and MADDPG models result in reduced ATHO values of 1.20 bps, 1.06 bps, 1.47 bps, 1.59 bps, and 1.66 bps, respectively. The proposed DNN was trained on an offline dataset of simulated UAV-aided mmWave. The parameters of the proposed algorithm were optimized to obtain the best learning performance. The training was conducted for 1000 epochs using Keras and Tensorflow on a Nvidia GTX 1060 GPU. The accuracy comparison of the proposed DNN was conducted against existing state-of-the-art algorithms. The results showed that the proposed IRA-AEODL technique achieved an average improvement in 11.5% over existing algorithms. This accuracy improvement was attributed to the stacked sparse autoencoder's ability to efficiently perform resource allocation and the AEO algorithm's ability to optimize the model.

Table 1. ATHRO analysis of the IRA-AEODL approach with other systems under varying UAVs.

	Average Throughput (bps)					
No of UAVs	Maximal Power	Random Power	MAB	DQL	MADDPG	IRA-AEODL
K = 2	0.71	0.72	1.43	1.50	1.57	1.62
K = 3	0.73	0.70	1.42	1.53	1.55	1.60
K = 4	1.11	1.04	1.41	1.55	1.66	1.72
K = 6	1.20	1.06	1.47	1.59	1.66	1.72

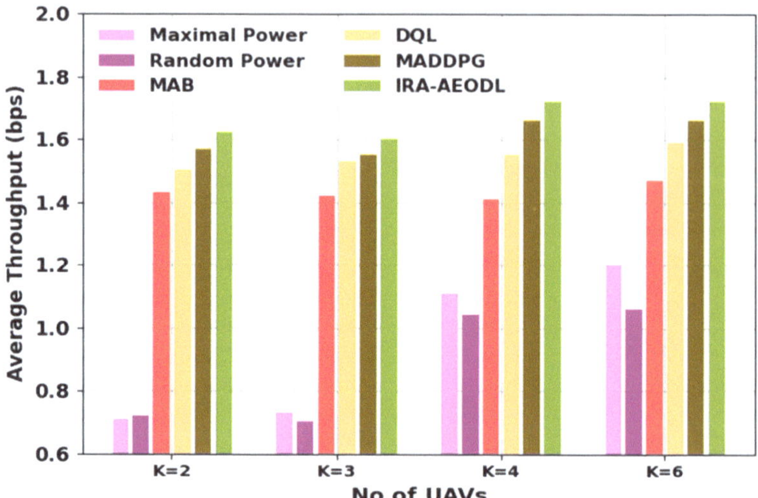

Figure 3. ATHRO analysis of the IRA-AEODL approach under varying UAVs.

The proposed model is a Deep Neural Network (DNN) model that has been trained on a dataset of images of different fruits. The DNN architecture uses convolutional layers to extract features from the images, followed by a densely connected set of layers to identify the classes of fruits. The training process will involve feeding the DNN model with labeled images of each of the desired fruit classes. The model will learn the features associated with each class and develop a set of weights that will allow it to recognize which fruits belong to which class. After training has been completed, the model can then be used to classify new images of fruits into their respective classes. Additionally, to improve accuracy, the model can also be fine-tuned using data augmentation techniques, such as randomly adjusting the size and orientation of the images as well as adjusting the brightness and contrast. This can help the model to better recognize the features in different images. Once training and fine-tuning is complete, the DNN can then be tested with a set of validation images to ensure that it is able to accurately classify the different types of fruits. Once satisfactory accuracy has been achieved, the model can then be deployed for use in applications.

Table 2 and Figure 4 demonstrate a comparative ATHRO study of the IRA-AEODL method with recent methods. The results represent the increasing ATHRO values of the IRA-AEODL technique under varying time slots. For 100 time slots, the IRA-AEODL method attains a maximum ATHO value of 1.84 bps whereas the MP, RP, MAB, DQL, and MADDPG methods attain decreased ATHO values of 0.97 bps, 0.92 bps, 1.47 bps, 1.58 bps, and 1.72 bps, respectively. Similarly, with 300-time slots, the IRA-AEODL method attains an increasing ATHO of 1.83 bps while the MP, RP, MAB, DQL, and MADDPG methods resulted in decreased ATHO values of 1.14 bps, 1.05 bps, 1.58 bps, 1.69 bps, and 1.75 bps, reseptively.

Table 2. ATHRO analysis of the IRA-AEODL approach with other systems under varying time slots.

Time Slots	Average Throughput (bps)					
	Maximal Power	Random Power	MAB	DQL	MADDPG	IRA-AEODL
100	0.97	0.92	1.47	1.58	1.72	1.84
200	1.02	0.99	1.48	1.66	1.73	1.83
300	1.14	1.05	1.58	1.69	1.75	1.83
400	1.34	1.28	1.58	1.70	1.75	1.84
500	1.47	1.38	1.63	1.77	1.85	1.92

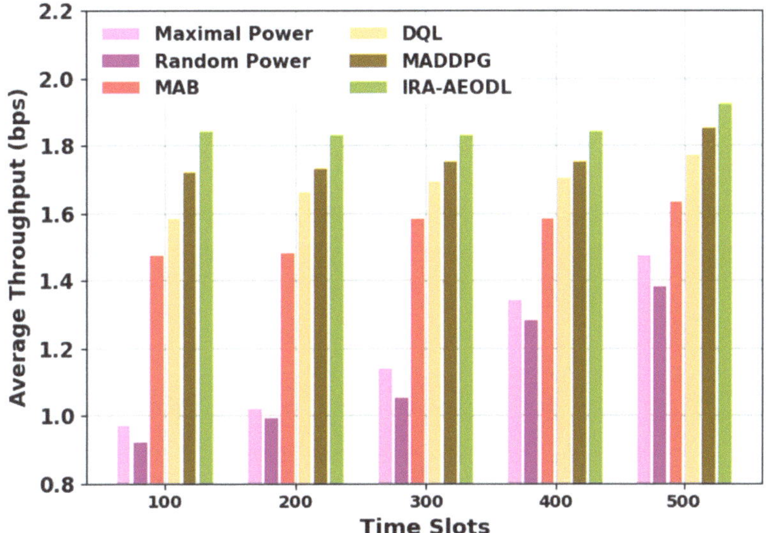

Figure 4. ATHRO analysis of the IRA-AEODL approach under varying time slots.

Table 3 and Figure 5 illustrate a comparative ATHRO study of the IRA-AEODL method with recent models. The results indicate the increasing ATHRO values of the IRA-AEODL technique under varying users. For 100 users, the IRA-AEODL technique obtains a higher ATHO value of 1.84 bps while the MP, RP, MAB, DQL, and MADDPG methods accomplish reduced ATHO values of 1.04 bps, 0.84 bps, 1.49 bps, 1.52 bps, and 1.70 bps, respectively. Similarly, with 300 users, the IRA-AEODL technique reaches an improving ATHO of 2.28 bps while the MP, RP, MAB, DQL, and MADDPG models resulted in reduced ATHO values of 1.43 bps, 1.36 bps, 1.79 bps, 1.91 bps, and 2.06 bps, respectively.

Table 3. ATHRO analysis of the IRA-AEODL approach with other systems under varying users.

No. of Users	Average Throughput (bps)					
	Maximal Power	Random Power	MAB	DQL	MADDPG	IRA-AEODL
100	1.04	0.84	1.49	1.52	1.70	1.84
200	1.23	1.20	1.63	1.78	1.89	2.08
300	1.43	1.36	1.79	1.91	2.06	2.28
400	1.65	1.55	1.95	2.13	2.33	2.55
500	1.70	1.73	2.10	2.24	2.57	2.61

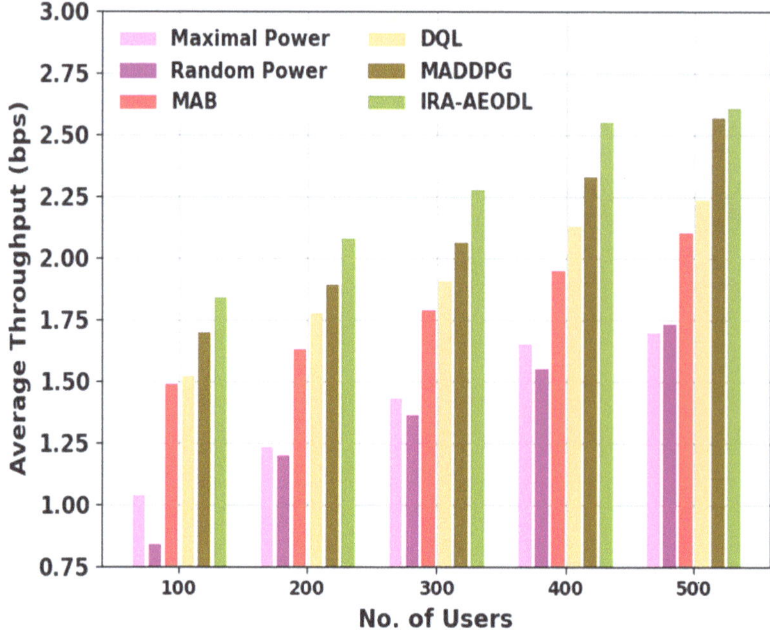

Figure 5. ATHRO analysis of the IRA-AEODL approach under varying users.

Table 4 and Figure 6 depict a comparative ATHRO study of the IRA-AEODL technique with recent models. The outcomes indicate the increasing ATHRO values of the IRA-AEODL technique under varying energy arrival E_{max}. For 80 energy arrival E_{max}, the IRA-AEODL technique attains a higher ATHO value of 1.73 bps while the MAB, DQL, and MADDPG methods obtain minimum ATHO values of 1.55 bps, 1.66 bps, and 1.71 bps respectively. Similarly, with the 160 energy arrival E_{max}, the IRA-AEODL technique reaches an improving ATHO of 1.85 bps while the MAB, DQL, and MADDPG models resulted in reduced ATHO values of 1.75 bps, 1.80 bps, and 1.83 bps, respectively.

Table 5 and Figure 7 demonstrate a comparative ATHRO study of the IRA-AEODL technique with recent methods. The results indicate the increasing ATHRO values of the IRA-AEODL technique under varying battery capacity (BC). For 3000 BC, the IRA-AEODL technique obtains a higher ATHO value of 1.74 bps while the MAB, DQL, and MADDPG methods accomplish reduced ATHO values of 1.55 bps, 1.63 bps, and 1.70 bps, respectively. Similarly, with 5000 BC, the IRA-AEODL technique reaches an improving ATHO of 1.79 bps while the MAB, DQL, and MADDPG models resulted in reduced ATHO values of 1.60 bps, 1.67 bps, and 1.79 bps, respectively.

Table 4. ATHRO analysis of the IRA-AEODL approach with other systems under varying energy arrival values E_{max}.

	Average Throughput (bps)			
Energy Arrival E_{max}	MAB	DQL	MADDPG	IRA-AEODL
80	1.55	1.66	1.71	1.73
100	1.64	1.71	1.79	1.81
120	1.65	1.74	1.80	1.83
140	1.69	1.76	1.82	1.85
160	1.75	1.80	1.83	1.85

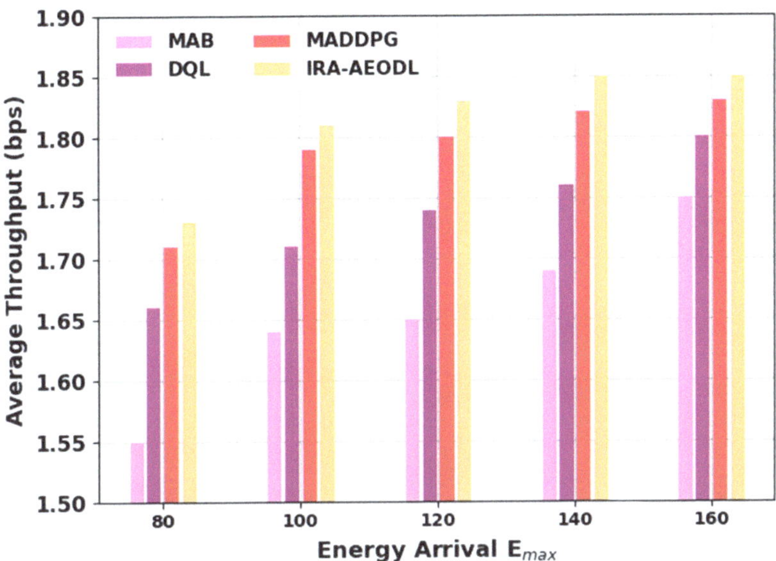

Figure 6. ATHRO analysis of the IRA-AEODL approach under varying energy arrival values E_{max}.

Table 5. ATHRO analysis of the IRA-AEODL approach with other systems under varying battery capacity.

Battery Capacity (C)	Average Throughput (bps)			
	MAB	DQL	MADDPG	IRA-AEODL
3000	1.55	1.63	1.70	1.74
3500	1.56	1.65	1.77	1.78
4000	1.58	1.66	1.77	1.78
4500	1.59	1.67	1.78	1.79
5000	1.60	1.67	1.79	1.79

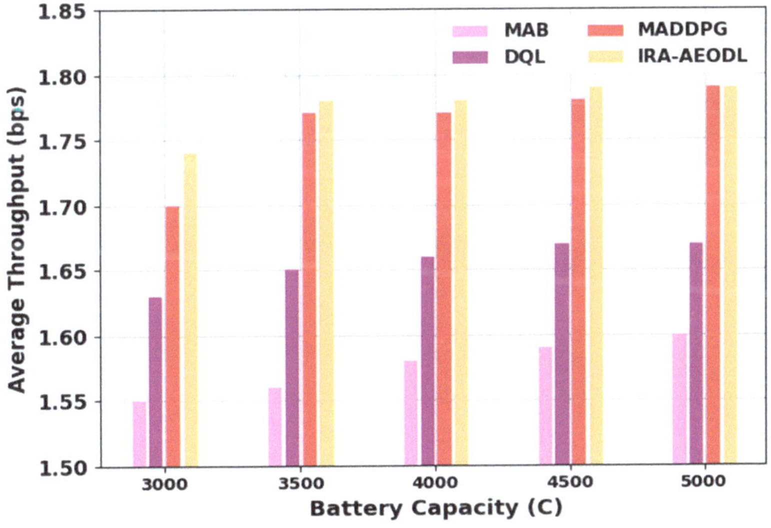

Figure 7. ATHRO analysis of the IRA-AEODL approach under varying battery capacity.

Table 6 and Figure 8 depict a comparative ATHRO study of the IRA-AEODL technique with recent models. The results indicate the increasing ATHRO values of the IRA-AEODL technique under varying Energy Transfer b/w Two UAVs (ETTUAV). For 3000 ETTUAV, the IRA-AEODL technique attains a higher ATHO value of 1.69 bps while the MAB, DQL, and MADDPG methods accomplish reduced ATHO values of 1.54 bps, 1.58 bps, and 1.65 bps, respectively. Similarly, with 5000 ETTUAV, the IRA-AEODL technique reaches an improving ATHO of 1.78 bps while the MAB, DQL, and MADDPG models resulted in reduced ATHO values of 1.65 bps, 1.70 bps, and 1.77 bps, respectively.

Table 6. ATHRO analysis of the IRA-AEODL approach with other systems under varying Energy Transfer b/w Two UAVs.

	Average Throughput (bps)			
Energy Transfer b/w Two UAVs	**MAB**	**DQL**	**MADDPG**	**IRA-AEODL**
3000	1.54	1.58	1.65	1.69
3500	1.58	1.63	1.68	1.73
4000	1.58	1.68	1.75	1.76
4500	1.64	1.69	1.76	1.77
5000	1.65	1.70	1.77	1.78

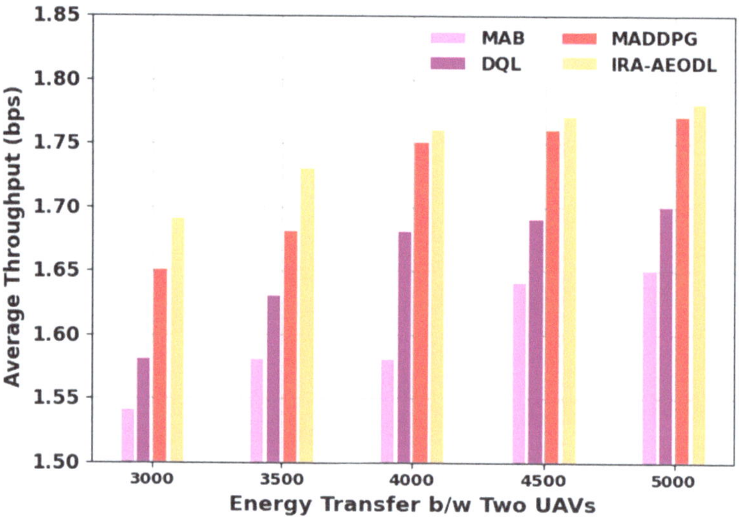

Figure 8. ATHRO analysis of the IRA-AEODL approach with other systems under varying ETTUAV.

Finally, the average reward examination of the IRA-AEODL technique with different models takes place in Table 7 and Figure 9. The results demonstrate that the IRA-AEODL technique gains increasing reward values over other models. For instance, with 200 episodes, the IRA-AEODL technique attains an increasing average reward of 1.41 while the MAB, DQL, and MADDPG techniques obtain reducing average rewards of 1.28, 1.34, and 1.29, respectively.

Table 7. Average reward analysis of the IRA-AEODL approach with other systems under varying pisodes.

	Average Reward			
Episodes	MAB	DQL	MADDPG	IRA-AEODL
0	1.13	1.18	1.29	1.33
200	1.28	1.34	1.29	1.41
400	1.33	1.36	1.34	1.44
600	1.31	1.39	1.39	1.43
800	1.31	1.38	1.43	1.44
1000	1.30	1.40	1.44	1.47
1200	1.31	1.40	1.44	1.50
1400	1.29	1.43	1.48	1.50
1600	1.29	1.43	1.47	1.54

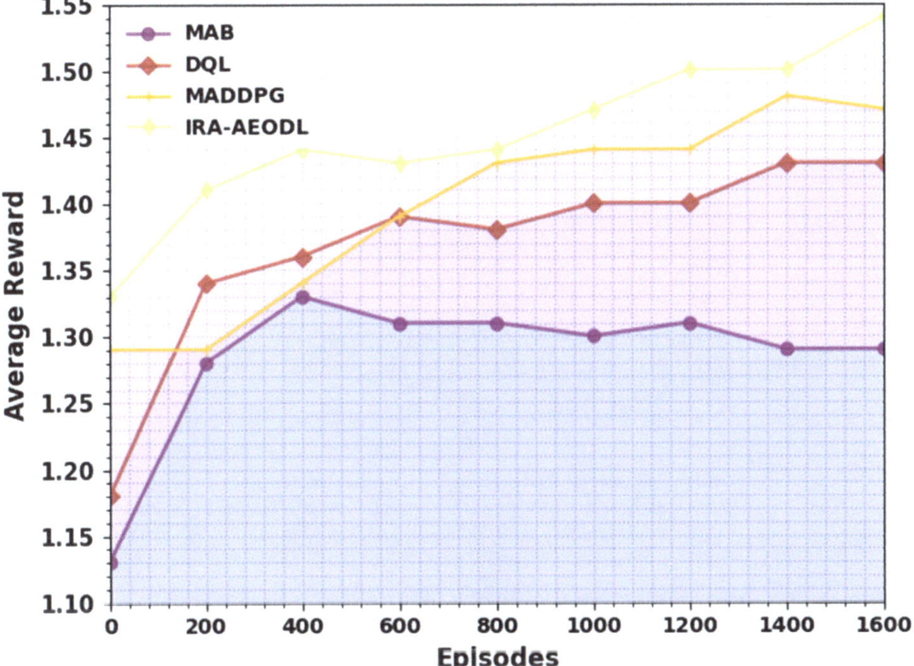

Figure 9. Average reward analysis of the IRA-AEODL approach under varying episodes.

Meanwhile, with 800 episodes, the IRA-AEODL technique attains an increasing average reward of 1.44 while the MAB, DQL, and MADDPG methods attain reducing average rewards of 1.31, 1.38, and 1.43, respectively. Eventually, with 1600 episodes, the IRA-AEODL technique attains an increasing average reward of 1.54 while the MAB, DQL, and MADDPG techniques obtain reducing average rewards of 1.29, 1.43, and 1.47, respectively. These results exhibited the superior performance of the IRA-AEODL technique over other existing models on the UAV networks.

5. Conclusions

In this article, we introduced a new IRA-AEODL technique for the optimal allocation of resources in UAV networks. The presented IRA-AEODL technique is intended for the effectual allocation of resources in wireless UAV networks. Here, the IRA-AEODL technique focused on the maximization of system utility over all users, combined trajectory design, user association, and energy scheduling. To optimally allocate the UAV policies, the SSAE model is used in the UAV networks. For the hyperparameter tuning process, the AEO algorithm is used to enhance the performance of the SSAE model. The experimental results of the IRA-AEODL technique are examined under different aspects and the outcomes stated the better performance of the IRA-AEODL approach over recent state of art approaches. In the future, the ensemble learning process can be included to improve the resource allocation performance of the IRA-AEODL technique. In comparison to other learning methods, the proposed algorithm has several advantages such as fast convergence, improved local optimization ability, and well-balanced global or local search ability. With the help of production, consumption, and decomposition operators, the proposed model is able to quickly explore the search space and find the optimal solution. Therefore, the proposed algorithm is vital for hyperparameter tuning as it ensures optimal results, sustainability, and robustness compared to other learning models.

Author Contributions: Conceptualization, R.A. and E.M.; methodology, M.S.A.M.; software, R.A.; validation, A.M. and E.M.; formal analysis, A.A.; investigation, M.S.A.M.; resources, A.R. and E.M.; data curation, A.A.; writing—original draft preparation, A.M. and M.S.A.M.; writing—review and editing, R.A. and E.M.; visualization, A.M.; supervision, A.M.; project administration, R.A.; funding acquisition, E.M. and R.A. All authors have read and agreed to the published version of the manuscript.

Funding: This work was supported by Princess Nourah bint Abdulrahman University Researchers Supporting Project number (PNURSP2023R323), Princess Nourah bint Abdulrahman University, Riyadh, Saudi Arabia, in part this paper is supported by postdoc fellowship granted by the Institute of Computer Technologies and Information Security, Southern Federal University, project № P.D./22-01-KT, and in part this paper has been supported by the RUDN University Strategic Academic Leadership Program (recipient Ammar Muthanna).

Institutional Review Board Statement: Not applicable.

Informed Consent Statement: Not applicable.

Data Availability Statement: The data are contained within the article and/or available from the corresponding author upon reasonable request.

Acknowledgments: The authors express their gratitude to Princess Nourah bint Abdulrahman University Researchers Supporting Project number (PNURSP2023R323), Princess Nourah bint Abdulrahman University, Riyadh, Saudi Arabia, and in part would like to acknowledge the RUDN University Strategic Academic Leadership Program (recipient Ammar Muthanna).

Conflicts of Interest: The authors declare no conflict of interest.

References

1. Seid, A.M.; Boateng, G.O.; Anokye, S.; Kwantwi, T.; Sun, G.; Liu, G. Collaborative computation offloading and resource allocation in multi-UAV-assisted IoT networks: A deep reinforcement learning approach. *IEEE Internet Things J.* **2021**, *8*, 12203–12218. [CrossRef]
2. Do, Q.V.; Pham, Q.-V.; Hwang, W.-J. Deep reinforcement learning for energy-efficient federated learning in UAV-enabled wireless powered networks. *IEEE Commun. Lett.* **2021**, *26*, 99–103. [CrossRef]
3. Dai, Z.; Zhang, Y.; Zhang, W.; Luo, X.; He, Z. A Multi-Agent Collaborative Environment Learning Method for UAV Deployment and Resource Allocation. *IEEE Trans. Signal Inf. Process. Netw.* **2022**, *8*, 120–130. [CrossRef]
4. Peng, H.; Shen, X. Multi-agent reinforcement learning based resource management in MEC- and UAV-assisted vehicular networks. *IEEE J. Sel. Areas Commun.* **2020**, *39*, 131–141. [CrossRef]
5. Hu, J.; Zhang, H.; Song, L.; Han, Z.; Poor, H.V. Reinforcement learning for a cellular internet of UAVs: Protocol design, trajectory control, and resource management. *IEEE Wirel. Commun.* **2020**, *27*, 116–123. [CrossRef]
6. Chen, X.; Liu, X.; Chen, Y.; Jiao, L.; Min, G. Deep Q-Network based resource allocation for UAV-assisted Ultra-Dense Networks. *Comput. Networks* **2021**, *196*, 108249. [CrossRef]
7. Munaye, Y.Y.; Juang, R.-T.; Lin, H.-P.; Tarekegn, G.B.; Lin, D.-B. Deep reinforcement learning based resource management in UAV-assisted IoT networks. *Appl. Sci.* **2021**, *11*, 2163. [CrossRef]

8. Liu, Y.; Yan, J.; Zhao, X. Deep reinforcement learning based latency minimization for mobile edge computing with virtualization in maritime UAV communication network. *IEEE Trans. Veh. Technol.* **2022**, *71*, 4225–4236. [CrossRef]
9. Qi, W.; Song, Q.; Guo, L.; Jamalipour, A. Energy-Efficient Resource Allocation for UAV-Assisted Vehicular Networks with Spectrum Sharing. *IEEE Trans. Veh. Technol.* **2022**, *71*, 7691–7702. [CrossRef]
10. Xu, Y.-H.; Sun, Q.-M.; Zhou, W.; Yu, G. Resource allocation for UAV-aided energy harvesting-powered D2D communications: A reinforcement learning-based scheme. *Ad Hoc Netw.* **2022**, *136*, 102973. [CrossRef]
11. Chang, Z.; Deng, H.; You, L.; Min, G.; Garg, S.; Kaddoum, G. Trajectory design and resource allocation for multi-UAV networks: Deep reinforcement learning approaches. *IEEE Trans. Netw. Sci. Eng.* **2022**, *10*, 2940–2951. [CrossRef]
12. Li, K.; Ni, W.; Dressler, F. LSTM-characterized deep reinforcement learning for continuous flight control and resource allocation in UAV-assisted sensor network. *IEEE Internet Things J.* **2021**, *9*, 4179–4189. [CrossRef]
13. Yin, S.; Yu, F.R. Resource allocation and trajectory design in UAV-aided cellular networks based on multiagent reinforcement learning. *IEEE Internet Things J.* **2021**, *9*, 2933–2943. [CrossRef]
14. Zhao, N.; Liu, Z.; Cheng, Y. Multi-agent deep reinforcement learning for trajectory design and power allocation in multi-UAV networks. *IEEE Access* **2020**, *8*, 139670–139679. [CrossRef]
15. Niu, Y.; Yan, X.; Wang, Y.; Niu, Y. An adaptive neighborhood-based search enhanced artificial ecosystem optimizer for UCAV path planning. *Expert Syst. Appl.* **2022**, *208*, 118047. [CrossRef]
16. Yin, Z.; Lin, Y.; Zhang, Y.; Qian, Y.; Shu, F.; Li, J. Collaborative Multiagent Reinforcement Learning Aided Resource Allocation for UAV Anti-Jamming Communication. *IEEE Internet Things J.* **2022**, *9*, 23995–24008. [CrossRef]
17. Nie, Y.; Zhao, J.; Gao, F.; Yu, F.R. Semi-distributed resource management in UAV-aided MEC systems: A multi-agent federated reinforcement learning approach. *IEEE Trans. Veh. Technol.* **2021**, *70*, 13162–13173. [CrossRef]
18. Zhang, S.; Zhu, Y.; Liu, J. Multi-UAV Enabled Aerial-Ground Integrated Networks: A Stochastic Geometry Analysis. *IEEE Trans. Commun.* **2022**, *70*, 7040–7054. [CrossRef]
19. Zhang, Y.; Satapathy, S.C.; Wang, S. Fruit category classification by fractional Fourier entropy with rotation angle vector grid and stacked sparse autoencoder. *Expert Syst.* **2022**, *39*, e12701. [CrossRef]
20. Izci, D.; Hekimoğlu, B.; Ekinci, S. A new artificial ecosystem-based optimization integrated with Nelder-Mead method for PID controller design of buck converter. *Alex. Eng. J.* **2022**, *61*, 2030–2044. [CrossRef]
21. Domingo, M.C. Power Allocation and Energy Cooperation for UAV-Enabled MmWave Networks: A Multi-Agent Deep Reinforcement Learning Approach. *Sensors* **2022**, *22*, 270. [CrossRef] [PubMed]

 *drones*

MDPI

Article

Hierarchical Matching Algorithm for Relay Selection in MEC-Aided Ultra-Dense UAV Networks

Wei Liang [1,2,*], Shaobo Ma [1], Siyuan Yang [3], Boxuan Zhang [1] and Ang Gao [1]

[1] School of Electronics and Information, Northwestern Polytechnical University, Xi'an 710060, China; 2022262290@mail.nwpu.edu.cn (S.M.); 18710723787@163.com (B.Z.); gaoang@nwpu.edu.cn (A.G.)
[2] Research & Development Institute, Northwestern Polytechnical University in Shenzhen, Shenzhen 518057, China
[3] Xian Aerospace & Computing Research Institution, Xi'an 710000, China; yangsiyuan@nwpu.edu.cn
* Correspondence: liangwei@nwpu.edu.cn

Abstract: With the rapid development of communication technology, unmanned aerial vehicle–mobile edge computing (UAV-MEC) networks have emerged with powerful capabilities. However, existing research studies have neglected the issues involving user grouping and relay selection structures under UAV cluster-assisted communication. Therefore, in this article, we present a comprehensive communication–computing resource allocation for UAV-MEC networks. In particular, ground users make stable user groups first, and then multiple UAVs act as relays in order to assist these user groups in simultaneously uploading their tasks to the terrestrial base station at the edge server. Moreover, in order to maximize the system's overall throughput, a more flexible and hierarchical matching relay selection algorithm is proposed in terms of matching the ground user groups and corresponding UAVs. For vulnerable users, we also propose a weighted relay selection algorithm to maximize the system performance. Furthermore, simulation results show that the proposed relay selection algorithm achieves a significant gain in comparison with the other benchmarks, and the stability of the proposed algorithms could be verified.

Keywords: MEC; relay selection; UAV; matching theory

Citation: Liang, W.; Ma, S.; Yang, S.; Zhang, B.; Gao, A. Hierarchical Matching Algorithm for Relay Selection in MEC-Aided Ultra-Dense UAV Networks. *Drones* **2023**, *7*, 579. https://doi.org/10.3390/drones7090579

Academic Editor: Petros Bithas

Received: 29 August 2023
Revised: 8 September 2023
Accepted: 12 September 2023
Published: 14 September 2023

1. Introduction

With the rapid development of communication technologies, the application scenarios of 6G networks often need to cope with massive access users and data volumes, which also bring about serious challenges in terms of network latency, quality of service (QoS), capacity, and other metrics. Traditional centralized cloud computing and fixed base station communication methods may suffer from high interaction latency and network congestion because they are far away from user terminals [1]. To address this problem, mobile edge computing (MEC) technology [2–5] sinks computing resources to wireless access networks closer to user terminals, and further extends it to non-user networks, such as Wi-Fi access. This effectively reduces the transmission delay and energy consumption and creates a service environment with high communication performance and transmission bandwidth. It could be used in vehicle networking, virtual reality, augmented reality, industrial control, autonomous driving, and other applications [6–10]. Unmanned aerial vehicles (UAVs) are highly autonomous and flexible, are able to transmit messages without signal occlusion and reduce fading caused by signal reflection, scattering, diffraction, and penetration. Therefore, UAVs have certain communication and computing capabilities and are often used as mission execution carriers, cellular network nodes, and transmission relays [11–15]. As a result, the UAV-MEC network was created. It combines the high autonomy and flexibility of UAVs with the benefits of MEC networks to provide users with flexible coverage, reliable communication connectivity, and powerful computing capabilities [16–18].

Most of the UAV-MEC research studies assume that a single UAV is applied to MEC networks [19–21]. However, in real scenarios, UAVs may need to face massive user access and data transmissions, where the energy and computational resources of a single UAV could not support efficient and continuous work. In contrast, UAV clusters cannot only reduce these problems, but also gain benefits through cluster collaborations, improving system scalability and performance. Therefore, it is important to consider how UAVs collaborate with each other. The authors of [22] proposed a decentralized deep reinforcement learning algorithm to enable UAV clusters to autonomously and distributedly learn dynamic coordination strategies by exploiting the deterministic state transfer property of the system, which effectively improves the system's task computational rate. The authors of [23] jointly optimize the system's computational resource allocation, power control, and user association to minimize the power of the system. In order to solve this non-convex optimization problem, the researchers proposed a centralized multi-intelligent body reinforcement learning algorithm and a semi-distributed federated reinforcement-learning algorithm, respectively, which effectively achieves the optimization of the system latency and power metrics. The authors of [24] modeled the optimization problem as a discrete Stackelberg game model with priorities to obtain network hierarchical characteristics. Also, they proved that the subgame at each priority level was an ordered potential game with the Nash equilibrium, and proposed a hierarchical learning algorithm that could achieve fast convergence of hierarchical grouping strategies for UAV clusters.

After summarizing and analyzing the existing research on UAV-MEC networks, we found the following limitations:

- **User grouping and relay selection:** In the case of UAV clusters as relays to assist the communication, the ground users should cooperate in groups and select different UAVs for transmission, but there is very limited research on the integration of the user grouping as well as the relay selection strategies.
- **Resource management and power allocation:** Existing research studies have been conducted on power allocation and resource management under specific grouping methods and relay selection structures, lacking a unified approach.

In this paper, we investigate the grouping method and relay selection structure problems in UAV cluster-assisted relay transmission. Specifically, we establish a communication model for UAV cluster-assisted ground users to offload computational tasks to the ground base station. Meanwhile, the system throughput is taken as the objective function under the constraints of communication delay and transmission power, and the problem is decoupled into two subproblems. More importantly, we propose a hierarchical matching relay selection algorithm and a weighted relay selection algorithm to maximize the throughput of the system, where simulation results demonstrate the effectiveness and superiority of the proposed algorithm.

The rest of the paper is organized as follows. The system model is elaborated on in Section 2. The proposed hierarchical matching relay selection algorithm and weighted relay selection algorithm are presented in Section 3. The simulation results are presented in Section 4. The conclusions and future work are presented in Section 5.

2. System Model

As shown in Figure 1, we consider a UAV cluster-assisted relay transmission model. The model has M ground users, N UAVs, and a ground base station with a deployed MEC server. The ground users and UAVs are assumed to be homogeneous, and both have the same constraint of transmission power, set as p_{max}^u and p_{max}^U, respectively. The computational ability of the MEC server is measured by the number of CPU cycles required by the MEC server to compute each bit of input data, set as f. The set of UAVs is defined as $\mathcal{N} = \{1, 2, \ldots, N\}$, and the set of ground users is defined as $\mathcal{M} = \{1, 2, \ldots, M\}$, where the number of ground users is much higher than the number of UAVs, $M \gg N$. The ground users are randomly distributed in a specific range of the horizontal space, and the 3D coordinate positions could be expressed as $\left(x_i^u, y_i^u, 0\right)$. The UAVs are randomly distributed in the three-dimensional space, and their coordinate positions could be expressed as $\left(x_i^U, y_i^U, h_i^U\right)$. Note that the UAVs need to keep a certain distance from each other to avoid mutual collision and interference. In addition, during the process of information transmission and relay forwarding, the UAVs maintain a hovering state with fixed three-dimensional coordinates to ensure the stability of the communication link.

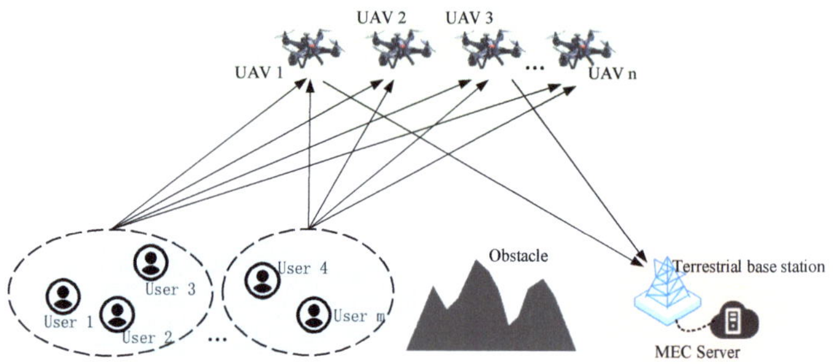

Figure 1. Transmission model of UAV-MEC networks.

In order to fully utilize the spectrum resources, M ground users will be divided into N groups with an unlimited number of ground users within each group. Assuming that the total system bandwidth is B, and the spectrum resources are equally divided into N orthogonal sub-channels, the bandwidth of the sub-channel assigned to the ith group is B_i, expressed as

$$B_i = \frac{B}{N}. \tag{1}$$

Specifically, ground users within the same group use uplink non-orthogonal multiple access (NOMA) to cooperate in offloading, and ground users between different groups use frequency division multiple access (FDMA) to transmit in different frequency bands. At the same time, different UAVs will also use FDMA to transmit information to the base station.

2.1. System Transmission Model

As shown in Figure 1, there may be obstacles in the communication environment that impede the communication. Thus, there is a line-of-sight (LOS) channel primary path component along with a non-line-of-sight (NLOS) channel multipath component in the communication link. Therefore, the Rician fading channel [25,26] is employed in this paper.

Additionally, $h_{i,j}^u$ denotes the channel gain between the ith UAV and the jth ground user. h_i^U denotes the channel gain between the ith UAV and the ground base station. Assuming that there are k ground users in the group assisted by the ith UAV, the received signal y_i of the ith UAV could be expressed as

$$y_i = \sum_{j=1}^{k} \sqrt{p_{i,j}^u} h_{i,j}^u x_{i,j} + n_0, \tag{2}$$

note that n_0 is the additive Gaussian white noise (AWGN), satisfying a mean of 0 and a variance of N_0. $p_{i,j}^u$ is the transmission power of the jth ground user in the ith UAV-assisted group, which could not exceed the upper limit of the ground user transmission power $p_{\max}^u$, which means

$$p_{i,j}^u \leq p_{\max}^u, \forall i \in \mathcal{N}, \forall j \in \mathcal{M}. \tag{3}$$

In the uplink NOMA mode, the decoding order at the receiving end is determined by the strength of each sub-signal. From Equation (2), the received strength is related to the signal power, so the decoding is done in descending order of the signal power. The inter-group interference of the jth ground user in the ith UAV-assisted group are computed as

$$I_{NOMA}^{i,j} = \sum_{t=j+1}^{N_i} p_{i,t}^u \left| h_{i,t}^u \right|^2,$$

(4)

where N_i is the number of users in the ith UAV-assisted user group; therefore, the signal-to-noise ratio (SNR) of the jth ground user in the ith UAV-assisted user group is given by

$$\gamma_i^j = \frac{p_{i,j}^u \left| h_{i,j}^u \right|^2}{I_{NOMA}^{i,j} + N_0} = \frac{p_{i,j}^u \left| h_{i,j}^u \right|^2}{\sum\limits_{t=j+1}^{N_i} p_{i,t}^u \left| h_{i,t}^u \right|^2 + N_0}.$$

(5)

According to the Shannon channel capacity, the information transmission rate $r_{i,j}^u$ could be expressed as

$$r_{i,j}^u = B_i log_2 \left(1 + \gamma_i^j \right),$$

$$= \frac{B}{N} log_2 \left(1 + \frac{p_{i,j}^u \left| h_{i,j}^u \right|^2}{\sum\limits_{t=j+1}^{N_i} p_{i,t}^u \left| h_{i,t}^u \right|^2 + N_0} \right),$$

(6)

therefore, the throughput from a user group to its corresponding UAV could be expressed as

$$C_{SR}^i = \sum_{j=1}^{N_i} r_{i,j}^u.$$

(7)

More specifically, the decode-and-forward (DF) technique is used for UAVs to relay the received signal. In this strategy, the UAV receives the signal from the user group and performs the decode–recode–forward operation on the signal, reducing the transmission process interference. Then, the UAV uses FDMA to transmit the signal to the ground base station. Therefore, the ith UAV information transmission rate r_i^U could be expressed as

$$r_i^U = \frac{B}{N} log_2 \left(1 + \frac{p_i^U \left| h_i^U \right|^2}{N_0} \right).$$

(8)

For the ith UAV, the transmission power could not exceed the constraint of the UAV transmission power p_{max}^U, which means

$$p_i^U \leq p_{max}^U, \forall i \in \mathcal{N},$$

(9)

similar to Equation (7), the throughput from the ith UAV to the ground base station could be given by

$$C_{RD}^i = r_i^U.$$

(10)

Under the DF method, the system throughput is limited by the minimum of the throughput from the source node to the relay node and the throughput from the relay node to the destination node. Therefore, the system throughput of the transmission process for the ith UAV-assisted user group would be expressed as

$$C_i = \min \left(C_{SR}^i, C_{RD}^i \right).$$

(11)

In this paper, the system throughput C is the sum of all user group throughputs, which could be expressed as

$$C = \sum_{i=1}^{N} C_i.$$

(12)

2.2. System Computation Model

In UAV-MEC networks, the computational task to be performed by the jth ground user could be defined as a binary group $D_j \triangleq (d_j, T_j)$, where d_j denotes the amount of data for the computational task to be performed by the jth ground user, and T_j denotes the maximum transmission delay tolerable for the computational task. These two parameters indicate that the computational task to be performed by the ground user is computationally intensive and time-sensitive. Assuming that the computational tasks are indivisible, they are transmitted to the ground base station by a UAV relay through a complete offloading method.

Note that time and energy consumption need to be considered during the offloading and transfer of computational tasks from the ground user to the MEC server. For the jth ground user in the NOMA group of the ith UAV-assisted relay, the computational task transmission time $t_{i,j}^{SR}$ and the energy consumption $E_{i,j}^{SR}$ from the source node to the relay node could be expressed as

$$t_{i,j}^{SR} = \frac{d_{i,j}}{B_i r_{i,j}^u} = \frac{d_{i,j}N}{Br_{i,j}^u},\tag{13}$$

$$E_{i,j}^{SR} = p_{i,j}^u t_{i,j}^{SR}.\tag{14}$$

Equation (13) utilizes the relationship between the total system bandwidth B and the sub-channel bandwidth B_i. From Equation (14), it could be seen that the energy consumption of the computational task is related to the ground user transmission power and the information transmission rate. Meanwhile, it could be seen from Equation (6) that the information transmission rate is also related to the ground user transmission power. Therefore, without loss of generality, this paper converts the transmission process energy constraint problem into a transmission power constraint problem.

Neglecting the UAV decoding–recoding time, for the ith UAV relay, the transmission time $t_{i,j}^{RD}$ of the computational task of the auxiliary NOMA group from the relay node to the destination node is the sum of the transmission times of all users in the group, denoted as

$$t_{i,j}^{RD} = \frac{\sum_{j=1}^{N_i} d_{i,j}}{B_i r_i^U} = \frac{N\sum_{j=1}^{N_i} d_{i,j}}{Br_i^U}.\tag{15}$$

For the jth ground user in the NOMA group of the ith UAV-assisted relay, the time of the computational task execution process at the MEC server $t_{i,j}^c$ could be expressed as

$$t_{i,j}^c = \frac{F_{i,j}d_{i,j}}{f},\tag{16}$$

where $F_{i,j}$ is the number of CPU cycles required by the jth ground user in the ith NOMA group to compute the task unit data volume. In this paper, ground users perform similar types of tasks, and without loss of generality, $F_{i,j}$ takes the same value for all ground users.

For the jth ground user in the NOMA group of the ith UAV-assisted relay, the energy consumption of the execution process of the computational task at the MEC server $E_{i,j}^c$ could be expressed as

$$E_{i,j}^c = P_{MEC}t_{i,j}^c,\tag{17}$$

where P_{MEC} is the computational power of the MEC server deployed at the ground base station. In this paper, the MEC server is deployed with high computational power and energy supply, and thus, without loss of generality, the energy consumption of the MEC server is negligible.

Therefore, for the jth ground user in the NOMA group of the ith UAV-assisted relay, the total task execution delay $t_{i,j}^o$ could be expressed as

$$t_{i,j}^o = t_{i,j}^{SR} + t_{i,j}^{RD} + t_{i,j}^c.\tag{18}$$

Since the computational tasks are delay-sensitive, for all UAV computational tasks, the total delay could not exceed the maximum tolerable transmission delay $T_{i,j}$, which means

$$t_{i,j}^o \leq T_{i,j}, \forall i \in \mathcal{N}, \forall j \in \mathcal{M}. \tag{19}$$

2.3. Problem Formulation

The system transmission and computation model show that ground users are required to perform computation-intensive and delay-sensitive tasks, which need to be transmitted by the UAV-assisted relay to the ground base station due to resource constraints. Therefore, our goal is to maximize the system throughput under the constraints of computational resources as well as transmission delay.

The system throughput of the proposed model is affected by various factors, including the ground user grouping situation, the UAV relay selection situation, and the power allocation of the ground users in the NOMA system. Among them, the two factors, ground user grouping and UAV relay selection, are correlated and together determine the actual communication grouping structure. Therefore, in this section, these two factors are unified as the relay selection structure. Thus, the system parameters include the relay selection structure and power allocation for ground users, where the system's objective function could be expressed as

$$\max_{\mathcal{A},\mathcal{P}} C$$

$$\begin{aligned}
s.t. \quad &C1 : t_{i,j}^o \leq T_{i,j}, \forall i \in \mathcal{N}, \forall j \in \mathcal{M}, \\
&C2 : p_{i,j}^u \leq p_{\max}^u, \forall i \in \mathcal{N}, \forall j \in \mathcal{M}, \\
&C3 : p_i^U \leq p_{\max}^U, \forall i \in \mathcal{N}, \\
&C4 : \bigcup_{i=1}^{N} \mathcal{A}_i = \mathcal{M}, \bigcap_{i=1}^{N} \mathcal{A}_i = \varnothing, \forall i \in \mathcal{N},
\end{aligned} \tag{20}$$

where parameter $\mathcal{A}$ is the set of relay selection structures and parameter $\mathcal{P}$ is the set of ground user power allocations. Constraint $C1$ indicates that the total delay of all ground users' computing tasks could not exceed the maximum tolerable transmission delay of computing tasks. Constraint $C2$ indicates that the transmission power of all ground users could not exceed the upper limit of ground users' transmission power. Constraint $C3$ indicates that the transmission power of all UAVs could not exceed the upper limit of the UAV transmission power constraint. In constraint $C4$, $\mathcal{A}_i$ indicates the relay selection structure of the ith UAV relay and its auxiliary NOMA user group. That means the concatenation set of all relay selection structures includes all ground users, and the intersection set of all relay selection structures is the empty set, which ensures the ground user completeness of the computational task transmission. Notably, we restrict the minimum distance between UAVs so that they cannot overlap or collide.

The objective function established in this paper is a joint problem of two parameters, the relay selection structure and ground user power allocation. To simplify the treatment, Equation (20) is decoupled into two subproblems—relay selection and power allocation—expressed as

$$P1 : \max_{\mathcal{A}} \sum_{i=1}^{N} \sum_{j=1}^{M} u_{i,j} \min \left(r_{i,j}^u, r_i^U \right),$$

$$s.t. \quad C1 : \bigcup_{i=1}^{N} \mathcal{A}_i = \mathcal{M}, \bigcap_{i=1}^{N} \mathcal{A}_i = \varnothing, \forall i \in \mathcal{N}, \tag{21}$$

$$C2 : \{u_{i,j}\} \in \{0, 1\}.$$

$$P2 : \max_{\mathcal{P}} C,$$

$$s.t. \quad C1 : t_{i,j}^o \leq T_{i,j}, \forall i \in \mathcal{N}, \forall j \in \mathcal{M},$$
$$C2 : p_{i,j}^u \leq p_{\max}^u, \forall i \in \mathcal{N}, \forall j \in \mathcal{M}, \tag{22}$$
$$C3 : p_i^U \leq p_{\max}^U, \forall i \in \mathcal{N}.$$

where $u_{i,j}$ is a binary offloaded variable that could be traversed over all ground users that reflecting their cooperation with each UAV. More specifically, subproblem $P2$ is the power allocation problem in the fixed relay selection structure, and in this paper, we deal with it using a fractional transmit power allocation (FTPA) algorithm. Subproblem $P1$ is a relay selection structure problem in the context of determining the ground user power allocation. To illustrate the nature of the problem, the objective function could be split into the sum of the throughputs of each NOMA group. However, subproblem $P1$ is non-convex because of the presence of discrete binary variables $u_{i,j}$, which is difficult to solve by traditional optimization methods. Therefore, this paper adopts the matching idea to solve the problem.

3. Relay Selection Algorithm

For the UAV-assisted relay transmission model developed in this paper, the bilateral matching participants consist of the set of ground users and the set of UAVs. The number of ground users is much more than the number of UAVs. Therefore, the bilateral matching problem in this paper is a many-to-one matching problem.

3.1. Hierarchical Matching Relay Selection

Ground users matched to the same UAV cooperate for information transmission with NOMA. However, after ground users and UAVs cooperate, the nature of the UAV set changes due to the case of intra-group interference. Specifically, the UAV that cooperates with the ground user changes its utility function value for other ground users. If the other ground users still follow the original preference relationship for the matching process, this does not reflect the actual situation of the system. As shown in Figure 2, a matching round is defined as the ground user set sends a cooperation request based on the preference relationship and the UAV set responds to the cooperation request based on the preference relationship. The process from the start of matching to the point where all participants have no intention to change the matching result is denoted as a matching round. After a matching round, UAVs that have cooperated with ground users exist, so the preference relationships of the participants need to be adapted. Specifically, the unmatched ground users are the match initiators, and the UAVs with these connected ground users are considered as match responders. Based on this, the preference relationship is updated and the next matching round begins until all ground users cooperate with the UAV. The proposed hierarchical matching relay selection algorithm is shown in Algorithm 1.

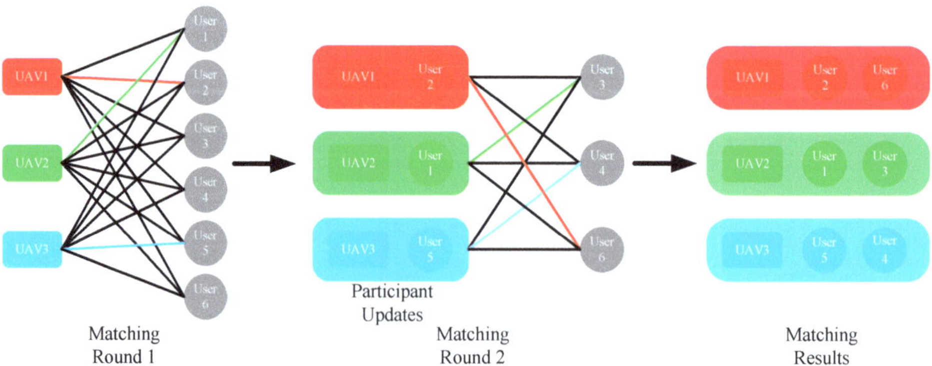

Figure 2. A matching round where different colors represent different groups.

Algorithm 1: Hierarchical matching relay selection algorithm

1 *Step 1: Initialization*
  (1)   Input parameters: ground users set M, UAVs set N, channel gain sets H.
  (2)   Initializing preference relationship sequences.

Step 2: Hierarchical matching
  (1)   While $M \neq \varnothing$:
  (2)   While $Pre_m \neq \varnothing$:
  (3)   Each unmatched ground user sends a match request to the highest-ranked UAV based on a preference relationship sequence;
  (4)   If UAV was not matched with ground users:
  (5)   Match successfully;
  (6)   Else:
  (7)   If the original matching user ranks higher:
  (8)   Refuse the new matching request;
  (9)   Else:
  (10)   Refuse the original matching request;
  (11)   Rejected users remove the selected UAV from the preference relationship sequence;
  (12)   End if;
  (13)   End if;
  (14)   Repeat (3);
  (15)   End while;
  (16)   Update the M, N, and preference relationship sequence set;
  (17)   Repeat (2);
  (18) End while.

In the first round, the priority during initialization is determined by the transmission process throughput in descending order. Note that the throughput from the source node to the relay node should be calculated as the OMA method because all users have not cooperated. During the matching process, each ground user checks the matching status and does not perform an operation if a match has already been realized with a UAV. Instead, the sequence of preference relationships is checked and the UAV with the highest ranking in the sequence of preference relationships is selected to issue a match request. Meanwhile, each UAV first checks the matching status after receiving the matching request sent by the ground user. If a match has not yet been realized with the ground user, it chooses to accept the match request. Instead, the UAV needs to make a choice based on the preference relationship sequence. If the ground user sending the match request has a higher preference relationship ranking, the UAV will end the cooperative relationship with the original ground user and it realizes a new match. On the contrary, the UAV will maintain a cooperative relationship with the original ground user and reject the new match request. In both cases, the rejected ground users are required to remove the UAV from the sequence of preference relationships. When all the ground users' matching requests are responded to by the UAVs, one matching round is over. Repeat the above steps until all unmatched ground users are no longer able to issue matching requests to the UAV, which means for all unmatched ground users, the preference relationship sequence is empty, represented by $Pre_{m_i} = \varnothing$. In this case, the set of ground users and the set of UAVs form N binary matching pairs (i, j), where i and j are the index of UAVs and ground users, respectively. With the number of ground users far exceeding the number of UAVs, there are still $(M - N)$ ground users that have not achieved a match. After the second round of matching, if there are still unmatched ground users, then $(M - 2N)$ of these users remain unmatched. The set of ground users and the set of UAVs then constitute N ternary matching pairs, denoted as (i, j, k). Subsequently, in the ith round, there are $(M - (i - 1)N)$ that remain unmatched.

These users, in combination with the set of UAVs, form N i-element matching pairs with the set of UAVs, which serves as a target for matching with the remaining ground users.

3.2. Weighted Relay Selection

Algorithm 1 considers the high-priority ground user and the UAV as a whole at the end of each matching round and acts as a matching responder for the next round, ensuring the accuracy of the preference relationship. However, the algorithm ignores a possible situation where vulnerable users are disadvantaged in this algorithm. Assuming that the preference relationship sequence and the function values for ground user a and ground user b are represented as

$$Pre_a = \{(n_1 : x) \succ_a (n_2 : 0.9x) \succ_a (n_3 : 0.6x)\}, \tag{23}$$

$$Pre_b = \{(n_1 : 0.8x) \succ_b (n_2 : 0.5x) \succ_b (n_3 : 0.3x)\}, \tag{24}$$

where Pre_a and Pre_b represent the preference relationships of users a and b, respectively, and $\succ$ means that the former has a higher utility value. In this case, a and b have identical preference relationship sequences but different utility function values. During the matching process, the vulnerable user b is rejected by UAV n_1 and is forced to choose to match with UAV n_2. However, user a always obtains a high utility value when matching with any UAV; vulnerable user b obtains a lower utility value when matched with the remaining UAVs, except n_1, and this situation leads to a lower total utility value of the system. Therefore, it is important to propose a new algorithm, which could compensate for vulnerable users.

The core idea of the improved algorithm is to weigh the value of the utility function of the vulnerable user to improve its chances of being selected by the UAVs. Specifically, in the improved algorithm, the preference relationship sequence should focus not only on the ranking order but also on the corresponding utility value. Thus, without loss of generality, the preference relationship sequence of any ground user m_j is re-expressed as

$$\left\{ \left(n_1 : v_1^j\right) \succ_{m_j} \left(n_2 : v_2^j\right) \succ_{m_j} \cdots \succ_{m_j} \left(n_n : v_n^j\right) \right\}, \tag{25}$$

where v_i^j is the value of the utility function when the ground user m_j is matched with the UAV n_i. Based on this, three parameters are designed: the trigger threshold μ, the weighted ratio θ, and the weighted round N_{lim}. In a matching round, when the ground user is rejected by the UAV to which it sends a matching request, the following operation is performed:

(1) Compare the utility values of the current UAV n_{cur} with a UAV that has the next highest preference ranking n_{next}. Calculate the percentage increase u_{real} in the utility of the ground user matching n_{cur} compared to matching n_{next}:

$$u_{real} = \frac{v_{cur}^j - v_{next}^j}{v_{next}^j} \tag{26}$$

(2) Compare u_{real} and μ. If $u_{real} < \mu$, we consider that the rejected user is not a vulnerable user and continue the normal matching process; if $u_{real} \geq \mu$, consider the rejected user is a vulnerable user.

(3) For vulnerable users, v_{cur}^i is weighted and a matching request is resent to UAV n_{cur} with the weighted utility function value. If the matching request is accepted, the algorithm ends; if the matching request is rejected, step (1) is repeated with an upper limit of N_{lim} number of repetitions.

The proposed weighted relay selection algorithm for vulnerable users is shown in Algorithm 2.

Algorithm 2: Weighted relay selection algorithm

---

1 *Step 1: Initialization*

   (1)   Input parameters: set of preference relationship sequences Pre_m, the trigger threshold μ, the weighted ratio θ, and the weighted round N_{lim}.

Step 2: Weighted matching

   (1)   While the ground user's matching request was rejected $\cap$ $N_{lim} \neq 0$:
   (2)   Calculate u_{real};
   (3)   If $u_{real} < \mu$:
   (4)   Stop;
   (5)   Else:
   (6) $v = v(1 + \theta)$, $N_{lim} = N_{lim} - 1$;
   (7)   Send a matching request to the UAV again;
   (8)   End if;
   (9)   End while.

3.3. Algorithm Stability

In order to understand the stability of the matching results, the definition of an impeded stable matching pair is given as follows.

If there exists a matching pair $(inviter, responder) \in \mathcal{M} \times \mathcal{N}$, and the matching pair does not exist in the set $\mathcal{R}$ of matches that have already appeared, which means $(inviter, responder) \notin \mathcal{R}$. However, for the participants of this matching pair, there exists a matching result $(inviter', responder) \in \mathcal{R}$, $(inviter, responder') \in \mathcal{R}$, which means $inviter \succ_{responder} inviter'$ and $responder \succ_{inviter} responder'$, which indicates that for the matching pair $(inviter, responder)$, this matching result destroys the original matching result, so it is called an impeded stable matching pair.

The algorithm proposed in this paper is essentially a one-to-one matching of multiple rounds, so as long as the stability of the one-to-one matching is understood, the stability of the algorithm would be understood.

For the final matching results $(inviter_{fin}, responder_{fin})$, assuming that there is an impeded stable matching pair $(inviter, responder)$, there are two possibilities: $inviter_{fin}$ sent a match request to the $responder$, or $inviter_{fin}$ did not send a match request to the $responder$. For the former, $(inviter, responder)$ does not exist, as the $responder$ would have received match requests from participants higher in their preference order. For the latter, $(inviter, responder)$ also does not exist because either a match request sent to $responder$ was declined, or there is another participant ranked higher than $responder$ in $inviter_{fin}$'s preference order.

4. Simulation Results and Analysis

This section verifies the performance simulation of the hierarchical matching relay selection algorithm and the weighted relay selection algorithm proposed in the above section. The simulation parameters are set as shown in Table 1.

The comparison algorithm we adopt is the classical relay selection strategy: max–SR [27], max–RD [28], and max–min [29]. The idea of the max–SR relay selection algorithm is to select the relay node with the largest instantaneous SNR between the S-R links, and the optimal relay under this algorithm could be expressed as

$$relay^* = \mathrm{argmax}\{ |h_{i,j}^u|^2 \}, \tag{27}$$

similarly, the max–RD algorithm finds the relay node with the largest instantaneous SNR of the R-D link, which could be given by

$$relay^* = \mathrm{argmax}\{ |h_i^U|^2 \}, \tag{28}$$

the core idea of the max–min algorithm is to consider the quality of both links simultaneously, which could be expressed as

$$relay^* = \operatorname{argmax}\{\min\{\left|h_{i,j}^u\right|^2, \left|h_i^U\right|^2\}\}, \tag{29}$$

note that the max–RD and max–min algorithms require a channel state feedback mechanism.

Table 1. Simulation parameter setting.

Parameters	Value
Radius of the horizontal distribution range of UAVs	800 m
Distance between UAVs	200 m
Height distribution range of UAVs	100–300 m
Radius of the horizontal distribution range of ground users	600 m
Sizes of task data	1–10 kbit
Maximum tolerable transmission delay for computing tasks	10–100 ms
Path loss index	0.8
Noise spectral density n_0	-90 dBm/Hz
Rice channel parameters	2
Weighted round limit N_{lim}	5
Average number of tests	1000

Figure 3 presents the system throughput using our proposed hierarchical matching relay selection algorithm with the comparison algorithm for a different number of ground users when the number of UAVs is fixed. Note that we set the number of UAVs to 4, $N = 4$, and the number of ground users ranges from 4 to 24. The parameter settings are based on the numerical relationship, $N \leq M$. This is because an increase in the number of users does not notably enhance the performance given the transmission power limitations of the UAVs. The simulation results show that with the increase in ground users, the system throughput with the max–RD algorithm varies only in a small range and tends to be smooth. This is because in the max–RD algorithm, the relay selection depends only on the channel gain of the R-D link, so there will be a situation where all the ground users select the same relay UAV, and the system throughput is less affected by the change in the number of users. In the max–min and max–SR algorithms, the system throughput gradually increases as the number of users increases, due to the fact that ground users have more UAVs to choose from. It is worth noting that in our proposed algorithm, the system throughput gradually increases and it is significantly higher than the comparison algorithm, which indicates that our proposed algorithm has better user scale adaptability. This is due to the advantages of NOMA and the fact that we always utilized all the UAVs available as relay nodes.

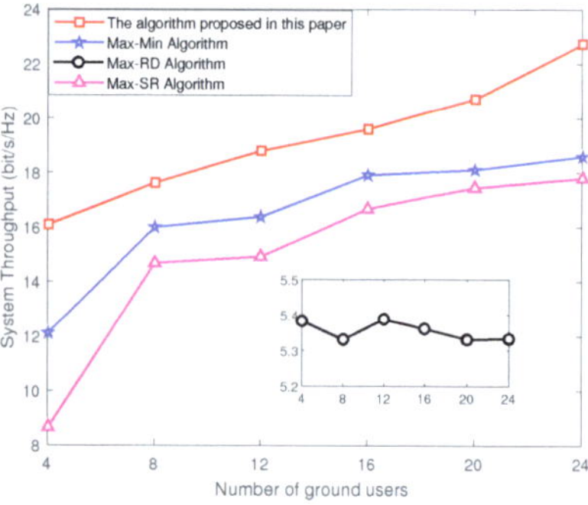

Figure 3. Variation of system throughput with the number of ground users for different relay selection algorithms.

Figure 4 presents the system throughput for different numbers of UAVs when the ground user numbers are fixed. Note that we set the number of ground users to 20, $M = 20$, and the number of UAVs to range from 2 to 7. The parameter settings are based on the numerical relationship between the two and the fact that more UAVs only have smaller performance gains with any algorithm under our tests. The simulation results show that as the available UAV numbers increase, the system throughput using the max–RD algorithm only changes within a small range and there is no upward trend. This is because in the case of the max–RD algorithm, the relay selection only depends on the channel gain between the UAVs and the destination node. Therefore, it may occur that ground users choose the same relay UAV, and the system throughput is less affected by changes in the number of UAVs. In the case of the max–min algorithm and max–SR algorithm, the communication links supporting simultaneous transmission increase as the UAV numbers increase. Therefore, the probability of UAVs with high channel gain increases, and ground users have more UAV choices, improving the system throughput. For the algorithm proposed in this paper, as the UAVs increase, the number of users in each NOMA group decreases, and user interference in the group decreases, which improves the system throughput. It could be seen that although increasing the number of UAVs increases the resource expenditure, it could significantly improve the system throughput.

Figure 5 presents the system throughput using our proposed hierarchical matching relay selection algorithm with the comparison algorithm for different user maximum transmission power scenarios, with the number of ground users $M = 20$ and the number of UAVs $N = 4$. The simulation results show that the system throughput with both the proposed algorithm and the comparison algorithm gradually increases as the maximum transmission power increases within a certain range. It is noteworthy that the system throughput is highest in the case of the proposed algorithm. This is due to the fact that when the maximum transmission power of the ground user increases, the SNR of the ground user's transmission also increases and, therefore, the system throughput increases. Note that the system throughput in the case of the proposed algorithm shows a decreasing growth rate tendency as the transmission power of the ground user increases, while the growth rate of the comparison algorithm remains almost constant. This is due to the fact that the comparison algorithm uses the OMA method for transmission and there is no intra-group user interference. However, the proposed algorithm uses the NOMA method for transmission, and as the ground user transmission power increases, the intra-group user interference also increases, and when it reaches a point where its effect on the SNR is close to that of the transmission power on the SNR, the growth rate of the system throughput decreases. This indicates that the proposed algorithm is sensitive to transmission power.

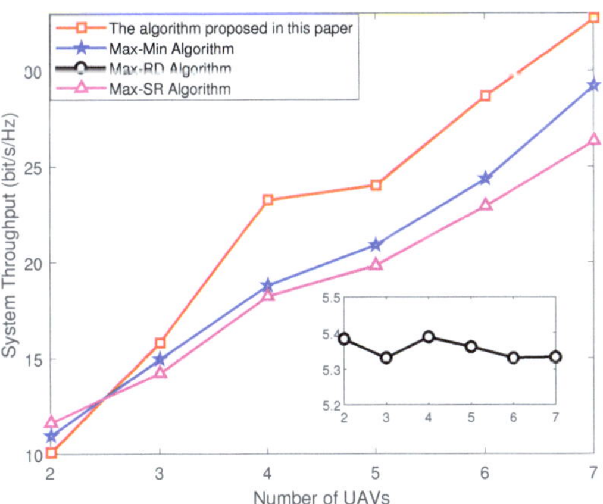

Figure 4. Variation of system throughput with the number of UAVs for different relay selection algorithms.

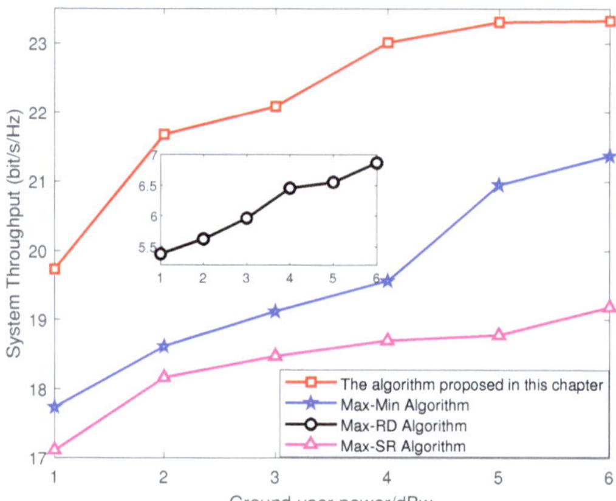

Figure 5. Variation of system throughput with maximum transmission power of terrestrial users for different relay selection algorithms.

Figure 6 presents the system throughput using the weighted relay selection algorithm for different algorithm parameters when the number of ground users is $M = 50$ and the number of UAVs is $N = 7$. The simulation results show that, within a certain parameter range, when the weighted ratio is fixed, the system throughput gradually increases as the trigger threshold decreases; when the trigger threshold is fixed, the system throughput gradually increases as the weighted ratio increases. Outside the parameter range, the system throughput remains stable and high. This is due to the fact that as the trigger threshold decreases and the weighted ratio increases, it makes the bias toward vulnerable users higher. Therefore, the vulnerable users are weighted to be re-matched in a way that improves the system throughput. After a certain range is reached, all possible vulnerable users have been involved in the proposed algorithm, so the system throughput remains stable at a higher value.

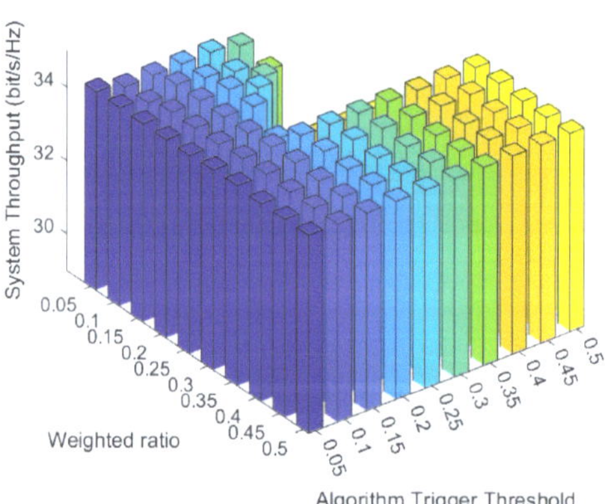

Figure 6. Performance of the weighted relay selection algorithm with different parameters.

Figure 7 presents the variation of the algorithm's runtime with the number of UAVs and the number of ground users for the weighted relay selection algorithm with the upper weighted round limit $N_{\lim} = 5$. The simulation results show that the algorithm running time rises gradually with the increase in the number of ground users and the growth rate increases gradually, which indicates that the ground users as the initiators of the matching have a greater impact on the algorithm running time. Note that the algorithm running time is less affected by the change in the number of UAVs. Based on this and Figure 4, we can consider that increasing the number of UAVs could significantly increase the system throughput while controlling the running time of the algorithm.

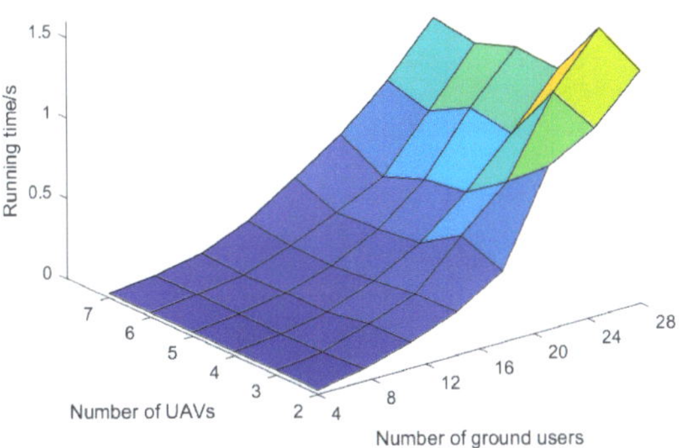

Figure 7. Variation in the algorithm running time with the number of matched participants.

5. Conclusions and Future Work

In this paper, a typical model of UAV-assisted relay transmission is established, and the transmission and computation process of the model is analyzed. Also, this paper jointly controls the transmission power and relay selection structure to optimize the objective function of the system throughput for resource-limited and delay-sensitive communication conditions. More importantly, we propose a hierarchical matching relay selection algorithm and a weighted relay selection algorithm for vulnerable users based on the matching idea, where the simulation results verify the superiority of the proposed algorithm. It should be noted that the weighted relay selection algorithm has better performance compared to the hierarchical matching relay selection algorithm, but there will be more resource investments and higher latency, which need to be selected according to the actual situation. For the power allocation problem in this scenario, please refer to our subsequent work.

Author Contributions: W.L. Conceptualization, Funding acquisition, Project administration, Supervision. S.M. Investigation, Software, Writing—review & editing, Writing—original draft. S.Y. Software, Writing—original draft. B.Z. Investigation, Validation. A.G. Resources, Supervision. All authors have read and agreed to the published version of the manuscript.

Funding: The work was supported by the Shenzhen Science and Technology program under Grant JCYJ20210324121006017 and in part by National Nature Science Foundation of China under Grants 62101450 and in part by Key R&D Plan of Shaan Xi Province Grants 2023YBGY037.

Data Availability Statement: Not applicable.

Conflicts of Interest: The authors declare no conflict of interest.

References

1. Linthicum, D.S. Connecting Fog and Cloud Computing. *IEEE Cloud Comput.* **2017**, *4*, 18–20. [CrossRef]
2. Taleb, T.; Samdanis, K.; Mada, B.; Flinck, H.; Dutta, S.; Sabella, D. On Multi-Access Edge Computing: A Survey of the Emerging 5G Network Edge Cloud Architecture and Orchestration. *IEEE Commun. Surv. Tutor.* **2017**, *19*, 1657–1681. [CrossRef]
3. Raeisi-Varzaneh, M.; Dakkak, O.; Habbal, A.; Kim, B.S. Resource Scheduling in Edge Computing: Architecture, Taxonomy, Open Issues and Future Research Directions. *IEEE Access* **2023**, *11*, 25329–25350. [CrossRef]
4. Li, B.; He, Q.; Cui, G.; Xia, X.; Chen, F.; Jin, H.; Yang, Y. READ: Robustness-Oriented Edge Application Deployment in Edge Computing Environment. *IEEE Trans. Serv. Comput.* **2022**, *15*, 1746–1759. [CrossRef]
5. Wang, D.; He, T.; Lou, Y.; Pang, L.; He, Y.; Chen, H.H. Double-edge Computation Offloading for Secure Integrated Space-air-aqua Networks. *IEEE Internet Things J.* **2023**, *10*, 15581–15593. [CrossRef]
6. Jiang, W.; Han, B.; Habibi, M.A.; Schotten, H.D. The Road Towards 6G: A Comprehensive Survey. *IEEE Open J. Commun. Soc.* **2021**, *2*, 334–366. [CrossRef]

7. Cao, K.; Hu, S.; Shi, Y.; Colombo, A.W.; Karnouskos, S.; Li, X. A Survey on Edge and Edge-Cloud Computing Assisted Cyber-Physical Systems. *IEEE Trans. Ind. Inform.* **2021**, *17*, 7806–7819. [CrossRef]
8. Zhang, J.; Tao, D. Empowering Things with Intelligence: A Survey of the Progress, Challenges, and Opportunities in Artificial Intelligence of Things. *IEEE Internet Things J.* **2021**, *8*, 7789–7817. [CrossRef]
9. Siriwardhana, Y.; Porambage, P.; Liyanage, M.; Ylianttila, M. A Survey on Mobile Augmented Reality with 5G Mobile Edge Computing: Architectures, Applications, and Technical Aspects. *IEEE Commun. Surv. Tutor.* **2021**, *23*, 1160–1192. [CrossRef]
10. Wang, D.; Wu, M.; Wei, Z.; Yu, K.; Min, L.; Mumtaz, S. Uplink secrecy performance of RIS-based RF/FSO three-dimension heterogeneous networks. *IEEE Trans. Wirel. Commun.* **2023**. [CrossRef]
11. Wu, Q.; Xu, J.; Zeng, Y.; Ng, D.W.K.; Al-Dhahir, N.; Schober, R.; Swindlehurst, A.L. A Comprehensive Overview on 5G-and-Beyond Networks With UAVs: From Communications to Sensing and Intelligence. *IEEE J. Sel. Areas Commun.* **2021**, *39*, 2912–2945. [CrossRef]
12. Liu, C.; Feng, W.; Chen, Y.; Wang, C.X.; Ge, N. Cell-Free Satellite-UAV Networks for 6G Wide-Area Internet of Things. *IEEE J. Sel. Areas Commun.* **2021**, *39*, 1116–1131. [CrossRef]
13. Peng, H.; Shen, X. Multi-Agent Reinforcement Learning Based Resource Management in MEC- and UAV-Assisted Vehicular Networks. *IEEE J. Sel. Areas Commun.* **2021**, *39*, 131–141. [CrossRef]
14. Jiang, F.; Wang, K.; Dong, L.; Pan, C.; Xu, W.; Yang, K. AI Driven Heterogeneous MEC System with UAV Assistance for Dynamic Environment: Challenges and Solutions. *IEEE Netw.* **2021**, *35*, 400–408. [CrossRef]
15. Wang, D.; He, T.; Zhou, F.; Cheng, J.; Zhang, R.; Wu, Q. Outage-driven link selection for secure buffer-aided networks. *Sci. China Inf. Sci.* **2022**, *65*, 182303. [CrossRef]
16. Zhang, P.; Wang, C.; Jiang, C.; Benslimane, A. UAV-Assisted Multi-Access Edge Computing: Technologies and Challenges. *IEEE Internet Things Mag.* **2021**, *4*, 12–17. [CrossRef]
17. Liu, Z.; Cao, Y.; Gao, P.; Hua, X.; Zhang, D.; Jiang, T. Multi-UAV network assisted intelligent edge computing: Challenges and opportunities. *China Commun.* **2022**, *19*, 258–278. [CrossRef]
18. Tun, Y.K.; Park, Y.M.; Tran, N.H.; Saad, W.; Pandey, S.R.; Hong, C.S. Energy-Efficient Resource Management in UAV-Assisted Mobile Edge Computing. *IEEE Commun. Lett.* **2021**, *25*, 249–253. [CrossRef]
19. Wang, D.; Tian, J.; Zhang, H.; Wu, D. Task Offloading and Trajectory Scheduling for UAV-Enabled MEC Networks: An Optimal Transport Theory Perspective. *IEEE Wirel. Commun. Lett.* **2022**, *11*, 150–154. [CrossRef]
20. Liu, B.; Wan, Y.; Zhou, F.; Wu, Q.; Hu, R.Q. Resource Allocation and Trajectory Design for MISO UAV-Assisted MEC Networks. *IEEE Trans. Veh. Technol.* **2022**, *71*, 4933–4948. [CrossRef]
21. Du, Y.; Yang, K.; Wang, K.; Zhang, G.; Zhao, Y.; Chen, D. Joint Resources and Workflow Scheduling in UAV-Enabled Wirelessly-Powered MEC for IoT Systems. *IEEE Trans. Veh. Technol.* **2019**, *68*, 10187–10200. [CrossRef]
22. Ye, Y.; Wei, W.; Geng, D.; He, X. Dynamic Coordination in UAV Swarm Assisted MEC via Decentralized Deep Reinforcement Learning. In Proceedings of the 2020 International Conference on Wireless Communications and Signal Processing (WCSP), Nanjing, China, 21–23 October 2020; pp. 1064–1069. [CrossRef]
23. Nie, Y.; Zhao, J.; Gao, F.; Yu, F.R. Semi-Distributed Resource Management in UAV-Aided MEC Systems: A Multi-Agent Federated Reinforcement Learning Approach. *IEEE Trans. Veh. Technol.* **2021**, *70*, 13162–13173. [CrossRef]
24. Chen, J.; Wu, Q.; Xu, Y.; Qi, N.; Dong, C. A Multi-leader Multi-follower Stackelberg Game for Coalition-based UAV MEC Networks. *IEEE Wirel. Commun. Lett.* **2021**, *10*, 2350–2354. [CrossRef]
25. Goddemeier, N.; Wietfeld, C. Investigation of Air-to-Air Channel Characteristics and a UAV Specific Extension to the Rice Model. In Proceedings of the 2015 IEEE Globecom Workshops (GC Wkshps), San Diego, CA, USA, 6–10 December 2015; pp. 1–5. [CrossRef]
26. Zhou, Y.; Hutu, F.; Villemaud, G. Analysis of a Spatial Modulation System over Time-varying Rician Fading Channel with a CSI Detector. In Proceedings of the 2020 IEEE Radio and Wireless Symposium (RWS), San Antonio, TX, USA, 26–29 January 2020; pp. 217–220. [CrossRef]
27. Lee, S.; Benevides da Costa, D.; Duong, T.Q. Outage probability of non-orthogonal multiple access schemes with partial relay selection. In Proceedings of the 2016 IEEE 27th Annual International Symposium on Personal, Indoor, and Mobile Radio Communications (PIMRC), Valencia, Spain, 4–8 September 2016; pp. 1–6. [CrossRef]
28. Lee, S.; Da Costa, D.B.; Vien, Q.T.; Duong, T.Q.; de Sousa, R.T., Jr. Non-orthogonal multiple access schemes with partial relay selection. *IET Commun.* **2017**, *11*, 846–854. [CrossRef]
29. Jing, Y.; Jafarkhani, H. Single and multiple relay selection schemes and their achievable diversity orders. *IEEE Trans. Wirel. Commun.* **2009**, *8*, 1414–1423. [CrossRef]

Article

LAP and IRS Enhanced Secure Transmissions for 6G-Oriented Vehicular IoT Services

Lingtong Min [1], Jiawei Li [1], Yixin He [2,*] and Qin Si [3]

1 School of Electronics and Information, Northwestern Polytechnical University, Xi'an 710072, China; minlingtong@nwpu.edu.cn (L.M.); jiawei-li@mail.nwpu.edu.cn (J.L.)
2 College of Information Science and Engineering, Jiaxing University, Jiaxing 314001, China
3 Changji Electric Power Supply Company, State Grid Xinjiang Electric Power Co., Ltd., Changji 831100, China; tbea_18599330971@163.com
* Correspondence: yixinhe@zjxu.edu.cn

Abstract: In 6G-oriented vehicular Internet of things (IoT) services, the integration of a low altitude platform (LAP) and intelligent reflecting surfaces (IRS) provides a promising solution to achieve seamless coverage and massive connections at low cost. However, due to the open nature of wireless channels, how to protect the transmission of privacy information in LAP-based IRS symbiotic vehicular networks remains a challenge. Motivated by the above, this paper investigates the LAP and IRS enhanced secure transmission problem in the presence of an eavesdropper. Specifically, we first deploy a fixed LAP equipped with IRS to overcome the blockages and introduce artificial noise against the eavesdropper. Next, we formulate a total secure channel capacity maximization problem by optimizing the phase shift, power distribution coefficient, and channel allocation. To effectively solve the formulated problem, we design an iterative algorithm with polynomial complexity, where the optimization variables are solved in turn. In addition, the complexity and convergence of the proposed iterative algorithm are analyzed theoretically. Finally, numerical results show that our proposed secure transmission scheme outperforms the comparison schemes in terms of the total secure channel capacity.

Keywords: intelligent reflecting surfaces (IRS); low-altitude platform (LAP); secure transmission; total secure channel capacity; vehicular Internet of things (IoT) services

Citation: Min, L.; Li, J.; He, Y.; Si, Q. LAP and IRS Enhanced Secure Transmissions for 6G-Oriented Vehicular IoT Services. *Drones* **2023**, *7*, 414. https://doi.org/10.3390/drones7070414

Academic Editor: Carlos Tavares Calafate

Received: 26 May 2023
Revised: 15 June 2023
Accepted: 20 June 2023
Published: 22 June 2023

1. Introduction

While the dense coverage of fifth-generation (5G) terrestrial networks can satisfy the demands of vehicular Internet of things (IoT) services in hotspots, people still have urgent requirements for ubiquitous connectivity with high data rates in remote areas [1]. Due to the inherent limitations of terrestrial networks, air-to-ground (A2G) communications are envisioned as a promising technique to serve sixth-generation (6G)-oriented vehicular IoT applications [2–4]. As the most representative A2G communications, low-altitude platform (LAP)-enhanced transmissions have lower path loss and higher line-of-sight (LoS) link probability, which can be deployed on demand via a levitation mode to provide seamless and flexible coverage [5–7]. On the other hand, intelligent reflecting surfaces (IRS) with low hardware cost and power consumption can be used for 6G-oriented vehicular IoT services by smartly reconfiguring wireless propagation environments [8–10].

Following the technological advancements of A2G communications, the combination of LAP and IRS has attracted a certain amount of attention [11]. Generally, this combination can be divided into two cases, i.e., mobile IRS schemes [12,13] and fixed IRS schemes [14–16]. However, in some practical vehicular network (VNet) scenarios (e.g., emergency rescues), mobile IRS schemes may be impractical. The reason is that the payload and flight time of LAPs with mobile capability are extremely limited. According to the above discussion, the authors in [14] derived the channel gain lower bound for LAP and IRS collaborative communications. Inspired by this work, the researchers in [15] investigated the sum rate maximization problem of LAP-aided IRS networks by optimizing the phase shift and LAP altitude. Moreover, by using quasi-stationary LAPs, the IRS-assisted multi-layer aerial architecture was proposed in [16], which pointed out a promising direction for 6G-oriented vehicular IoT services. Furthermore, in order to improve the channel capacity, more works focused on the network optimization problems, including beamforming, resource (e.g., power and spectrum) allocation, and energy efficiency optimization [17–20]. We have summarized these works in Table 1.

Table 1. Summary of key contributions and limitations of existing works on UAV-aided RIS-assisted IoT networks.

Reference	Key Contributions	Limitation
[14]	The channel gain lower bound for LAP and IRS collaborative communications was derived.	These works make an implicit assumption that LAP-based IRS symbiotic vehicular networks (VNets) are secure. In LAP-based IRS symbiotic VNets, the privacy information is susceptible to eavesdropping due to the open nature of A2G channels.
[15]	The sum rate maximization problem of LAP-aided IRS networks was investigated, where the phase shift and LAP altitude were optimized.	
[16]	The IRS-assisted multi-layer aerial architecture was proposed.	
[17–20]	By considering the beamforming, resource allocation, and energy efficiency, the channel capacity was improved.	

Although the above works present optimization policies and models of LAP and IRS enhanced transmissions, these works make an implicit assumption that LAP-based IRS symbiotic vehicular networks (VNets) are secure. In LAP-based IRS symbiotic VNets, the privacy information is susceptible to eavesdropping due to the open nature of A2G channels [21]. Traditionally, the network security is protected by upper-layer encryption methods. However, such encryption algorithms and key allocation strategies will significantly improve the complexity of the system [22]. Faced with the above challenges, by using the wireless channel characterizations, the physical layer security (PLS) technique can be regarded as a promising alternative technique, which can be widely applied to 6G-oriented vehicular IoT services to ensure information security [23]. Therefore, under the constraints of network security, how to improve the total secure channel capacity of LAP and IRS enhanced transmissions is a key technical difficulty.

Motivated by the above, this paper investigates the secure transmission problem in LAP-based IRS symbiotic VNets in the presence of an eavesdropper. First, we deploy a fixed LAP equipped with IRS to overcome the blockages and exploit artificial noise (AN) to interfere with the eavesdropper. Next, aiming to maximize the total secure channel capacity, we formulate this problem as a mixed-integer and non-convex program. To effectively solve the formulated problem, an iterative algorithm with polynomial complexity is proposed, where the phase shift, power distribution coefficient, and channel allocation are optimized in turn. Then, we theoretically analyze the complexity and convergence of the proposed iterative algorithm. Finally, numerical results show that the proposed secure transmission scheme significantly improves the total secure channel capacity against the current works [2,23] and baseline scheme. In addition, the influence of the number of reflection elements is discussed. The above results are a meaningful guide for improving the quality of service (QoS) of 6G-oriented vehicular IoT services.

The rest of this article is organized as follows. Section 2 introduces the network model and presents the total secure channel capacity maximization problem. Then, in Section 3, we design an iterative algorithm with polynomial complexity to solve the formulated problem. Simulation results are presented in Section 4. Finally, Section 5 concludes the paper.

2. Network Model and Problem Formulation

Figure 1 illustrates the considered LAP-based IRS symbiotic VNet, which consists of a remote base station (RBS), a fixed LAP equipped with IRS, U legitimate vehicle users, and an eavesdropper. The set of legitimate vehicle users is denoted as $\mathcal{U} = \{1, 2, ..., U\}$. We assume that there is no direct communication link between the RBS and the legitimate vehicle user/eavesdropper due to obstacles [8]. Under this condition, we adopt the LAP equipped with IRS to enhance transmissions. The IRS can be controlled by an intelligent controller.

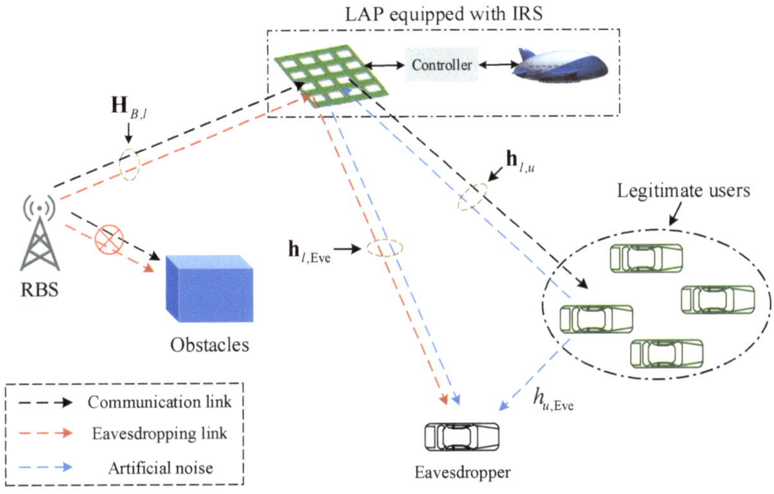

Figure 1. LAP-based IRS symbiotic VNets.

It is assumed that the IRS has G_h horizontal reflection elements and G_v vertical reflection elements, denoted as $\mathcal{G} = \{1, 2, ..., G\}$, where $G = G_h \times G_v$. Moreover, the RBS has N antennas and K channels, denoted as $\mathcal{K} = \{1, 2, ..., K\}$, where $K \geq U$. Let $\mathbf{K} = \{k_u | \forall k \in \mathcal{K}, \forall u \in \mathcal{U}\}$ denote the channel allocation policy. If the u-th ($\forall u \in \mathcal{U}$) legitimate vehicle user occupies the k-th ($\forall k \in \mathcal{K}$) channel, $k_u = 1$; otherwise, $k_u = 0$. Furthermore, each legitimate vehicle user with self-interference cancellation capability has two antennas that can implement full-duplex communication. Meanwhile, we assume that the AN emitted by the legitimate vehicle users will not affect the received signals, and the eavesdropper is equipped with a single antenna [24]. Since the total power $P_u^{\max}$ of the system is limited, the RBS and the u-th legitimate vehicle user need to negotiate to decide the transmitted power P_u^{down} of RBS (downlink) and the transmitted power P_u^{up} of AN (uplink). Especially, as discussed in [21], the channel is assumed to have reciprocity. Likewise, it is assumed that the channel state information (CSI) associated with the eavesdropper/IRS is available. The reason is that even for a passive eavesdropper, it can also estimate its CSI through local oscillator power inadvertently leaked from the eavesdropper's receiver radio frequency frontend [25]. Since the investigated scenario is highly dynamic, imperfect estimation of the reflection phases and phase errors are possible with respect to the link between LAP and ground nodes. In this situation, the CSI of the LAP vehicle links needs to be periodically reported to the RBS with a feedback period. According to [26], the first-order Gauss–Markov process can be utilized to estimate the CSI of LAP-vehicle links.

According to the above description, the received signal y_u of the u-th legitimate vehicle user can be expressed as

$$y_u = \mathbf{h}_{l,u}^{\text{H}} \mathbf{\Phi} \mathbf{H}_{B,l} P_u^{\text{down}} s_u + \eta_u, \tag{1}$$

where $\mathbf{h}_{l,u}^{\mathrm{H}}$ is the channel from IRS to the u-th legitimate vehicle user, $\mathbf{h}_{l,u} \in \mathbb{C}^{G \times 1}$; $\mathbf{\Phi}$ is the phase shift matrix, and $\mathbf{\Phi} = \mathrm{diag}\left\{e^{jX_g}\right\}$, $X_g \in [0, 2\pi)$, where X_g is the phase shift of the g-th ($\forall g \in \mathcal{G}$) reflection element; $\mathbf{H}_{B,l}$ is the channel from the RBS to IRS, $\mathbf{H}_{B,l} \in \mathbb{C}^{G \times N}$; s_u is the transmitted signal from the RBS for the u-th legitimate vehicle user with zero mean and normalized power; η_u is the noise received by the u-th legitimate vehicle user, $\eta_u \sim \mathcal{CN}\left(0, \sigma_u^2\right)$, where σ_u^2 is the noise power of the u-th legitimate vehicle user.

Similarly, the received signal y_u^{Eve} of the eavesdropper is

$$
\begin{aligned}
y_u^{\mathrm{Eve}} = {} & \mathbf{h}_{l,\mathrm{Eve}}^{\mathrm{H}} \mathbf{\Phi} \mathbf{H}_{B,l} P_u^{\mathrm{down}} s_u \\
& + \mathbf{h}_{l,\mathrm{Eve}}^{\mathrm{H}} \mathbf{\Phi} \mathbf{h}_{l,u} P_u^{\mathrm{up}} a_u + h_{u,\mathrm{Eve}} P_u^{\mathrm{up}} a_u \\
& + \eta_{\mathrm{Eve}},
\end{aligned}
\tag{2}
$$

where $\mathbf{h}_{l,\mathrm{Eve}}^{\mathrm{H}}$ is the channel from IRS to the eavesdropper, $\mathbf{h}_{l,\mathrm{Eve}} \in \mathbb{C}^{G \times 1}$; a_u is the AN signal emitted by the u-th legitimate vehicle user with zero mean and normalized power; $h_{u,\mathrm{Eve}}$ is the channel from the u-th legitimate vehicle user to the eavesdropper, $h_{u,\mathrm{Eve}} \in \mathbb{C}$; η_{Eve} is the noise received by the eavesdropper, $\eta_{\mathrm{Eve}} \sim \mathcal{CN}\left(0, \sigma_{\mathrm{Eve}}^2\right)$, where σ_{Eve}^2 is the noise power of the eavesdropper.

According to (1), the information rate $R_{B,u}\left(\mathbf{\Phi}, P_u^{\mathrm{down}}, k_u\right)$ of the u-th legitimate vehicle user is given by

$$
R_{B,u}\left(\mathbf{\Phi}, P_u^{\mathrm{down}}, k_u\right) = \sum_{k=1}^{K} B_u k_u \log_2(1 + \mathrm{SINR}_{B,u}),
\tag{3}
$$

where B_u is the channel bandwidth of the u-th legitimate vehicle user, and $\mathrm{SINR}_{B,u}$ can be expressed as

$$
\mathrm{SINR}_{B,u} = \frac{P_u^{\mathrm{down}} \left| \mathbf{h}_{l,u}^{\mathrm{H}} \mathbf{\Phi} \mathbf{H}_{B,l} \right|^2}{\sigma_u^2}.
\tag{4}
$$

The information rate $R_{u,\mathrm{Eve}}\left(\mathbf{\Phi}, P_u^{\mathrm{down}}, P_u^{\mathrm{up}}, k_u\right)$ of the eavesdropper is given by

$$
R_{u,\mathrm{Eve}}\left(\mathbf{\Phi}, P_u^{\mathrm{down}}, P_u^{\mathrm{up}}, k_u\right) = \sum_{k=1}^{K} B_{\mathrm{Eve}} k_u \log_2(1 + \mathrm{SINR}_{u,\mathrm{Eve}}),
\tag{5}
$$

where $\mathrm{SINR}_{u,\mathrm{Eve}}$ can be expressed as

$$
\mathrm{SINR}_{u,\mathrm{Eve}} = \frac{P_u^{\mathrm{down}} \left| \mathbf{h}_{l,\mathrm{Eve}}^{\mathrm{H}} \mathbf{\Phi} \mathbf{H}_{B,l} \right|^2}{P_u^{\mathrm{up}} \left| \mathbf{h}_{l,\mathrm{Eve}}^{\mathrm{H}} \mathbf{\Phi} \mathbf{h}_{u,l} \right|^2 + P_u^{\mathrm{up}} |h_{u,\mathrm{Eve}}|^2 + \sigma_{\mathrm{Eve}}^2}.
\tag{6}
$$

For notational simplicity, we define Ψ_u as the power distribution coefficient of the u-th legitimate vehicle user. Since $P_u^{\mathrm{max}} = P_u^{\mathrm{down}} + P_u^{\mathrm{up}}$, we have $P_u^{\mathrm{down}} = \Psi_u P_u^{\mathrm{max}}$ and $P_u^{\mathrm{up}} = (1 - \Psi_u) P_u^{\mathrm{max}}$. According to (3) and (5), in LAP-based IRS symbiotic VNets, the secure channel capacity R_u^{sec} of the u-th legitimate vehicle user is

$$
R_u^{\mathrm{sec}}(\mathbf{\Phi}, \Psi_u, k_u) = [R_{B,u}(\mathbf{\Phi}, \Psi_u, k_u) - R_{u,\mathrm{Eve}}(\mathbf{\Phi}, \Psi_u, k_u)]^+,
\tag{7}
$$

where $[\cdot]^+$ represents $\max\{\cdot, 0\}$.

Therefore, the total secure channel capacity $R_{\mathrm{tot}}^{\mathrm{sec}}(\mathbf{\Phi}, \mathbf{\Psi}, \mathbf{K})$ can be expressed as

$$
R_{\mathrm{tot}}^{\mathrm{sec}}(\mathbf{\Phi}, \mathbf{\Psi}, \mathbf{K}) = \sum_{u=1}^{U} R_u^{\mathrm{sec}}(\mathbf{\Phi}, \Psi_u, k_u),
\tag{8}
$$

where $\mathbf{\Psi} = \{\Psi_u | \forall u \in \mathcal{U}\}$.

By optimizing the power distribution coefficient $\mathbf{\Psi}$, phase shift $\mathbf{\Phi}$, and channel allocation policy $\mathbf{K}$, we aim to maximize the total secure channel capacity $R_{\text{tot}}^{\text{sec}}(\mathbf{\Phi}, \mathbf{\Psi}, \mathbf{K})$. The total secure channel capacity maximization problem can be mathematically formulated as

$$\mathbf{P1}: \max_{\mathbf{\Phi}, \mathbf{\Psi}, \mathbf{K}} \quad R_{\text{tot}}^{\text{sec}}(\mathbf{\Phi}, \mathbf{\Psi}, \mathbf{K}) \tag{9a}$$

$$\text{s.t.} \quad 0 < \Psi_u \leq 1, \forall u, \tag{9b}$$

$$\sum_{u=1}^{U} P_u^{\max} = P_{\text{tot}}, \tag{9c}$$

$$\mathbf{\Phi} = \text{diag}\left\{e^{jX_g}\right\}, \forall g, \tag{9d}$$

$$\left|e^{jX_g}\right| = 1, \ X_g \in [0, 2\pi), \forall g, \tag{9e}$$

$$k_u \in \{0,1\}, \sum_{k=1}^{K} k_u = 1, \sum_{u=1}^{U} k_u \leq 1, \forall k, u, \tag{9f}$$

where P_{tot} is the total power of the system.

The main notations are summarized in Table 2.

Table 2. Definition of parameters.

Parameter	Definition
U	Number of legitimate vehicle users
G	Number of reflection elements
N	Number of antennas
K	Number of channels
$P_u^{\max}$	Total power
P_u^{down}	Transmitted power of the RBS
P_u^{up}	Transmitted power of AN
y_u	Received signal of the u-th legitimate vehicle user
$\mathbf{h}_{l,u}^{H}$	Channel from IRS to the u-th legitimate vehicle user
$\mathbf{\Phi}$	Phase shift matrix
$\mathbf{H}_{B,l}$	Channel from the RBS to IRS
s_u	Transmitted signal from the RBS for the u-th legitimate vehicle user
$\mathbf{h}_{l,\text{Eve}}^{H}$	Channel from IRS to the eavesdropper
a_u	AN signal emitted by the u-th legitimate vehicle user
$h_{u,\text{Eve}}$	Channel from the u-th legitimate vehicle user to the eavesdropper
η_{Eve}	Noise received by the eavesdropper
$R_{R,u}$	Information rate of the u-th legitimate vehicle user
B_u	Channel bandwidth of the u-th legitimate vehicle user
$R_{u,\text{Eve}}$	Information rate of the eavesdropper
Ψ_u	Power distribution coefficient of the u-th legitimate vehicle user
R_u^{sec}	Secure channel capacity of the u-th legitimate vehicle user
$R_{\text{tot}}^{\text{sec}}$	Total secure channel capacity
P_{tot}	Total power of the system

In P1, (9b) and (9c) together limit the transmitted power of the RBS and legitimate vehicle users; (9d) and (9e) constrain the IRS phase shift; (9f) defines the channel allocation mode of multiple legitimate vehicle users. Since $\left|e^{jX_g}\right| = 1$ and $k_u \in \{0,1\}$, P1 is a mixed-integer and non-convex program. It is hard to obtain a global optimal solution for P1. Therefore, in Section 3, we propose an iterative algorithm, where $\mathbf{\Psi}$, $\mathbf{\Phi}$, and $\mathbf{K}$ are solved in turn.

3. Total Secure Channel Capacity Maximization Scheme

3.1. Phase Shift Optimization

In this stage, given $\mathbf{\Psi}$ and $\mathbf{K}$, the phase shift optimization problem P2 is given by

$$\textbf{P2}: \max_{\mathbf{\Phi}} \ R_{\text{tot}}^{\text{sec}}(\mathbf{\Phi}) = \sum_{u=1}^{U} R_{u}^{\text{sec}}(\mathbf{\Phi}) \tag{10a}$$

$$\text{s.t.} \ \ \mathbf{\Phi} = \text{diag}\left\{ e^{jX_g} \right\}, \forall g, \tag{10b}$$

$$\left| e^{jX_g} \right| = 1, \ X_g \in [0, 2\pi), \forall g. \tag{10c}$$

Next, an intermediate variable $\mathbf{X}$ is introduced, where $\mathbf{X} = \left[e^{jX_1}, \ldots, e^{jX_G} \right]^{\text{H}}$. We have $\mathbf{\Phi} = \text{diag}\left\{ \mathbf{X}^{\text{H}} \right\}$. Let $\mathbf{A}_{l,u} = \text{diag}\left\{ \mathbf{h}_{l,u}^{\text{H}} \right\}$ and $\mathbf{B}_{l,\text{Eve}} = \text{diag}\left\{ \mathbf{h}_{l,\text{Eve}}^{\text{H}} \right\}$. Based on the property of matrix transformation (i.e., $\mathbf{a}^{\text{H}}\mathbf{\Phi}\mathbf{b} = \mathbf{X}^{\text{H}}\text{diag}\{\mathbf{a}^{\text{H}}\}\mathbf{b}$), $\text{SINR}_{u}^{\text{sec}}(X)$ can be recast as

$$\text{SINR}_{u}^{\text{sec}}(X) = \frac{\mathbf{X}^{\text{H}} w_1 \mathbf{X}}{\mathbf{X}^{\text{H}}(w_2 + w_3 + w_4)\mathbf{X}} \times \mathbf{X}^{\text{H}}(w_5 + w_6)\mathbf{X}, \tag{11}$$

where $R_{u}^{\text{sec}}(\mathbf{\Phi}) = \log_2[1 + \text{SINR}_{u}^{\text{sec}}(X)]$, and $\mathbf{I}_G$ is the unit matrix. In addition, we have

$$w_1 = \left(\frac{1}{G}\right)\mathbf{I}_G + \frac{\mathbf{\Psi}_u P_u^{\max}\left(\mathbf{A}_{l,u}\mathbf{H}_{B,l}\mathbf{H}_{B,l}^{\text{H}}\mathbf{A}_{l,u}^{\text{H}}\right)}{\sigma_u^2}, \tag{12}$$

$$w_2 = \mathbf{\Psi}_u P_u^{\max}\left(\mathbf{B}_{l,\text{Eve}}\mathbf{H}_{B,l}\mathbf{H}_{B,l}^{\text{H}}\mathbf{B}_{l,\text{Eve}}^{\text{H}}\right), \tag{13}$$

$$w_3 = (1 - \mathbf{\Psi}_u)P_u^{\max}\left(\mathbf{B}_{l,\text{Eve}}\mathbf{h}_{l,u}\mathbf{h}_{l,u}^{\text{H}}\mathbf{B}_{l,\text{Eve}}^{\text{H}}\right), \tag{14}$$

$$w_4 = \left[\frac{(1 - \mathbf{\Psi}_u)P_u^{\max}|h_{u,\text{Eve}}|^2 + \sigma_{\text{Eve}}^2}{G}\right]\mathbf{I}_G, \tag{15}$$

$$w_5 = (1 - \mathbf{\Psi}_u)P_u^{\max}\left(\mathbf{B}_{l,\text{Eve}}\mathbf{h}_{l,u}\mathbf{h}_{l,u}^{\text{H}}\mathbf{B}_{l,\text{Eve}}^{\text{H}}\right), \tag{16}$$

and

$$w_6 = \frac{\left[(1 - \mathbf{\Psi}_u)P_u^{\max}|h_{u,\text{Eve}}|^2 + \sigma_{\text{Eve}}^2\right]}{G}\mathbf{I}_G. \tag{17}$$

To tackle P2, we further introduce three intermediate variables (α, β, and χ), which can be respectively expressed as

$$\alpha = \left(\frac{1}{G}\right)\mathbf{I}_G + \frac{\mathbf{\Psi}_u P_u^{\max}\left(\mathbf{A}_{l,u}\mathbf{H}_{B,l}\mathbf{H}_{B,l}^{\text{H}}\mathbf{A}_{l,u}^{\text{H}}\right)}{\sigma_u^2}, \tag{18}$$

$$\beta = \mathbf{B}_{l,\text{Eve}}\mathbf{H}_{B,l}\mathbf{H}_{B,l}^{\text{H}}\mathbf{B}_{l,\text{Eve}}^{\text{H}}, \tag{19}$$

and

$$\chi = (1 - \mathbf{\Psi}_u)P_u^{\max}\left(\mathbf{B}_{l,\text{Eve}}\mathbf{h}_{l,u}\mathbf{h}_{l,u}^{\text{H}}\mathbf{B}_{l,\text{Eve}}^{\text{H}}\right) \tag{20}$$

$$+ \left[(1 - \mathbf{\Psi}_u)P_u^{\max}|h_{u,\text{Eve}}|^2 + \sigma_{\text{Eve}}^2\right]G^{-1}\mathbf{I}_G.$$

Then, we simplify (11), and $\mathrm{SINR}_u^{\mathrm{sec}}(X)$ can be rewritten as

$$\mathrm{SINR}_u^{\mathrm{sec}}(X) = \frac{\mathrm{tr}\left(\alpha \mathbf{X}\mathbf{X}^{\mathrm{H}}\right)\mathrm{tr}\left(\beta \mathbf{X}\mathbf{X}^{\mathrm{H}}\right)}{\mathrm{tr}(\chi \mathbf{X}\mathbf{X}^{\mathrm{H}})}, \tag{21}$$

where $\mathrm{tr}(\cdot)$ is the trace of matrix.

To satisfy (10b) and (10c), we have

$$\begin{cases} \mathrm{rank}\left(\mathbf{X}\mathbf{X}^{\mathrm{H}}\right) = 1, \\ \left(\mathbf{X}\mathbf{X}^{\mathrm{H}}\right)_{g,g} = 1, \ \forall g \in \mathcal{G}. \end{cases} \tag{22}$$

Afterward, a slack variable $\Im$ is introduced. By using $\Im$, P2 can be rewritten as

$$\mathbf{P3}: \min_{\mathbf{X}} \ \Im \tag{23a}$$

$$\text{s.t.} \ e^{\log_2\left[\mathrm{tr}\left(\chi \mathbf{X}\mathbf{X}^{\mathrm{H}}\right)\right] - \log_2\left[\mathrm{tr}\left(\beta \mathbf{X}\mathbf{X}^{\mathrm{H}}\right)\right] - \log_2\left[\mathrm{tr}\left(\alpha \mathbf{X}\mathbf{X}^{\mathrm{H}}\right)\right]} - \Im \leq 0, \tag{23b}$$

$$\mathrm{tr}\left(\alpha \mathbf{X}\mathbf{X}^{\mathrm{H}}\right) \geq e^{\log_2\left[\mathrm{tr}\left(\alpha \mathbf{X}\mathbf{X}^{\mathrm{H}}\right)\right]}, \tag{23c}$$

$$\mathrm{tr}\left(\beta \mathbf{X}\mathbf{X}^{\mathrm{H}}\right) \geq e^{\log_2\left[\mathrm{tr}\left(\beta \mathbf{X}\mathbf{X}^{\mathrm{H}}\right)\right]}, \tag{23d}$$

$$\mathrm{tr}\left(\chi \mathbf{X}\mathbf{X}^{\mathrm{H}}\right) \leq e^{\log_2\left[\mathrm{tr}\left(\chi \mathbf{X}\mathbf{X}^{\mathrm{H}}\right)\right]}, \tag{23e}$$

$$\mathrm{rank}\left(\mathbf{X}\mathbf{X}^{\mathrm{H}}\right) = 1, \tag{23f}$$

$$\left(\mathbf{X}\mathbf{X}^{\mathrm{H}}\right)_{g,g} = 1, \ \forall g. \tag{23g}$$

By using the sequential convex approximation (SCA) method, we take the first-order Taylor expansion of (23e), which can be expressed as

$$\begin{aligned} & e^{\log_2\left[\mathrm{tr}\left(\chi \mathbf{X}\mathbf{X}^{\mathrm{H}}\right)\right] - \Delta} + e^{\log_2\left[\mathrm{tr}\left(\chi \mathbf{X}\mathbf{X}^{\mathrm{H}}\right)\right] - \Delta} \ln[e(\Delta)] \\ & \leq e^{\log_2\left[\mathrm{tr}\left(\chi \mathbf{X}\mathbf{X}^{\mathrm{H}}\right)\right]} \Rightarrow \mathrm{tr}\left(\chi \mathbf{X}\mathbf{X}^{\mathrm{H}}\right) \\ & \leq e^{\log_2\left[\mathrm{tr}\left(\chi \mathbf{X}\mathbf{X}^{\mathrm{H}}\right)\right] - \Delta}(1 + \Delta), \end{aligned} \tag{24}$$

where Δ is a minuscule negative value. Therefore, $\left\{\log_2\left[\mathrm{tr}\left(\chi \mathbf{X}\mathbf{X}^{\mathrm{H}}\right)\right] - \Delta\right\}$ can be considered an approximation of $\log_2\left[\mathrm{tr}\left(\chi \mathbf{X}\mathbf{X}^{\mathrm{H}}\right)\right]$.

According to (24), we adopt the semi-definite relaxation (SDR) method to relax (23f). Under this condition, P3 can be relaxed as

$$\mathbf{P4}: \min_{\mathbf{X}} \ \Im \tag{25a}$$

$$\text{s.t.} \ (23b) - (23d), (23g) \tag{25b}$$

$$\mathrm{tr}\left(\chi \mathbf{X}\mathbf{X}^{\mathrm{H}}\right) \leq e^{\log_2\left[\mathrm{tr}\left(\chi \mathbf{X}\mathbf{X}^{\mathrm{H}}\right)\right] - \Delta}(1 + \Delta). \tag{25c}$$

Obviously, P4 is a convex optimization problem, which can be solved by the convex problem solver. However, since the SDR method is used to relax (23f), the obtained phase shift cannot always satisfy $\mathrm{rank}\left(\mathbf{X}\mathbf{X}^{\mathrm{H}}\right) = 1$ [27]. Therefore, the Gaussian random process is employed to acquire the approximate solution, which satisfies rank-one, i.e., $\mathrm{rank}\left(\mathbf{X}\mathbf{X}^{\mathrm{H}}\right) = 1$.

3.2. Power Distribution Coefficient Optimization

In this stage, since it is assumed that $\mathbf{\Phi}$ and $\mathbf{K}$ have been determined, the power distribution problem can be expressed as

$$\mathbf{P5} : \max_{\boldsymbol{\Psi}} \ R_{\text{tot}}^{\text{sec}}(\boldsymbol{\Psi}) = \sum_{u=1}^{U} R_u^{\text{sec}}(\boldsymbol{\Psi}_u) \tag{26a}$$

$$\text{s.t.} \ 0 < \Psi_u \leq 1, \forall u, \tag{26b}$$

$$\sum_{u=1}^{U} P_u^{\max} = P_{\text{tot}}. \tag{26c}$$

In P5, $\text{SINR}_u^{\text{sec}}(\Psi_u)$ can be rewritten as

$$\text{SINR}_u^{\text{sec}}(\Psi_u) = \frac{f_1 f_2}{P_u^{\max}(f_3 - f_4) + \sigma_{\text{Eve}}^2}, \tag{27}$$

where $R_u^{\text{sec}}(\Psi_u) = \log_2[1 + \text{SINR}_u^{\text{sec}}(\Psi_u)]$. In addition, we can obtain

$$f_1 = \frac{1 + \Psi_u P_u^{\max}\left(\mathbf{H}_{B,l}^{\text{H}}\boldsymbol{\Phi}^{\text{H}}\mathbf{h}_{l,u}\mathbf{h}_{l,u}^{\text{H}}\boldsymbol{\Phi}\mathbf{H}_{B,l}\right)}{\sigma_u^2}, \tag{28}$$

$$f_2 = (1 - \Psi_u)\left(P_u^{\max}\left|\mathbf{h}_{l,\text{Eve}}^{\text{H}}\boldsymbol{\Phi}\mathbf{h}_{l,u}\right|^2 + P_u^{\max}|h_{u,\text{Eve}}|^2\right) + \sigma_{\text{Eve}}^2, \tag{29}$$

$$f_3 = \left|\mathbf{h}_{l,\text{Eve}}^{\text{H}}\boldsymbol{\Phi}\mathbf{h}_{l,u}\right|^2 + |h_{u,\text{Eve}}|^2, \tag{30}$$

and

$$f_4 = \Psi_u\left[\left|\mathbf{h}_{l,\text{Eve}}^{\text{H}}\boldsymbol{\Phi}\mathbf{h}_{l,u}\right|^2 + |h_{u,\text{Eve}}|^2 - \left(\mathbf{H}_{B,l}^{\text{H}}\boldsymbol{\Phi}^{\text{H}}\mathbf{h}_{l,\text{Eve}}\mathbf{h}_{l,\text{Eve}}^{\text{H}}\boldsymbol{\Phi}\mathbf{H}_{B,l}\right)\right]. \tag{31}$$

Lemma 1. *The objective function $\text{SINR}_u^{\text{sec}}(\Psi_u)$ is a convex function.*

Proof of Lemma 1. The first-order derivative of $\text{SINR}_u^{\text{sec}}(\Psi_u)$ with respect to Ψ_u is derived as

$$\frac{\partial \text{SINR}_u^{\text{sec}}(\Psi_u)}{\partial \Psi_u} = (\Psi_u)^2 \times (y_1 - y_2)$$

$$-\frac{2\left(\mathbf{H}_{B,l}^{\text{H}}\boldsymbol{\Phi}^{\text{H}}\mathbf{h}_{l,u}\mathbf{h}_{l,u}^{\text{H}}\boldsymbol{\Phi}\mathbf{H}_{B,l}\right)\left[P_u^{\max}\left(\left|\mathbf{h}_{l,\text{Eve}}^{\text{H}}\boldsymbol{\Phi}\mathbf{h}_{l,u}\right|^2 + |h_{u,\text{Eve}}|^2\right) + \sigma_{\text{Eve}}^2\right]}{\sigma_u^2(P_u^{\max})^{-2}\left[\left(\left|\mathbf{h}_{l,\text{Eve}}^{\text{H}}\boldsymbol{\Phi}\mathbf{h}_{l,u}\right|^2 + |h_{u,\text{Eve}}|^2\right)\right]^{-1}(\Psi_u)^{-1}}$$

$$+\frac{P_u^{\max}\left(\mathbf{H}_{B,l}^{\text{H}}\boldsymbol{\Phi}^{\text{H}}\mathbf{h}_{l,u}\mathbf{h}_{l,u}^{\text{H}}\boldsymbol{\Phi}\mathbf{H}_{B,l}\right)}{\sigma_u^2\left[P_u^{\max}\left(\left|\mathbf{h}_{l,\text{Eve}}^{\text{H}}\boldsymbol{\Phi}\mathbf{h}_{l,u}\right|^2 + |h_{u,\text{Eve}}|^2\right) + \sigma_{\text{Eve}}^2\right]^{-2}}$$

$$- P_u^{\max}\left(\mathbf{H}_{B,l}^{\text{H}}\boldsymbol{\Phi}^{\text{H}}\mathbf{h}_{l,\text{Eve}}\mathbf{h}_{l,\text{Eve}}^{\text{H}}\boldsymbol{\Phi}\mathbf{H}_{B,l}\right)\left[P_u^{\max}\left(\left|\mathbf{h}_{l,\text{Eve}}^{\text{H}}\boldsymbol{\Phi}\mathbf{h}_{l,u}\right|^2 + |h_{u,\text{Eve}}|^2\right) + \sigma_{\text{Eve}}^2\right], \tag{32}$$

where

$$y_1 = \frac{(P_u^{\max})^3\left(\mathbf{H}_{B,l}^{\text{H}}\boldsymbol{\Phi}^{\text{H}}\mathbf{h}_{l,u}\mathbf{h}_{l,u}^{\text{H}}\boldsymbol{\Phi}\mathbf{H}_{B,l}\right)}{\sigma_u^2\left(\left|\mathbf{h}_{l,\text{Eve}}^{\text{H}}\boldsymbol{\Phi}\mathbf{h}_{l,u}\right|^2 + |h_{u,\text{Eve}}|^2\right)^{-2}}, \tag{33}$$

and

$$y_2 = \frac{(P_u^{\max})^3 \left(\mathbf{H}_{B,l}^{H} \mathbf{\Phi}^{H} \mathbf{h}_{l,u} \mathbf{h}_{l,u}^{H} \mathbf{\Phi} \mathbf{H}_{B,l} \right) \left(\left| \mathbf{h}_{l,\mathrm{Eve}}^{H} \mathbf{\Phi} \mathbf{h}_{l,u} \right|^2 + |h_{u,\mathrm{Eve}}|^2 \right)}{\sigma_u^2 \left(\mathbf{H}_{B,l}^{H} \mathbf{\Phi}^{H} \mathbf{h}_{l,\mathrm{Eve}} \mathbf{h}_{l,\mathrm{Eve}}^{H} \mathbf{\Phi} \mathbf{H}_{B,l} \right)^{-1}}. \tag{34}$$

The second-order derivative of $\mathrm{SINR}_u^{\mathrm{sec}}(\Psi_u)$ with respect to Ψ_u is derived as

$$\frac{\partial^2 \mathrm{SINR}_u^{\mathrm{sec}}(\Psi_u)}{\partial(\Psi_u)^2} = \frac{-2g_1 g_2}{(\Psi_u g_3 + g_4)^3}, \tag{35}$$

where

$$g_1 = P_u^{\max} \left(\mathbf{H}_{B,l}^{H} \mathbf{\Phi}^{H} \mathbf{h}_{l,\mathrm{Eve}} \mathbf{h}_{l,\mathrm{Eve}}^{H} \mathbf{\Phi} \mathbf{H}_{B,l} \right) \tag{36}$$
$$\times \left[P_u^{\max} \left(\left| \mathbf{h}_{l,\mathrm{Eve}}^{H} \mathbf{\Phi} \mathbf{h}_{l,u} \right|^2 + |h_{u,\mathrm{Eve}}|^2 \right) + \sigma_{\mathrm{Eve}}^2 \right],$$

$$g_2 = P_u^{\max} \left(\left| \mathbf{h}_{l,\mathrm{Eve}}^{H} \mathbf{\Phi} \mathbf{h}_{l,u} \right|^2 + |h_{u,\mathrm{Eve}}|^2 \right)$$
$$- P_u^{\max} \left(\mathbf{H}_{B,l}^{H} \mathbf{\Phi}^{H} \mathbf{h}_{l,\mathrm{Eve}} \mathbf{h}_{l,\mathrm{Eve}}^{H} \mathbf{\Phi} \mathbf{H}_{B,l} \right) \tag{37}$$
$$- \frac{P_u^{\max} \left(\mathbf{H}_{B,l}^{H} \mathbf{\Phi}^{H} \mathbf{h}_{l,u} \mathbf{h}_{l,u}^{H} \mathbf{\Phi} \mathbf{H}_{B,l} \right)}{\sigma_u^2 \left(P_u^{\max} \left(\left| \mathbf{h}_{l,\mathrm{Eve}}^{H} \mathbf{\Phi} \mathbf{h}_{l,u} \right|^2 + |h_{u,\mathrm{Eve}}|^2 \right) + \sigma_{\mathrm{Eve}}^2 \right)^{-1}},$$

$$g_3 = P_u^{\max} \left(\mathbf{H}_{B,l}^{H} \mathbf{\Phi}^{H} \mathbf{h}_{l,\mathrm{Eve}} \mathbf{h}_{l,\mathrm{Eve}}^{H} \mathbf{\Phi} \mathbf{H}_{B,l} \right)$$
$$- P_u^{\max} \left(\left| \mathbf{h}_{l,\mathrm{Eve}}^{H} \mathbf{\Phi} \mathbf{h}_{l,u} \right|^2 + |h_{u,\mathrm{Eve}}|^2 \right), \tag{38}$$

and

$$g_4 = P_u^{\max} \left(\left| \mathbf{h}_{l,\mathrm{Eve}}^{H} \mathbf{\Phi} \mathbf{h}_{l,u} \right|^2 + |h_{u,\mathrm{Eve}}|^2 \right) + \sigma_{\mathrm{Eve}}^2. \tag{39}$$

We can obtain $g_1 > 0$, $g_2 > 0$, and $\Psi_u g_3 + g_4 > 0$. Therefore, we have $\frac{-2g_1 g_2}{(\Psi_u g_3 + g_4)^3} < 0$, i.e., $\frac{\partial^2 \mathrm{SINR}_u^{\mathrm{sec}}(\Psi_u)}{\partial(\Psi_u)^2} < 0$. In this case, the objective function $R_u^{\mathrm{sec}}(\Psi_u)$ can be regarded as a convex function, thus proving Lemma 1. $\square$

According to Lemma 1, when $\frac{\partial^2 \mathrm{SINR}_u^{\mathrm{sec}}(\Psi_u)}{\partial(\Psi_u)^2} = 0$, we can obtain the maximum of $R_u^{\mathrm{sec}}(\Psi_u)$. As can be seen from (32), $\frac{\partial^2 \mathrm{SINR}_u^{\mathrm{sec}}(\Psi_u)}{\partial(\Psi_u)^2}$ is a quadratic function with respect to Ψ_u. Therefore, $(\Psi_u)^*$ is derived as

$$(\Psi_u)^* = -\frac{g_4}{g_3} \pm \frac{\sqrt{g_1 g_2 g_5}}{g_3 g_5}, \tag{40}$$

where

$$g_5 = \frac{(P_u^{\max})^2 \left(\mathbf{H}_{B,l}^{H} \mathbf{\Phi}^{H} \mathbf{h}_{l,u} \mathbf{h}_{l,u}^{H} \mathbf{\Phi} \mathbf{H}_{B,l} \right)}{\sigma_u^2}$$
$$\times \left(\left| \mathbf{h}_{l,\mathrm{Eve}}^{H} \mathbf{\Phi} \mathbf{h}_{l,u} \right|^2 + |h_{u,\mathrm{Eve}}|^2 \right). \tag{41}$$

However, for $(\Psi_u)^* = -\frac{g_4}{g_3} + \frac{\sqrt{g_1 g_2 g_5}}{g_3 g_5}$, we have

$$
\begin{aligned}
(\Psi_u)^* &= -\frac{g_4}{g_3} + \frac{\sqrt{g_1 g_2 g_5}}{g_3 g_5} \geq -\frac{g_4}{g_3} \\[2mm]
&= \frac{P_u^{\max}\left(\left|\mathbf{h}_{l,\text{Eve}}^{\text{H}}\mathbf{\Phi}\mathbf{h}_{l,u}\right|^2 + |h_{u,\text{Eve}}|^2\right) + \sigma_{\text{Eve}}^2}{P_u^{\max}\left[\left(\left|\mathbf{h}_{l,\text{Eve}}^{\text{H}}\mathbf{\Phi}\mathbf{h}_{l,u}\right|^2 + |h_{u,\text{Eve}}|^2\right) - \left(\mathbf{H}_{B,l}^{\text{H}}\mathbf{\Phi}^{\text{H}}\mathbf{h}_{l,\text{Eve}}\mathbf{h}_{l,\text{Eve}}^{\text{H}}\mathbf{\Phi}\mathbf{H}_{B,l}\right)\right]} \\[2mm]
&\geq \frac{P_u^{\max}\left(\left|\mathbf{h}_{l,\text{Eve}}^{\text{H}}\mathbf{\Phi}\mathbf{h}_{l,u}\right|^2 + |h_{u,\text{Eve}}|^2\right) + \sigma_{\text{Eve}}^2}{P_u^{\max}\left(\left|\mathbf{h}_{l,\text{Eve}}^{\text{H}}\mathbf{\Phi}\mathbf{h}_{l,u}\right|^2 + |h_{u,\text{Eve}}|^2\right)} \\[2mm]
&= 1 + \frac{\sigma_{\text{Eve}}^2}{P_u^{\max}\left(\left|\mathbf{h}_{l,\text{Eve}}^{\text{H}}\mathbf{\Phi}\mathbf{h}_{l,u}\right|^2 + |h_{u,\text{Eve}}|^2\right)} > 1.
\end{aligned}
\tag{42}
$$

According to (42), we know that $(\Psi_u)^* = -\frac{g_4}{g_3} + \frac{\sqrt{g_1 g_2 g_5}}{g_3 g_5}$ cannot satisfy (26b), i.e., $0 < \Psi_u \leq 1$. Under this condition, the optimal power distribution coefficient $(\Psi_u)^*$ is

$$
(\Psi_u)^* \begin{cases} -\left(\frac{g_4}{g_3} + \frac{\sqrt{g_1 g_2 g_5}}{g_3 g_5}\right), & 0 < \Psi_u \leq 1, \\ 1, & \text{else.} \end{cases}
\tag{43}
$$

3.3. Channel Allocation

Similarly, we assume that $\mathbf{\Psi}$ and $\mathbf{\Phi}$ have been given in advance. The channel allocation problem takes the form

$$
\mathbf{P6}: \max_{\mathbf{K}} \ R_{\text{tot}}^{\text{sec}}(\mathbf{K}) = \sum_{u=1}^{U} R_u^{\text{sec}}(k_u)
\tag{44a}
$$

$$
\text{s.t. } k_u \in \{0,1\}, \sum_{k=1}^{K} k_u = 1, \sum_{u=1}^{U} k_u \leq 1, \forall k, u.
\tag{44b}
$$

As discussed in [23], P6 turns out to be a maximum weight bipartite matching (MWBM) problem. In polynomial time, the MWBM problem can be solved by the Hungarian algorithm. Based on above analysis, we can obtain the optimal channel allocation policy by using Algorithm 1.

Algorithm 1 Optimal channel allocation algorithm for P6

1: **for** $k = 1 : K$ **do**
2:     **for** $u = 1 : U$ **do**
3:         According to the SCA and SDR methods, as well as Gaussian random process, we can obtain the optimal phase shift $(\mathbf{\Phi})^*$.
4:         According to (43), we can acquire the optimal power distribution coefficient $(\mathbf{\Psi})^*$.
5:         We substitute $(\mathbf{\Phi})^*$ and $(\mathbf{\Psi})^*$ into (7) to obtain $R_u^{\text{sec}}(k_u)$.
6:     **end for**
7: **end for**
8: The Hungarian algorithm is adopted to solve P6.
9: Output the optimal channel allocation policy $(\mathbf{K})^*$.

3.4. Overall Algorithmic Framework

In this paper, we design a total secure channel capacity maximization scheme for LAP and IRS enhanced transmissions, where the phase shift $\mathbf{\Phi}$, power distribution coefficient $\mathbf{\Psi}$, and channel allocation $\mathbf{K}$ are optimized. Figure 2 shows the overall algorithmic framework, where $\mathbf{\Phi}$, $\mathbf{\Psi}$, and $\mathbf{K}$ are solved iteratively. Specifically, by using the SCA and SDR methods, we can solve the formulated phase shift optimization problem P2, based on which the optimal phase shift $(\mathbf{\Phi})^*$ satisfying the rank-one constraint can be obtained by adopting the Gaussian random process. Next, according to (43), we can obtain the closed-form expression of optimal power distribution coefficient $(\mathbf{\Psi})^*$. Then, Algorithm 1 employs the Hungarian algorithm to acquire $(\mathbf{K})^*$. Finally, the above processes are repeated until satisfying the termination condition.

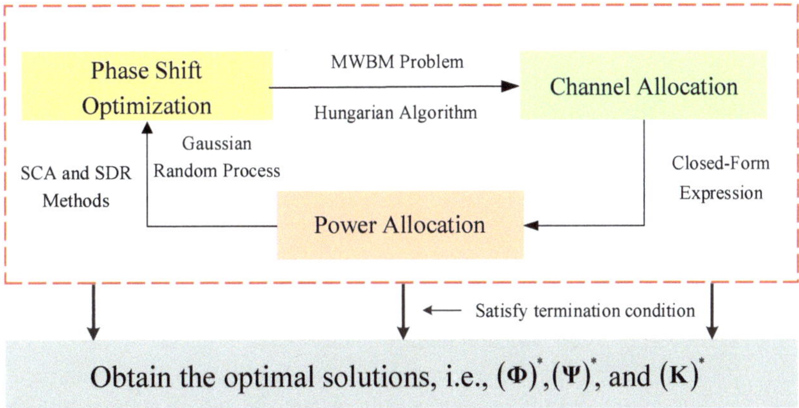

Figure 2. Overall algorithmic framework.

The complexity of the total secure channel capacity maximization scheme is mainly composed of three parts: (1) phase shift optimization; (2) power distribution coefficient optimization; (3) channel allocation. For the first part, since the SCA and SDR methods are used, the complexity of this part is $O(G^{3.5})$. Moreover, for the second part, we can derive the closed-form of the optimal power distribution coefficient; thus, the complexity of this part is $O(1)$. Furthermore, for the third part, the complexity of the channel allocation policy using the Hungarian algorithm is $O\left((U+K)^3\right)$. To summarize, the total computational complexity of solving P1 is $O\left(I_{\text{tot}}G^{3.5}\right) + O(I_{\text{tot}}) + O\left(I_{\text{tot}}(U+K)^3\right)$, where I_{tot} is the total number of iterations.

Discussion (Convergence Analysis): In this paper, the total secure channel capacity is maximized by iterative optimization. Therefore, the convergence needs to be analyzed. First, we present a simple scenario, which consists of an RBS, a fixed LAP equipped with IRS, a legitimate vehicle user, and an eavesdropper. In this case, $(\mathbf{K})^*$ can be obtained by using the enumeration method. As shown in Figure 3, for a given $(\mathbf{K})^*$, we iteratively optimize $(\mathbf{\Phi})^*$ and $(\mathbf{\Psi})^*$ based on the coordinated polling method. The objective function value (i.e., the total secure channel capacity) is improved partly after each iteration. Since the objective function value of P1 is bounded, our designed iterative algorithm can always converge to the optimal value or some certain values after finite iterations.

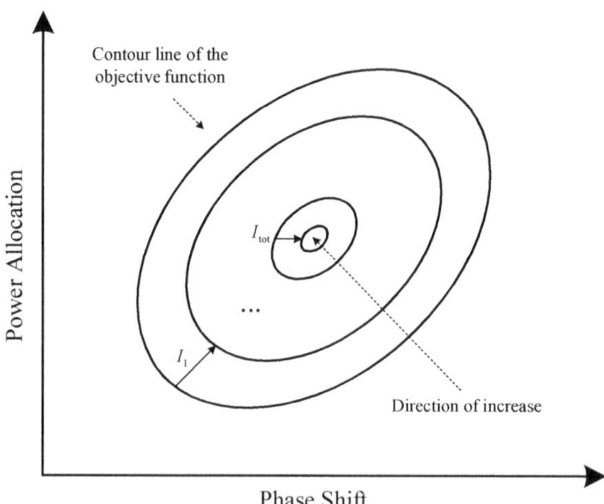

Figure 3. Coordinate polling method.

4. Performance Evaluation

In this section, simulation experiments are conducted to evaluate the performance of the proposed total secure channel capacity maximization scheme. Specifically, the comparison schemes are as follows. (a) Scheme 1 (LAP-PLS-CPO) [23]: This work uses the LAP to relay signals of the RBS and the PLS technique to ensure information security. In addition, the channel and power are optimized. (b) Scheme 2 (LAP-SPHO) [2]: This work adopts the LAP-enabled relay method to improve the data rate, based on which the spectrum, power, and LAP height are optimized. (c) Scheme 3 (LAP-RIS-CPO): In this scheme, $\mathbf{\Phi}$ is initialized by random value, and then, $\mathbf{K}$ and $\mathbf{\Psi}$ are optimized by the Algorithm 1 and (43), respectively.

In our simulations, we consider a scenario, where $U = [10, 55]$, $P_u^{\max} = 33$ dBm, $N = 32$, $K = [15, 60]$, $G = 64$, and $\sigma_u^2 = \sigma_{\text{Eve}}^2 = -174$ dBm/Hz. In order to analyze conveniently, a Cartesian coordinate is established in Figure 1, where the RBS is located at $(0, 0, 0)$ m, the fixed LAP equipped with IRS is located at $(800, 0, 200)$ m, the eavesdropper is located at $(650, 300, 0)$ m, and the cell radius of RBS is 1000 m. Moreover, the A2G channel model is $32.44 + 20 \lg[d(\text{km})] + 20 \lg[f_c(\text{MHz})]$. We model the fast fading channels as independent and identically distributed (i.i.d.) Rayleigh fading channels. As shown in Figure 1, the fast fading channels can be regarded as Rayleigh fading channels, taking into account the rich reflections and diffractions from surface-based obstacles.

Figure 4 illustrates the comparison of the total secure channel capacity with respect to the number of legitimate vehicle users under the different schemes. It is obvious that our proposed total secure channel capacity maximization scheme outperforms other comparison schemes. The reason is that the LAP and IRS enhanced transmissions are adopted in the considered scenario, based on which the phase shift, power distribution coefficient, and channel allocation are optimized. Compared to Scheme 1 (LAP-PLS-CPO), Scheme 2 (LAP-SPHO), and Scheme 3 (LAP-RIS-CPO), the total secure channel capacity can be increased by 67.56%, 141.3%, and 31.94%, respectively. Especially, for Scheme 2 (LAP-SPHO), since the PLS technique is not adopted, security cannot be satisfied, resulting in the lowest total secure channel capacity. In addition, even when the number of legitimate vehicle users is large, our designed scheme can still achieve relatively high information security rates.

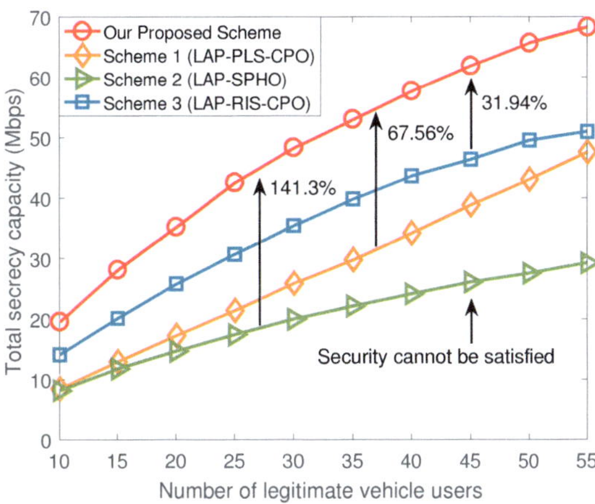

Figure 4. The total secure channel capacity versus the number of legitimate vehicle users.

Figure 5 shows the comparison of the total secure channel capacity with respect to the maximum transmitted power under the different schemes. We can observe that the total secure channel capacity increases monotonously with the increase in the maximum transmitted power $P_u^{\max}$. In addition, the larger the $P_u^{\max}$, the faster the growth of the total secure channel capacity. This is because, in this case, more power is allocated to AN to jam the eavesdropper, which can protect the security of 6G-oriented vehicular IoT services.

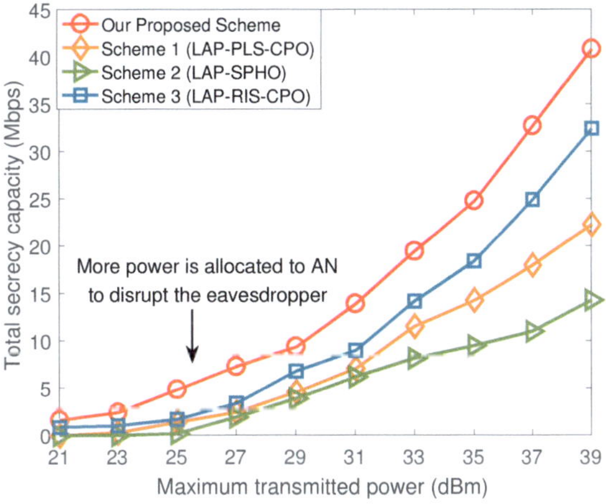

Figure 5. The total secure channel capacity versus the maximum transmitted power.

Next, we investigate the impact of the number of reflection elements on the performance of the proposed scheme. In Figure 6, we plot the comparison of the total secure channel capacity under different numbers of reflection elements. It is observed that the total secure channel capacity increases with the number of reflection elements. This phenomenon is more obvious when the number of legitimate vehicle users is small. This is because more reflection elements can better improve the channel quality. However, as discussed in Section 3.4, since the complexity of solving P2 is $O\left(G^{3.5}\right)$, adding reflection elements will significantly increase the algorithm complexity. Therefore, there is a tradeoff between the total secure channel capacity and the algorithm complexity in terms of the number of reflection elements. The total secure channel capacity maximization by jointly considering the above two factors is a meaningful problem for future research.

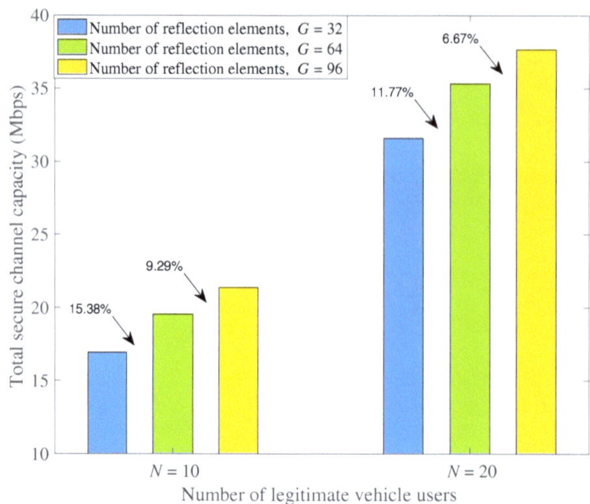

Figure 6. Comparison of the total secure channel capacity under different numbers of reflection elements.

As shown in Figure 7, we investigate the impact of the LAP's altitude on the total secure channel capacity. We can find that with the increase in the LAP's altitude, the total secure channel capacity decreases. The reason is that increasing the LAP's altitude will lead to an increase in the path loss, thereby reducing the total secure channel capacity. However, there is a minimum altitude limit for using this A2G channel model. For altitudes below 100 m, we need to change the large-scale fading model.

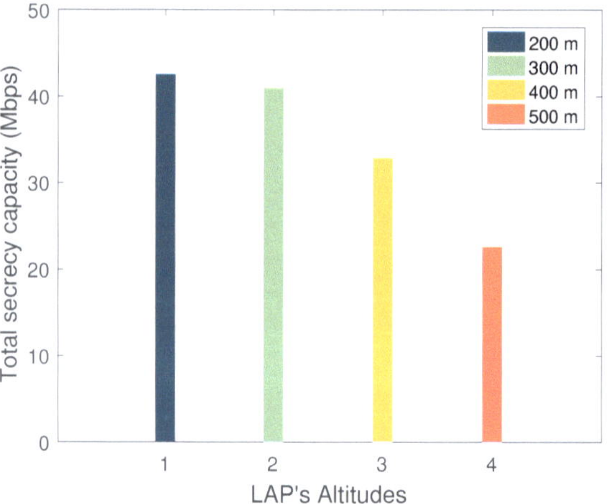

Figure 7. Comparison of the total secure channel capacity under different LAP altitudes.

As shown in Figure 8, we investigate the impact of the distance on the total secure channel capacity, where the LAP's X-axis positions are changed. It can be observed that the total secure channel capacity increases first and then decreases. Similarly, this is because the LAP's position will affect the path loss, thereby influencing the total secure channel capacity. Therefore, optimizing the LAP's deployment is an interesting topic that deserves further study.

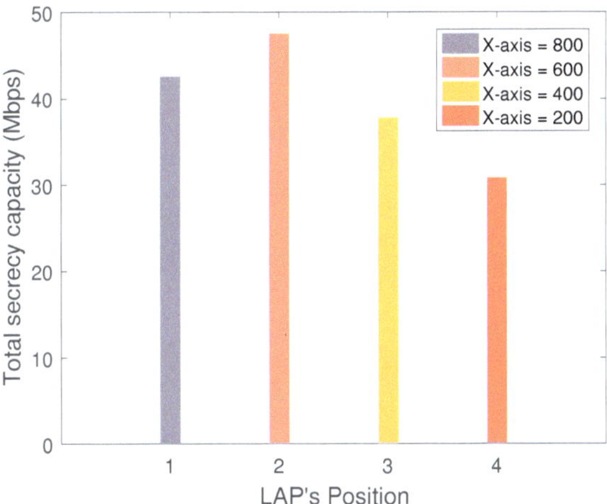

Figure 8. Comparison of the total secure channel capacity under different distances.

5. Conclusions

In order to improve the QoS of 6G-oriented vehicular IoT services, this paper used LAP equipped with IRS to overcome blockages, based on which the secure transmission problem was investigated. First, we introduced AN to enhance the security performance, which could prevent the eavesdropper from receiving privacy information. Next, by jointly considering the phase shift and power distribution coefficient optimization as well as channel allocation, we formulated a total secure channel capacity maximization problem for the LAP-based IRS symbiotic VNets. Then, to deal with this intractable problem, we devised an iterative algorithm, based on which the convergence and the complexity were analyzed. Finally, numerical results demonstrated that the proposed scheme significantly outperformed the comparison schemes in terms of the total secure channel capacity. Furthermore, the joint optimization of the LAP location and network resources with imperfect CSI to maximize the total secure channel capacity is worth investigating and is challenging, which will be our future work.

Author Contributions: L.M.: Conceptualization, Methodology, Software, Writing—Original Draft Preparation, Visualization. J.L.: Conceptualization, Resources, Methodology, Software, Writing—Review. Y.H.: Conceptualization, Resources, Writing—Review and Editing, Supervision, Project Administration, Funding Acquisition. Q.S.: Conceptualization, Resources, Writing—Review. All authors have read and agreed to the published version of the manuscript.

Funding: This work was supported by the National Natural Science Foundation of China, under grant 62206221, and in part by the Fundamental Research Funds for the Central Universities.

Institutional Review Board Statement: Not applicable.

Informed Consent Statement: Not applicable.

Data Availability Statement: Not applicable.

Conflicts of Interest: The authors declare no conflict of interest for publishing in this journal.

References

1. Wang, C.; Li, Z.; Zhang, H.; Ng, D.W.K.; Al-Dhahir, N. Achieving covertness and security in broadcast channels with finite blocklength. *IEEE Trans. Wireless Commun.* **2022**, *21*, 7624–7640.
2. Zhai, D.; Li, H.; Tang, X.; Zhang, R.; Ding, Z.; Yu, F.R. Height optimization and resource allocation for NOMA enhanced UAV-aided relay networks. *IEEE Trans. Commun.* **2021**, *69*, 962–975.
3. Wang, C.; Li, Z.; Shi, J.; Ng, D.W.K. Intelligent reflecting surface-assisted multi-antenna covert communications: Joint active and passive beamforming optimization. *IEEE Trans. Commun.* **2021**, *69*, 3984–4000.

4. Fawaz, W. Effect of non-cooperative vehicles on path connectivity in vehicular networks: A theoretical analysis and UAV-based remedy. *Veh. Commun.* **2018**, *11*, 12–19. [CrossRef]
5. Fotouhi, A.; Qiang, H.; Ding, M.; Hasson, M.; Giordano, L.G.; Garcia-Rodriguez, A.; Yuan, J. Survey on UAV cellular communications: Practical aspects standardization advancements regulation and security challenges. *IEEE Commun. Surveys Tuts.* **2019**, *21*, 3417–3442.
6. Schweiger, K.; Preis, L. Urban air mobility: Systematic review of scientific publications and regulations for vertiport design and operation. *Drones* **2022**, *6*, 179.
7. Alsamhi, S.H.; Shvetsov, A.V.; Kumar, S.; Hassan, J.; Alhartomi, M.A.; Shvetsova, S.V.; Sahal, R.; Hawbani, A. Computing in the sky: A survey on intelligent ubiquitous computing for UAV-assisted 6G networks and industry 4.0/5.0. *Drones* **2022**, *6*, 177.
8. Cao, Y.; Lv, T.; Ni, W. Intelligent reflecting surface aided multi-user mmWave communications for coverage enhancement. In Proceedings of the 2020 IEEE 31st Annual International Symposium on Personal, Indoor and Mobile Radio Communications, London, UK, 31 August–3 September 2020; pp. 1–6
9. Pan, Q.; Wu, J.; Nebhen, J.; Bashir, A.K.; Su, Y.; Li, J. Artificial intelligence-based energy efficient communication system for intelligent reflecting surface-driven VANETs. *IEEE Trans. Intell. Transp. Syst.* **2022**, *23*, 19714–19726.
10. Zhi, K.; Pan, C.; Ren, H.; Wang, K. Power scaling law analysis and phase shift optimization of RIS-aided massive MIMO systems with statistical CSI. *IEEE Trans. Commun.* **2022**, *70*, 3558–3574.
11. You, C.; Kang, Z.; Zeng, Y.; Zhang, R. Enabling smart reflection in integrated air-ground wireless network: IRS meets UAV. *IEEE Wireless Commun.* **2021**, *28*, 138–144. [CrossRef]
12. Shafique, T.; Tabassum, H.; Hossain, E. Optimization of wireless relaying with flexible UAV-borne reflecting surfaces. *IEEE Trans. Commun.* **2021**, *69*, 309–325. [CrossRef]
13. Samir, M.; Elhattab, M.; Assi, C.; Sharafeddine, S.; Ghrayeb, A. Optimizing age of information through aerial reconfigurable intelligent surfaces: A deep reinforcement learning approach. *IEEE Trans. Veh. Technol.* **2021**, *70*, 3978–3983. [CrossRef]
14. Iacovelli, G.; Coluccia, A.; Grieco, L.A. Channel gain lower bound for IRS-assisted UAV-aided communications. *IEEE Commun. Lett.* **2021**, *25*, 3805–3809. [CrossRef]
15. Li, Y.; Zhang, H.; Long, K.; Nallanathan, A. Exploring sum rate maximization in UAV-based multi-IRS networks: IRS association, UAV altitude, and phase shift design. *IEEE Trans. Commun.* **2022**, *70*, 7764–7774. [CrossRef]
16. Al-Jarrah, M.; Al-Dweik, A.; Alsusa, E.; Iraqi, Y.; Alouini, M.-S. On the performance of IRS-assisted multi-layer UAV communications with imperfect phase compensation. *IEEE Trans. Commun.* **2021**, *69*, 8551–8568. [CrossRef]
17. Su, Y.; Pang, X.; Chen, S.; Jiang, X.; Zhao, N.; Yu, F.R. Spectrum and energy efficiency optimization in IRS-assisted UAV networks. *IEEE Trans. Commun.* **2022**, *70*, 6489–6502. [CrossRef]
18. Zhang, X.; Wang, J.; Poor, H.V. Joint optimization of IRS and UAV-trajectory: For supporting statistical delay and error-rate bounded QoS over mURLLC-driven 6G mobile wireless networks using FBC. *IEEE Veh. Technol. Mag.* **2022**, *17*, 55–63. [CrossRef]
19. Ji, Z.; Yang, W.; Guan, X.; Zhao, X.; Li, G.; Wu, Q. Trajectory and transmit power optimization for IRS-assisted UAV communication under malicious jamming. *IEEE Trans. Veh. Technol.* **2022**, *71*, 11262–11266. [CrossRef]
20. Wang, D.; He, T.; Zhou, F.; Cheng, J.; Zhang, R.; Wu, Q. Outage-driven link selection for secure buffer-aided networks. *Sci. China Inf. Sci.* **2022**, *65*, 182303. [CrossRef]
21. Sun, G.; Tao, X.; Li, N.; Xu, J. Intelligent reflecting surface and UAV assisted secrecy communication in millimeter-wave networks. *IEEE Trans. Veh. Technol.* **2022**, *70*, 11949–11961. [CrossRef]
22. Wang, D.; Wu, M.; He, Y.; Pang, L.; Xu, Q.; Zhang, R. An HAP and UAVs collaboration framework for uplink secure rate maximization in NOMA-enabled IoT networks. *Remote Sens.* **2022**, *14*, 4501. [CrossRef]
23. He, Y.; Nie, L.; Guo, T.; Kaur, K.; Hassan, M.M.; Yu, K. A NOMA-enabled framework for relay deployment and network optimization in double-layer airborne access VANETs. *IEEE Trans. Intell. Transp. Syst.* **2022**, *23*, 22452–22466. [CrossRef]
24. Ding, X.; Song, T.; Zou, Y.; Chen, X.; Hanzo, L. Security-reliability tradeoff analysis of artificial noise aided two-way opportunistic relay selection. *IEEE Trans. Veh. Technol.* **2017**, *66*, 3930–3941. [CrossRef]
25. Xu, S.; Liu, J.; Cao, Y. Intelligent reflecting surface empowered physical-layer security: Signal cancellation or jamming? *IEEE Internet Things J.* **2022**, *9*, 1265–1275. [CrossRef]
26. He, Y.; Wang, D.; Huang, F.; Zhang, R.; Gu, X.; Pan, J. A V2I and V2V collaboration framework to support emergency communications in ABS-aided Internet of Vehicles. *IEEE Trans. Green Commun. Netw.* 2023 , *Early access* . [CrossRef]
27. Luo, Z.; Ma, W.; So, A.M.; Ye, Y.; Zhang, S. Semidefinite relaxation of quadratic optimization problems. *IEEE Signal Process. Mag.* **2010**, *27*, 20–34. [CrossRef]

drones

Article

A Multi-Subsampling Self-Attention Network for Unmanned Aerial Vehicle-to-Ground Automatic Modulation Recognition System

Yongjian Shen [1,*], Hao Yuan [2], Pengyu Zhang [2], Yuheng Li [2], Minkang Cai [2] and Jingwen Li [1]

[1] Beijing University of Aeronautics and Astronautics, Beijing 100191, China; lijingwen@buaa.edu.cn

[2] Beijing Research Institute of Telemetry, Beijing 100076, China; haoyuan_1@stu.xidian.edu.cn (H.Y.); pengpengdezahuodian@mail.nwpu.edu.cn (P.Z.); liyuheng@mail.nwpu.edu.cn (Y.L.); mkcai@stu.xidian.edu.cn (M.C.)

* Correspondence: shenyongshen@buaa.edu.cn

Abstract: In this paper, we investigate the deep learning applications of radio automatic modulation recognition (AMR) applications in unmanned aerial vehicle (UAV)-to-ground AMR systems. The integration of deep learning in a UAV-aided signal processing terminal can recognize the modulation mode without the provision of parameters. However, the layers used in current models have a small data processing range, and their low noise resistance is another disadvantage. Most importantly, large numbers of parameters and high amounts of computation will burden terminals in the system. We propose a multi-subsampling self-attention (MSSA) network for UAV-to-ground AMR systems, for which we devise a residual dilated module containing ordinary and dilated convolution to expand the data processing range, followed by a self-attention module to improve the classification, even in the presence of noise interference. We subsample the signals to reduce the number of parameters and amount of calculation. We also propose three model sizes, namely large, medium, and small, and the smaller the model, the more suitable it will be for UAV-to-ground AMR systems. We conduct ablation experiments with state-of-the-art and baseline models on the common AMR and radio machine learning (RML) 2018.01a datasets. The proposed method achieves the highest accuracy of 97.00% at a 30 dB signal-to-noise ratio (SNR). The weight file of the small MSSA is only 642 KB.

Keywords: automatic modulation recognition (AMR); deep learning; self-attention mechanism

Citation: Shen, Y.; Yuan, H.; Zhang, P.; Li, Y.; Cai, M.; Li, J. A Multi-Subsampling Self-Attention Network for Unmanned Aerial Vehicle-to-Ground Automatic Modulation Recognition System. *Drones* **2023**, *7*, 376. https://doi.org/10.3390/drones7060376

Academic Editors: Dawei Wang and Ruonan Zhang

Received: 23 May 2023
Accepted: 31 May 2023
Published: 4 June 2023

1. Introduction

With the promotion of unmanned combat concepts, the enhancement of drone-mounting capabilities, and the increase in flight time, unmanned aerial vehicle-(UAV)-to-ground automatic modulation recognition (AMR) systems are increasingly used in modern, social, especially in emergency communications. UAVs carrying communication reconnaissance payloads can perform the reconnaissance of general radio stations, with strong concealment and long detection distances, and are widely used. The aim of communication reconnaissance is to analyze various parameters of intercepted radio signals and find suitable methods to demodulate them.

However, in the actual transmission process, a signal not only is affected by the antagonistic factors of the transmitter (such as the radio frequency chain), but also changes due to the type of interference and propagation environment, increasing the difficulty of the signal communication and analysis. To simplify the transmission process, AMR methods are proposed, which receive the modulated signal and recognize the modulation mode without the provision of parameters.

Signal modulation recognition can provide essential modulation information for the received radio signals, especially non-cooperative radio signals, which contain cognitive radio, spectrum sensing, signal surveillance, and interference identification [1–4]. Traditional methods face difficulty when coping with the complex and growing types of transmitters. Given that AMR serves as a bridge between signal detection and demodulation, a simple and feasible approach that can be deployed on terminals outfitted on UAV platforms is sorely needed.

1.1. Related Work

1.1.1. Traditional AMR Methods

Traditional AMR can be categorized as likelihood theory-based AMR (LB-AMR) [5,6] or feature-based AMR (FB-AMR) [7]. TB-AMR methods recognize modulation schemes by Bayesian estimation and have high computational complexity. FB-AMR methods analyze a large number of signals, extract interesting features, and determine the category of modulation methods using instantaneous time-domain [8], transform domain [9], and statistical [10,11] features. The AMR task is a regression problem with multiple dimensions. Traditional machine learning methods, such as decision tree [12] and support vector machine (SVM) [13], are easily realized. However, the performance of such methods is reduced when addressing complex or multiple-modulation schemes.

1.1.2. DL-Based AMR Methods

FB-AMR must pre-train on a large set of signals and learn the features of each modulation scheme. Compared with LB-AMR, FB-AMR methods can approximate the ground truth with lower computational complexity and can fit various modulation schemes. Deep learning AMR (DL-AMR) methods are highly dependent on prior knowledge.

Since the first open access AMR dataset, Radio Machine Learning (RML) 2016.04C, was proposed, the convolutional neural network (CNN) has been introduced in the AMR task [14]. There are many methods based on deep learning, which can be divided into three groups, depending on their research content. Some researchers are concerned with enhancing the capability of models. Methods include long short-term memory (LSTM) [15], ResNet [16], the gated recurrent unit (GRU) [17], and deep learning blocks such as temporal attention [18]. LSTM and GRUs extract features by comparing information in the time dimension. They are more suitable for tasks with sequential data inputs, but a large amount of computation limits their performance in mobile terminals. Generally, to avoid overfitting, just two blocks are used in a model. Therefore, CNN-based networks [14,16] have been introduced in the AMR task. These models have a moderate number of parameters and moderate structure complexity for the same performance as LSTM and GRUs. However, the current models all have a small processing range, and the signal data are too long for the models to process efficiently.

A second group of researchers have attempted to increase the channel's number of inputs to improve the classification accuracy. Normally, the inputs of models are signals with two channels, namely in-phase (I) and quadrature-phase (Q) channels, which are supported in public datasets. Researchers use signal processing methods to expand the number of input channels. The amplitude and phase, which can be easily calculated from I and Q, are frequently used. Networks that expand the channels of inputs [19–21] greatly increase the computational cost, and additional factors that increase the requirement of inputs betray the initial goal of easily realizing the AMR task. In preprocessing inputs before these are input to the models, some methods are too specialized and complex to be applied to AMR tasks on a large scale. The models have more parameters, with a deeper and wider network structure.

The last improvement direction of the AMR task is converting the signal inputs into images. The most representative methods use three-channel constellation images as inputs [22,23] and use AlexNet [24] and GoogLeNet [25] for image classification. However, these methods make the AMR task more complex because the input is usually a signal rather than an image.

Comprehensively considering the advantages and disadvantages of the three groups of methods, this paper follows the idea of the first group. These methods have a small processing range in each layer but have weaknesses in terms of noise resistance. However, noise in signal transmission is unavoidable. Moreover, even with LSTM and recurrent neural networks (RNNs), which are common in the analysis of sequence data, either the computation or the number of parameters is too much for deployment on mobile terminals mounted on UAVs, which means that they are unsuitable for UAV-to-ground AMR systems.

1.2. Contributions

The performance of uniform subsampling is equal to that of no subsampling when the subsampling rate is 2 [26]. By reducing the input signal length by subsampling, the parameters and computation in models are greatly decreased. We propose a multi-subsampling self-attention network (MSSA) that can be deployed on a terminal in our UAV-to-ground AMR system. Our main contributions are summarized as follows:

- We design an information integration module with ordinary convolution and dilated convolution branches. Dilated convolution has a larger receptive field than ordinary convolution and is more suitable for global information extraction. The sum of the two branches provides more detailed information.
- To enhance the noise resistance, we introduce a self-attention module with a strong feature extraction capability. The module can dynamically adjust the weights of parameters to amplify the influence of those that are beneficial for modulation recognition and diminish the influence of invalid parameters during the recognition process.
- We subsample the signal into multiple signals with two branches, I and Q, and concatenate them channel-wise. We finesse the model architecture to prevent overfitting. We propose MSSAs in large, medium, and small sizes, with fewer parameters and faster speeds, which are more suitable for our UAV-to-ground AMR system.
- Ablation experiments on a common dataset with current models show the ability of the proposed method in AMR. MSSA has the best performance on RML 2018.01a and 97.00% accuracy when the signal-to-noise ratio (SNR) is 30 dB. Different sizes of MSSA each have their advantages in terms of accuracy, speed, and parameters. The weight file of MSSA(S) is only 652 KB.

1.3. Organization

The remainder of this paper is organized as follows. Section 2 presents the system model. Section 3 analyzes the structure and theory of the proposed method. Section 4 discusses experiments on the dataset, including the comparison of current models and different hyperparameters. Section 5 provides our conclusions.

2. System Model

The emergence of unmanned aerial vehicles (UAVs) has revolutionized the field of remote sensing and other aerial applications by providing both low-altitude and high-altitude platforms for data collection [27]. Depending on the payloads mounted, a UAV can serve as either a computational server or a relay [28], thus enabling diverse ranges of applications in mapping, wildlife conservation, and emergency communications [29], among others.

Our UAV-to-ground AMR system consists of a reconnaissance drone and a ground control terminal, as shown in Figure 1. The drone conducts reconnaissance on communication links such as air-to-air, air-to-ground, and ground-to-ground. The reconnaissance equipment includes various types of airborne radios, radars, vehicle radios, and handheld radios. The ground control terminal, functioning as either a high-capacity computer or server cluster, can receive brief reconnaissance results in real-time and analyze reconnaissance data after the drone returns. AMR methods are typically implemented on the ground control terminal due to the heavy computational burden involved. The information fed back by the drone includes the number of reconnaissance signals, as well as the information such as the direction, time, frequency, modulation method, power, and possible transmission source type of each signal. The processing of reconnaissance data by ground control terminals can achieve signal demodulation and analysis. Nevertheless, due to the low computational cost of our proposed AMR methods, the UAV-mounted payloads with embedded microprocessors can also perform signal demodulation and analysis. The ground terminal solely performs signal behavior analysis and transmits relevant instructions to the drone.

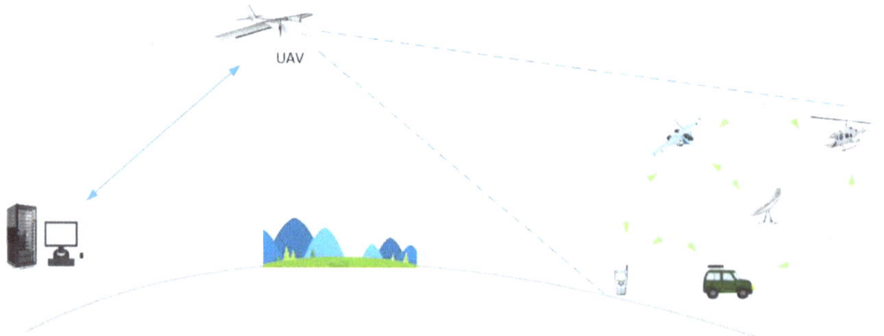

Figure 1. Schematic diagram of a UAV-to-ground AMR system.

Our system can also be utilized as a UAV-assisted mobile edge computing (MEC) architecture, in which the UAV and ground control terminal jointly demodulate and analyze the received signal before transmitting it to end users. Additionally, a flying ad hoc network (FANET) comprising multiple UAVs could be integrated into our system to reduce response delays and increase response probabilities [30]. By receiving signals from various transmitter sources, the UAVs can also support signal modulation analysis, significantly contributing to emergency communications.

3. Design and Implementation of Multi-Subsampling Self-Attention Network

3.1. Architecture

Although current DL-AMR methods have high accuracy, the complexity is high. Except for expanding the number of input channels by preprocessing, many complex modules are used in DL-AMR. The optional expanding channel can obtain the amplitude and phase through the provided I/Q signal [15], and the models used for DL-AMR include LSTM, DAE [31], and GRU [32]. The amplitude A and phase ϕ, which can be easily calculated from I and Q, with the formula:

$$x = \left[\begin{array}{c} A \\ \phi \end{array} \right] = \left[\begin{array}{c} \sqrt{I^2 + Q^2} \\ \arctan 2(Q/I) \end{array} \right]. \tag{1}$$

However, the received signal in common applications is typically represented as I and Q components, which requires specialized knowledge to process the expanded channels, including the amplitude, phase [19–21], and constellation image [22,23], as used in these modulation schemes. In recent years, the structure of deep learning has become deeper and more complex, with higher computational costs. For signal data, the parameters can be limited using simple CNN networks and are sufficient for the modulation classification task. We selected a CNN as our main framework.

Our network aims to achieve a more balanced architecture that minimizes the computational cost, accelerates the training speed, and requires less prior knowledge. Our experiments show that the original input with a length of 1024 is excessively long for this task, which can be attributed to the large number of parameters in the models used. Thus, before the CNN module, we reshape the $(1024, 2)$ inputs into $(4, 256, 2)$ or $(2, 512, 2)$ to reduce the length, through uniform subsampling with two subsampling rates. This expands the channels of inputs without changing the original data, and the decrease in input length reduces the number of parameters. It will greatly reduce the inference speed when embedded on the terminal. We chose to utilize multiple cascaded residual modules following the first layer of our network. The residual mechanism allows for the preservation of original information during the feature transmission process. Then, we use dilated convolution, which can extract features from greater global information, and an ordinary convolution layer for local information.

With extracted local and global information, the addition of two branches facilitates the subsequent self-attention task. The self-attention mechanism has a strong capability to extract interesting information, which will affect the results of category classification. The traditional fully connected layer generally has more parameters than the convolution layer and benefits convergence. We put the fully connected layer at the two last layers to obtain the features of the signal, and a softmax function distributes the probabilities of each category. The number of dense units should match the number of modulation schemes used. By utilizing subsampled inputs, we can significantly reduce the number of trainable parameters in the last two layers of our model, which tend to be the main source of computational cost. Figure 2 shows the main framework of our model with a subsampling rate of 4. Figure 3 depicts the primary architecture of ResNet as presented in [16]. Our method employs fewer kernels in each convolutional layer compared to ResNet. Additionally, the input size of the first fully connected layer in ResNet is $(32 \times 16 \times 1)$, while our model's input size is $(16 \times 8 \times 1)$. Consequently, the number of trainable parameters in our approach has been reduced by a factor of four.

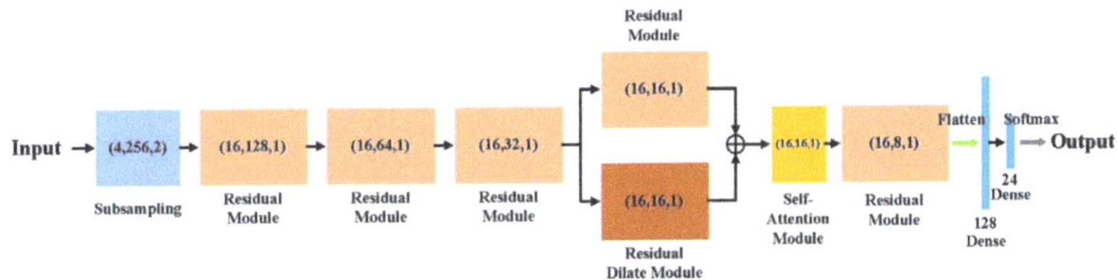

Figure 2. Structure of MSSA(M) when the subsampling rate is 4. The number of last dense units is the same as the number of types of modulation schemes.

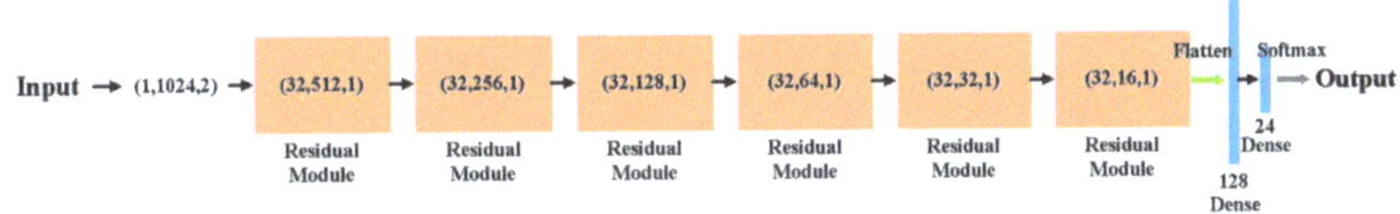

Figure 3. Structure of ResNet [16]. Our method has fewer kernels in each convolutional layer.

We selected suitable hyperparameters to fine-tune the model, including the number of convolution layers, kernel size, learning rate, and the sample size of each batch. The key point in DL-AMR is to solve the overfitting of models. The latest model, especially one with a self-attention module, could easily be subject to overfitting because of the higher computation cost. We conducted experiments on different hyperparameters and selected the best.

3.2. Methodology

3.2.1. Enhanced Processing Range via Dilated Residual Connections

As the signal has fewer dimensions, the length of the signal data is greater than that of the image data. The receptive field is defined as the region in the input space that a particular CNN feature is looking at. The input of the common AMR dataset has a length of 1024. However, ordinary convolution always has fewer receptive fields, which are 3 with a 3 kernel size, and successive convolution layers have 5 receptive fields. Let k be the kernel size, and s be the stride of the convolution layer. Then, the receptive field of the convolution layer CF can be calculated as:

$$CF_i = F_{i-1} + ((k-1) * \prod_{j=1}^{i-1} s_j), \tag{2}$$

where CF_i refers to the receptive field of the current convolution layer, and F_{i-1} is the previous level. Hence, we serially connect four convolution layers as the base module. The stride and kernel sizes are set to 1 and 3, respectively. The final receptive field is 9, which cannot cover the entire length of the signal. The number of convolution layers in the residual modules needs to be controlled to avoid overfitting, but this number has already reached the limit of the modules.

Unlike ordinary convolution, dilated convolution conducts a convolution operation whose kernel has holes. Figure 4 shows the different kernels of two convolutions. By filling kernels with holes, the receptive field of dilated convolution DCF is increased,

$$DCF_i = F_{i-1} + ((k * r - 1) * \prod_{j=1}^{i-1} s_j), \tag{3}$$

where r is the dilated rate of dilated convolution. Then, the whole receptive field of the residual dilated module is 11 when $r = 2$ and $k = 3$. The expanded receptive field provides more global information. In contrast to the pooling layer, the dilated convolution layer does not remove elements with smaller values, which can retain the most information of inputs. Dilated convolution operates on data at equally spaced intervals, effectively performing a form of specialized subsampling. Consequently, the features extracted from dilated convolution will differ from the features extracted by parallel residual modules. In addition, to maintain gradient stability, we set a bridge between every two layers. The structure is shown in Figure 5. The residual block output is:

$$H_{Res} = x + W_i \times (W_{i-1} \times x + b_{i-1}) + b_i, \tag{4}$$

where $W \times x + b$ represents convolution, and x is the input of the one residual block. The gradient of this module is:

$$\begin{aligned}
\frac{\partial H_{Res}}{\partial x} &= 1 + \frac{\partial W_i \times (W_{i-1} \times x + b_{i-1}) + b_i}{\partial x} \\
&= 1 + W_i \times W_{i-1},
\end{aligned} \tag{5}$$

where 1 keeps the gradient in a controllable range. This can avoid the explosion and disappearance of the gradient. We combine one convolution with a $(1,1)$ kernel size, two residual blocks, and one max-pooling layer. The $(1,1)$ convolution layer extracts the channel-wise information and the $(3,1)$ layer extracts the height and channel dimensions. The $(2,1)$ max-pooling layer reduces the width of the inputs. Due to the $(channel, length, 2)$ input shape, the kernel size in the first residual module is set as $(3,2)$ to consider the information in both I/Q signals. Similarly, the kernel size of the max pooling layer in the first residual module is set as $(2,2)$ to reduce the dimension in width. We replace the $(3,1)$ convolution layer with a $(3,1)$ dilated convolution layer in the residual dilated module.

The larger receptive field of residual dilated convolution can cover information that ordinary convolution cannot, and the features extracted by dilated convolution can be regarded as information from larger-size inputs. Ordinary convolution and dilated convolution branches in the proposed model provide features extracted from different scales to the next self-attention module.

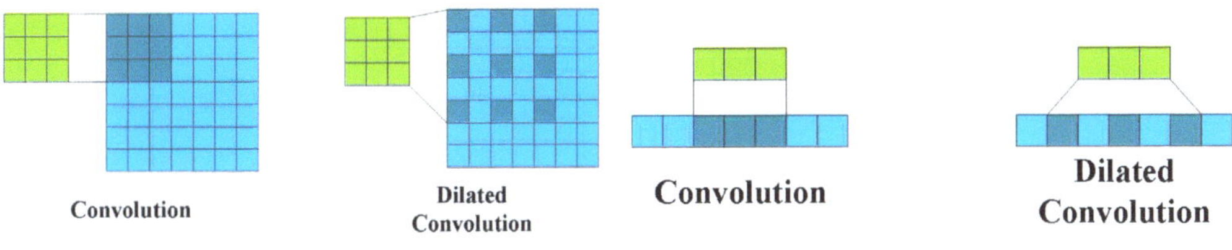

(a) 2D convolution   (b) 1D convolution

Figure 4. Two types of convolutions with data of different dimensions. By filling kernels with holes, the receptive field of dilated convolution is increased.

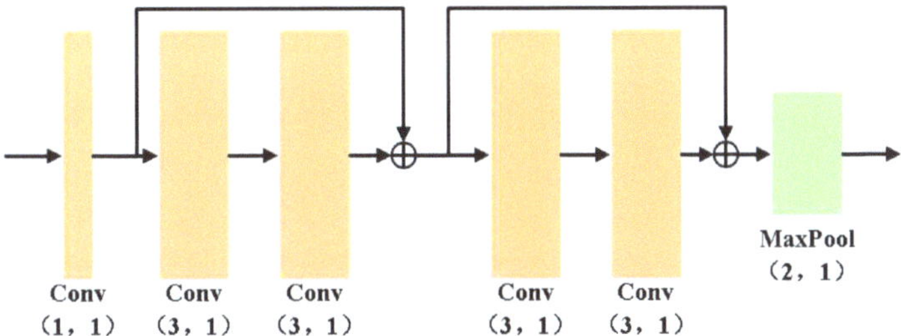

Figure 5. Structure of the residual module. Each block has two convolution layers with a kernel size of $(3, 1)$ and one max-pooling layer with a kernel size of $(2, 1)$; however, the sizes of these layers are $(3, 2)$ and $(2, 2)$, respectively, when in the first residual module of the MSSA.

3.2.2. Enhanced Robustness of Attention Models against Noise

After the two branches of convolution and dilated convolution, a module that can integrate global and local information to extract interesting features is needed. Therefore, a self-attention module is introduced. The attention model can filter ineffective information [33], which can be noise generated in the transmission process or information repeated in a periodic signal. The output of attention module H_A can be formulated as:

$$H_A = \sigma(Mask(x)) \odot \phi(Trunk(x)), \tag{6}$$

where $Mask(x)$ represents the gate controlled by the output of $Trunk(x)$. $Mask(x)$ and $Trunk(x)$ can be any type of structure and are the outputs of the mask and trunk branch. The mechanism assumes the weights for elements in each location, and the trainable weights can dynamically extract interesting features to enhance the classification ability of models.

Comparing AMR and image classification, the AMR task is relatively simple. The number of parameters and layers should be controlled to avoid wasting computational resources. We use the attention mechanism once in the proposed model. We adjust the structure of the attention module, i.e., the self-attention module [34],

$$
\begin{aligned}
H_{SA} &= P_{QK}(x) \times V(x) \\
&= softmax(\frac{Q(x) \times K(x)}{\sqrt{d_K}}) \times V(x),
\end{aligned}
\tag{7}
$$

where $Q(x), K(x)$, and Vx refer to the extracted features in three branches, and d_K is the number of channels. In this module, $Q(x) \times K(x)$ strengthens the interesting elements, and the softmax function assumes the probabilities of elements in each location. Then, the weights $P_{QK}(x)$ generated by $softmax(\frac{Q(x) \times K(x)}{\sqrt{d_K}})$ select the features in $V(x)$ as the output of the self-attention module. In the calculation of DL-AMR, the features can be a matrix with shape $(batch, length, 1, channels)$, with a dimension of 1. Then, the result of $P_{QK}(x)$ in

H_{SA} would have the shape $(batch, 1, 1, channels)$. The probabilities for $V(x)$ are constant values that are inefficient for the classification task. Therefore, we use the element-wise product $\odot$ to replace matrix multiplication $\times$, as in Equation (6), with the formula:

$$
\begin{aligned}
H'_{SA} &= x + Relu(P'_{QK}(x) \odot V(x)) \\
&= x + Relu(softmax(Q(x) \odot K(x)) \odot V(x)).
\end{aligned}
\tag{8}
$$

The kernel size in Q is $(1,1)$, and the others are $(3,1)$. The Q branch learns the characteristics channel-wise. Then, the impact factors in the channel and location are $Q(x) \odot K(x)$. The probabilities can be obtained by the softmax function. Activated by the ReLU function, $Relu(P'_{QK}(x) \odot V(x))$ contributes to extracting key features in both local and global information. According to the gradient formula of parameter W_V in $V(x)$,

$$
\begin{aligned}
\frac{\partial H'_{SA}(x')}{\partial W_V} &= Relu'(P'_{QK}(x') \odot V(x')) \frac{\partial P'_{QK}(x') \odot V(x')}{\partial W_V} \\
&= Relu'(P'_{QK}(x') \odot V(x')) P'_{QK}(x') \odot \frac{\partial V(x')}{\partial W_V} \\
&= Relu'(P'_{QK}(x') \odot V(x')) P'_{QK}(x') \odot x',
\end{aligned}
\tag{9}
$$

where x', $P'_{QK}(x')$, and $Relu'(P'_{QK}(x') \odot V(x'))$ are constants in the parameter update process, which could be controlled by $Q(x)$ and $K(x)$. Similarly, the parameters in Q, K, and V are all learnable and trainable, and they can influence each other. Following extensive training, the value of these valid parameters can be adjusted, resulting in a significant amplification of their impact on modulation recognition. Additionally, dynamic parameters can filter out noisy data and place greater emphasis on valid data. The information from the addition of the convolution and dilated convolution layers is effectively utilized. Like the residual module, the addition operation is used before the output of the self-attention module.

As a result, the gradient of this module becomes controllable and the noise resistance is increased. It avoids overfitting because of the addition with input x. The structure of our self-attention module is shown in Figure 6. Ablation experiments demonstrate the capability of the self-attention mechanism in the AMR task. However, self-attention increases computation, which could affects the deployment on a mobile terminal.

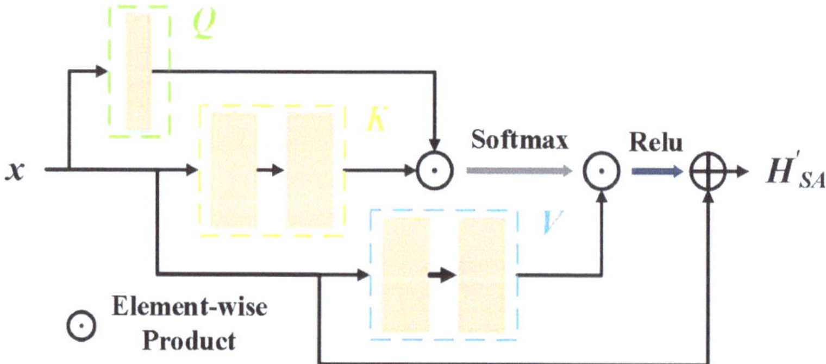

Figure 6. Self-attention module. Kernel size in Q is $(1,1)$; others are $(3,1)$. We add input x and output H'_{SA} to prevent the gradient exploding and disappearing.

3.2.3. Streamlined Modeling with Subsampling Layer

To simplify the models and decrease the number of parameters, a subsampling layer is added at the start of our network. Ramjee et al. conducted experiments on uniform, random, and magnitude rank subsampling [26]. Uniform subsampling had an equal performance with no subsampling when the subsampling rate was 2, while the others had no improvement.

Hence, we subsample the data with a constant interval and concatenate them channel-wise. Then, the length of inputs will be reduced, and the number of channels increased. However, the original network is too complex to learn the features in shorter data, which will cause overfitting in the training process. Therefore, the hyperparameters in the other three sizes of MSSA are adjusted to be suitable for shorter data after subsampling, which include the number of residual modules, residual module filters, and fully connected layers (dense layers) units. This greatly reduces the number of trainable parameters and increases the training speed. Table 1 shows the details of each model size.

Table 1. MSSA with different hyperparameters.

Hyperparameters	Inputs Shape	Residual Modules	Residual Module Filters	Dense Units
MSSA(XL)	(1, 1024, 2)	6	32	128
MSSA(L)	(2, 512, 2)	6	32	128
MSSA(M)	(4, 256, 2)	5	16	128
MSSA(S)	(4, 256, 2)	5	16	64

Finally, we proposed four different models featuring two subsampling rates. The utilization of dilated convolution provided an additional specialized subsampling result for each model. We compared our methods with current models on a common public dataset, showing good performance. The proposed method has a 60.90% mean classification accuracy on SNR from −20 dB to 30 dB. Compared with current models, MSSA(S) has the fewest parameters and the fastest training speed, but slightly less accuracy. It is more suitable for signal detection and recognition systems.

3.3. Equipment and Facilities

Ablation experiments were conducted on a 64-bit Linux system equipped with an Nvidia GeForce RTX 2080 Ti graphics card with 12 GB memory. All models were trained with TensorFlow v1.14, CUDNN v7.4, and CUDA v10.0. We used the Adam optimizer with a learning rate of $1 \times e^{-4}$. The batch size was 1000 and iterations were limited to 100, except that the iterations of the other three MSSAs were 200. We saved the best model at each epoch. The loss function used the category cross-entropy function.

Figure 7 shows the deployment details of our UAV-to-ground AMR system. We selected an Nvidia Jetson TX2 as the signal processing terminal and a MicroPhase ANTSDR E310 transmitter. We leveraged TensorRT on the terminal to accelerate the inference process of (MSSA(S)).

(a) Signal processing terminal

(b) Transmitter

Figure 7. Configuration of Nvidia Jetson TX2 signal processing terminal and MicroPhase ANTSDR E310 transmitter.

4. Experiments and Results

For evaluation, we selected the RML 2018.01a dataset [16], which is a common AMR dataset. This dataset contains 24 types of modulation schemes: OOK, 4ASK, 8ASK, BPSK, QPSK, 8PSK, 16PSK, 32PSK, 16APSK, 32APSK, 64APSK, 128APSK, 16QAM, 32QAM, 64QAM, 128QAM, 256QAM, AM-SSB-WC, AM-SSB-SC, AM-DSB-WC, AM-DSB-SC, FM, GMASK, and OQPSK. There were 2555904 sample signals with a shape of $(2, 1024)$. The signal-to-noise ratio (SNR) of signals in the dataset ranged from -20 dB to 30 dB with an increment of 2. Each modulation scheme had 26 sets of signals with different SNRs, and each SNR set had 4096 signals. Compared with other public AMR datasets, this dataset has more samples and more kinds of modulation schemes. Table 2 compares the properties of the main AMR datasets. The signals have two branches, I and Q. We trained the models with all kinds of modulation schemes, conducted ablation experiments on each hyperparameter, and proposed the best couple of hyperparameters. The dataset was divided into training and test sets at a 7:3 ratio. There were 1,789,128 signals in the training set and 766,776 in the test set.

Table 2. Comparison of open AMR datasets.

Dataset	Number of Modulation Schemes	Sample Dimension	Dataset Size	SNR Range (dB)
RML 2016.04c	11	2×128	162,060	$-20:2:18$
RML 2016.10a	11	2×128	220,000	$-20:2:18$
RML 2016.10b	10	2×128	1,200,000	$-20:2:18$
RML 2018.01a	24	2×1024	2,555,904	$-20:2:30$

LSTM had the best performance when tested on the RML 2018.01a dataset and was the main comparison network [1]. In addition, MSSA is an improvement network based on ResNet, and a CNN is the baseline of the AMR task. It is necessary to compare MSSA with CNN and ResNet.

Our experiment had two main purposes, the more important one being to demonstrate the modulation recognition ability of the proposed model compared with current or basic models, and the other being to provide a faster and simpler network while retaining most of the performance. We display the results of comparisons with other models in Section 4.1, and Section 4.2 shows the improvement of our model in terms of adjusting the structure and hyperparameters.

4.1. Experimental Comparison for AMR Task

To fairly test the ability of models in radio modulation recognition, no data enhancement methods were used, not even random shuffling, because this would affect the gradient updates when the number of samples was too large. The inputs in this experiment were with the shape $(1024, 2)$, which was reshaped to a $(1, 1024, 2)$ tensor by reshaping the layer with the data format "channels first". The network was fed one of the types of modulation schemes and estimated the probability of the 24 classes. The prediction of a model is the greatest value in the confidence matrix. Figure 8 shows the recognition accuracy curves of each class of models in each SNR. The proposed model had a higher accuracy in most SNRs, except SNRs in the range of $-5 \sim 8$ dB. The CNN in [14] performed the worst among the four models.

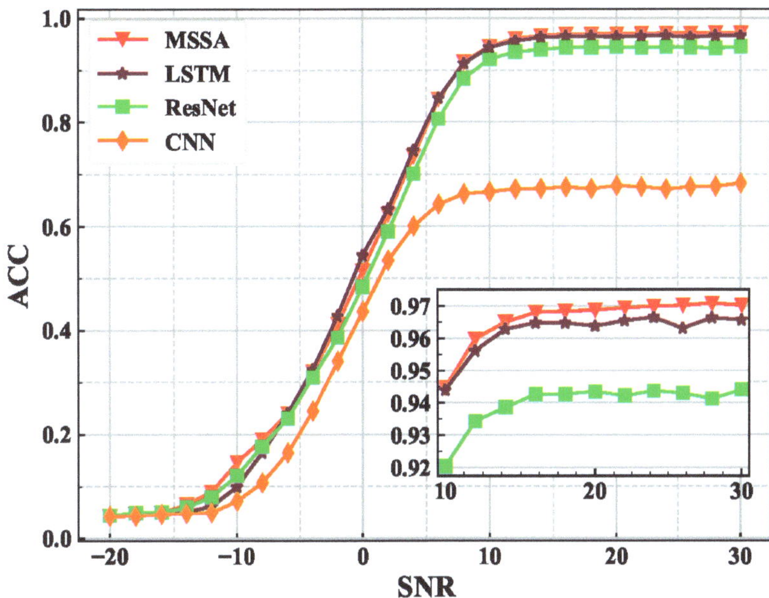

Figure 8. Recognition accuracy of models on the RML 2018.01a dataset. The proposed MSSA has the highest accuracy when SNR is above 10.

The other evaluation index of AMR is the confidence confusion matrix of models, which will reflect the quality and performance of the classifier. It demonstrates the certainty and correctness of the classifier's classification results for each category during the prediction phase. By observing the confidence confusion matrix, we can identify which categories have more accurate classification results, which categories are often confused, and the performance of the classifier when classifying samples with high uncertainty. We tested samples with different SNRs and calculated the mean confidence indexes of each class when inputting a signal. Then, confusion results are visualized in Figure 9. The vertical axis on each matrix denotes the true labels, and the horizontal axis denotes the predicted labels. As shown in Figure 8, MSSA performed the best in the confidence confusion matrix. MSSA had one modulation scheme, 16QAM, which was difficult to recognize, while LSTM and ResNet had two modulation schemes.

For the SNR of signals that are almost above 6 dB or 8 dB in the realistic AMR task, this paper shows the accuracy of models when the SNR is 6, 14, 22, and 30 dB. We calculated the average classification recognition accuracy without distinguishing the SNR and classification. The mean class accuracy values of models are shown in Table 3. The proposed model had the highest accuracy when the SNR was above 14 dB, and it had the highest mean accuracy. From ResNet to MSSA, the average accuracy increased by 2.09%. This experiment shows the capability of our models in the AMR task.

Table 3. Comparison of models in terms of classification recognition accuracy (%).

Acc (%) in SNR (dB)	6	14	22	30	Mean (−20:2:30)
CNN	64.16	67.14	67.43	68.12	43.89
ResNet	80.54	93.85	94.20	94.39	58.81
LSTM	**84.65**	96.28	96.54	96.59	60.22
MSSA	84.38	**96.49**	**96.93**	**97.00**	**60.90**

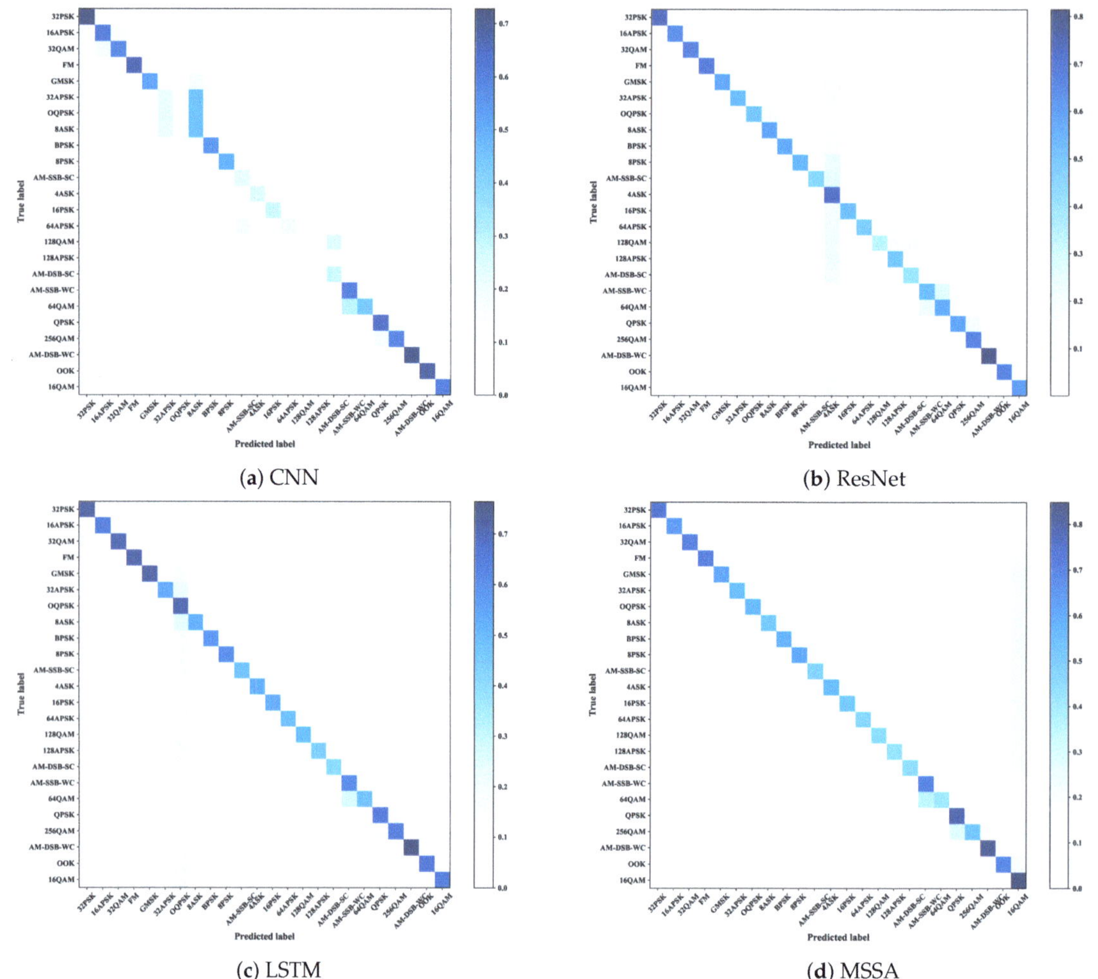

(a) CNN

(b) ResNet

(c) LSTM

(d) MSSA

Figure 9. Confidence confusion matrices of the models. MSSA has one modulation scheme, 16QAM, that is difficult to recognize.

4.2. Experimental Comparison on Hyperparameters

This paper adjusted the structure of the model with a subsampling rate of 4, as shown in Table 1, because the length of the data is too short to achieve a higher mean classification accuracy when trained on the original model. The accuracy values of models are shown in Table 4, where the indicator values are shown for the convenient comparison of the training speed, parameters, and classification recognition accuracy. The number of trainable parameters in a model is strongly correlated with the utilization of computational and memory resources. Increasing the number of parameters may lead to a higher computational load on the terminal or embedded microprocessor, thus reducing both the training and inference speeds. MSSA(L) and MSSA(S) had the fastest speeds. MSSA(S) used the fewest parameters, while the accuracy of MSSA(L) was higher. While MSSA(L) had more parameters, it had the same training speed as ResNet and higher accuracy. Figure 10 shows the accuracy curves of models in each SNR.

Table 4. Comparison of models on size, complexity, and classification recognition accuracy (%).

	Time (Second/Epoch)	Parameters	SNR = 6 (dB) Acc (%)	SNR = 30 (dB) Acc (%)	Mean Acc (%)
CNN	367	13,064,524	64.16	68.12	43.89
ResNet	171	139,192	80.54	94.39	58.81
LSTM	1242	202,766	**84.65**	96.59	60.22
MSSA(XL)	283	218,200	84.38	**97.00**	**60.90**
MSSA(L)	171	152,696	82.09	95.78	59.70
MSSA(M)	**99**	54,632	75.80	93.48	57.01
MSSA(S)	**99**	**36,648**	72.43	90.50	55.25

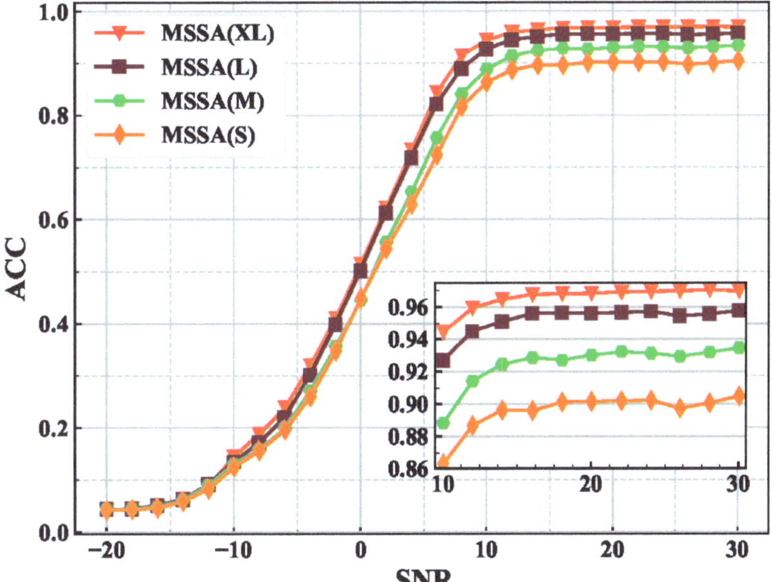

Figure 10. Recognition accuracy of MSSA with different hyperparameters on the RML 2018.01a dataset. As the subsampling rate increases, the model has fewer parameters, and the accuracy decreases by approximately 2.5%.

The capabilities of models decreased as the number of parameters decreased, but the speeds increased. These models with different sizes all outperformed the CNN, and even at the same training speed, MSSA(L) was better than ResNet. These four MSSAs in different sizes could be applied to different situations with different requirements, such as lower parameters or faster speeds. Figure 11 shows the confusion matrices. The OOK modulation scheme was difficult for MSSA in the M or S size. Although they have a problem recognizing signals with OOK modulation, the confidence indexes of the OOK scheme are high (above 0.8).

We proposed three sizes of MSSA models, with faster speeds and fewer parameters. They have their own advantages. MSSA(XL) performed the best among current models, with a few more parameters and a slower training speed. The L-sized MSSA(L) was the most recommended model, with a moderate performance in terms of accuracy, training speed, and parameters. MSSA(M) was more suitable when speed was required. MSSA(S) was more suitable when the parameter requirement was very strict.

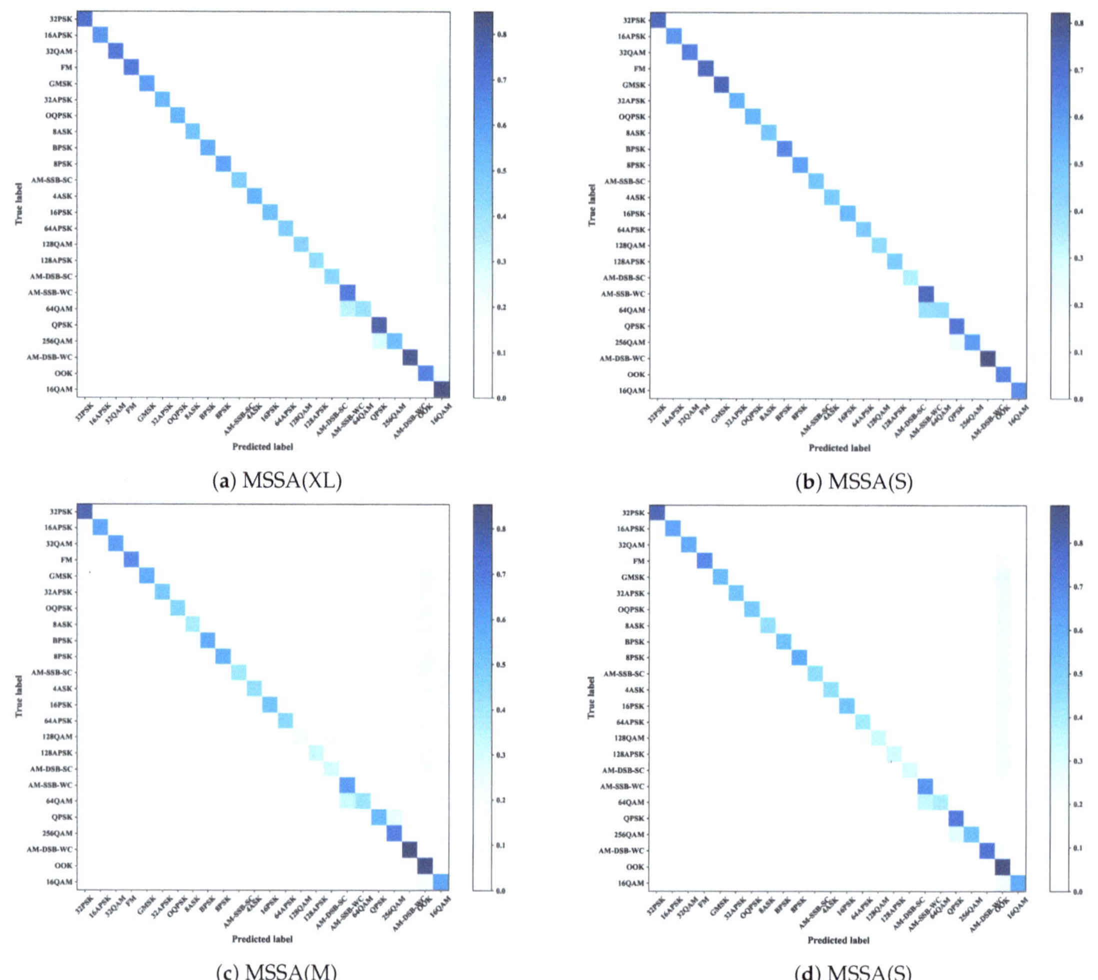

(a) MSSA(XL)   (b) MSSA(S)

(c) MSSA(M)   (d) MSSA(S)

Figure 11. Confidence in the confusion matrices of MSSA in different sizes. OOK modulation scheme is still a difficult task for MSSA in M or S size.

5. Conclusions

Current AMR methods with large numbers of parameters and high computational complexity are difficult to employ on UAV-to-ground AMR systems. The limited data processing range and low noise resistance also restrict the performances of deep learning methods. Therefore, we proposed MSSA, with fewer parameters, for drone–ground signal processing platforms. We proposed a residual dilated module with a larger receptive field to expand the data processing range and a self-attention module to dynamically acquire information from either local or global contexts, which provides strong noise resistance. Finally, we adjusted the structure of MSSA with different subsampling rates and proposed large, medium, and small MSSA models, which all performed well on the AMR task and had different advantages. The L-sized MSSA(L) was most recommended, with a moderate performance in terms of accuracy, training speed, and the number of parameters. The default MSSA model had the highest accuracy among current models, with a moderately slower training speed than ResNet. Compared with the LSTM, ResNet, and CNN models, MSSA had fewer parameters, making it suitable and scalable for practical applications in drone-based AMR systems with limited computing resources.

As illustrated in Figure 11, the OOK modulation scheme posed the greatest challenge for the M- and S-sized MSSA models, while the MSSA(XL) model had issues recognizing 16QAM modulation. This could possibly be attributed to the subsampling rate of the input samples. Hence, combining data with different subsampling rates may be an effective solution for the AMR task. Future work will explore how to implement this combination and select appropriate subsampling rates.

The signal pattern recognition algorithm proposed in this study demonstrated superior performance in computer simulations and showed promise for deployment on resource-constrained UAV platforms to enable real-time signal analysis. Further verification through physical experiments remains a priority in future research.

Author Contributions: Y.S.: conceptualization, methodology, software, writing—original draft preparation, and visualization; H.Y.: conceptualization, resources, writing—original draft, supervision, project administration, and methodology; P.Z.: data curation, and software; Y.L.: validation and resources; M.C.: investigation and formal analysis; J.L.: writing—review and editing. All authors have read and agreed to the published version of the manuscript.

Funding: This research received no external funding.

Data Availability Statement: Not applicable.

Conflicts of Interest: The authors declare no conflict of interest.

References

1. Zhang, F.; Luo, C.; Xu, J.; Luo, Y.; Zheng, F. Deep Learning Based Automatic Modulation Recognition: Models, Datasets, and Challenges. *Digit. Signal Process.* **2022**, *129*, 103650. [CrossRef]
2. Bhatti, F.A.; Khan, M.J.; Selim, A.; Paisana, F. Shared Spectrum Monitoring Using Deep Learning. *IEEE Trans. Cogn. Commun. Netw.* **2021**, *7*, 1171–1185. [CrossRef]
3. Dobre, O.; Abdi, A.; Bar-Ness, Y.; Su, W. Survey of automatic modulation classification techniques: Classical approaches and new trends. *IET Commun.* **2007**, *1*, 137–156. [CrossRef]
4. Zhang, W.; Feng, M.; Krunz, M.; Hossein Yazdani Abyaneh, A. Signal Detection and Classification in Shared Spectrum: A Deep Learning Approach. In Proceedings of the IEEE Conference on Computer Communications (IEEE INFOCOM 2021), Vancouver, BC, Canada, 10–13 May 2021; IEEE Press: Piscataway, NJ, USA, 2021; pp. 1–10. [CrossRef]
5. Dulek, B. Online Hybrid Likelihood Based Modulation Classification Using Multiple Sensors. *IEEE Trans. Wirel. Commun.* **2017**, *16*, 4984–5000. [CrossRef]
6. Xu, J.L.; Su, W.; Zhou, M. Likelihood-Ratio Approaches to Automatic Modulation Classification. *IEEE Trans. Syst. Man Cybern. Part C Appl. Rev.* **2011**, *41*, 455–469. [CrossRef]
7. Hazza, A.; Shoaib, M.; Alshebeili, S.A.; Fahad, A. An overview of feature-based methods for digital modulation classification. In Proceedings of the 2013 1st International Conference on Communications, Signal Processing, and Their Applications (ICCSPA), Sharjah, United Arab Emirates, 12–14 February 2013; pp. 1–6. [CrossRef]
8. Chan, Y.; Gadbois, L.; Yansouni, P. Identification of the modulation type of a signal. In Proceedings of the IEEE International Conference on Acoustics, Speech, and Signal Processing (ICASSP '85), Tampa, FL, USA, 26–29 April 1985; Volume 10, pp. 838–841. [CrossRef]
9. Hong, L.; Ho, K. Identification of digital modulation types using the wavelet transform. In Proceedings of the IEEE Military Communications Conference (MILCOM 1999), Atlantic City, NJ, USA, 31 October–3 November 1999; Cat. No.99CH36341, Volume 1, pp. 427–431. [CrossRef]
10. Liu, L.; Xu, J. A Novel Modulation Classification Method Based on High Order Cumulants. In Proceedings of the 2006 International Conference on Wireless Communications, Networking and Mobile Computing, Wuhan, China, 22–24 September 2006; pp. 1–5. [CrossRef]
11. Swami, A.; Sadler, B. Hierarchical digital modulation classification using cumulants. *IEEE Trans. Commun.* **2000**, *48*, 416–429. [CrossRef]
12. Yuan, J.; Zhao-Yang, Z.; Pei-Liang, Q. Modulation classification of communication signals. In Proceedings of the Military Communications Conference (IEEE MILCOM 2004), Monterey, CA, USA, 31 October–3 November 2004; Volume 3, pp. 1470–1476. [CrossRef]
13. Park, C.S.; Choi, J.H.; Nah, S.P.; Jang, W.; Kim, D.Y. Automatic Modulation Recognition of Digital Signals using Wavelet Features and SVM. In Proceedings of the 2008 10th International Conference on Advanced Communication Technology, Gangwon, Republic of Korea, 17–20 February 2008; Volume 1, pp. 387–390. [CrossRef]
14. O'Shea, T.J.; Corgan, J.; Clancy, T.C. Convolutional Radio Modulation Recognition Networks. In Proceedings of the 17th International Conference (EANN 2016), Aberdeen, UK, 2–5 September 2016.

15. Rajendran, S.; Meert, W.; Giustiniano, D.; Lenders, V.; Pollin, S. Deep Learning Models for Wireless Signal Classification with Distributed Low-Cost Spectrum Sensors. *IEEE Trans. Cogn. Commun. Netw.* **2018**, *4*, 433–445. [CrossRef]
16. O'Shea, T.J.; Roy, T.; Clancy, T.C. Over-the-Air Deep Learning Based Radio Signal Classification. *IEEE J. Sel. Top. Signal Process.* **2018**, *12*, 168–179. [CrossRef]
17. Njoku, J.N.; Morocho-Cayamcela, M.E.; Lim, W. CGDNet: Efficient Hybrid Deep Learning Model for Robust Automatic Modulation Recognition. *IEEE Netw. Lett.* **2021**, *3*, 47–51. [CrossRef]
18. Hu, S.; Pei, Y.; Liang, P.P.; Liang, Y.C. Deep Neural Network for Robust Modulation Classification Under Uncertain Noise Conditions. *IEEE Trans. Veh. Technol.* **2020**, *69*, 564–577. [CrossRef]
19. Zhang, M.; Zeng, Y.; Han, Z.; Gong, Y. Automatic Modulation Recognition Using Deep Learning Architectures. In Proceedings of the 2018 IEEE 19th International Workshop on Signal Processing Advances in Wireless Communications (SPAWC), Kalamata, Greece, 25–28 June 2018; pp. 1–5. [CrossRef]
20. Ghasemzadeh, P.; Hempel, M.; Sharif, H. GS-QRNN: A High-Efficiency Automatic Modulation Classifier for Cognitive Radio IoT. *IEEE Internet Things J.* **2022**, *9*, 9467–9477. [CrossRef]
21. Wang, T.; Hou, Y.; Zhang, H.; Guo, Z. Deep Learning Based Modulation Recognition With Multi-Cue Fusion. *IEEE Wirel. Commun. Lett.* **2021**, *10*, 1757–1760. [CrossRef]
22. Li, Y.; Shao, G.; Wang, B. Automatic Modulation Classification Based on Bispectrum and CNN. In Proceedings of the 2019 IEEE 8th Joint International Information Technology and Artificial Intelligence Conference (ITAIC), Chongqing, China, 24–26 May 2019; pp. 311–316. [CrossRef]
23. Peng, S.; Jiang, H.; Wang, H.; Alwageed, H.; Zhou, Y.; Sebdani, M.M.; Yao, Y.D. Modulation Classification Based on Signal Constellation Diagrams and Deep Learning. *IEEE Trans. Neural Netw. Learn. Syst.* **2019**, *30*, 718–727. [CrossRef] [PubMed]
24. Krizhevsky, A.; Sutskever, I.; Hinton, G.E. ImageNet classification with deep convolutional neural networks. *Commun. ACM* **2012**, *60*, 84–90. [CrossRef]
25. Szegedy, C.; Liu, W.; Jia, Y.; Sermanet, P.; Reed, S.; Anguelov, D.; Erhan, D.; Vanhoucke, V.; Rabinovich, A. Going Deeper with Convolutions. *arXiv* **2014**, arXiv:1409.4842.
26. Ramjee, S.; Ju, S.; Yang, D.; Liu, X.; Gamal, A.E.; Eldar, Y.C. Fast Deep Learning for Automatic Modulation Classification. *arXiv* **2019**, arXiv:1901.05850.
27. Alsamhi, S.H.; Shvetsov, A.V.; Kumar, S.; Hassan, J.; Alhartomi, M.A.; Shvetsova, S.V.; Sahal, R.; Hawbani, A. Computing in the Sky: A Survey on Intelligent Ubiquitous Computing for UAV-Assisted 6G Networks and Industry 4.0/5.0. *Drones* **2022**, *6*, 177. [CrossRef]
28. Hu, X.; Wong, K.K.; Yang, K.; Zheng, Z. UAV-Assisted Relaying and Edge Computing: Scheduling and Trajectory Optimization. *IEEE Trans. Wirel. Commun.* **2019**, *18*, 4738–4752. [CrossRef]
29. He, Y.; Wang, D.; Huang, F.; Zhang, R.; Gu, X.; Pan, J. A V2I and V2V Collaboration Framework to Support Emergency Communications in ABS-Aided Internet of Vehicles. *IEEE Trans. Green Commun. Netw.* 2023, *early access*. [CrossRef]
30. Shi, J.; Zhao, L.; Wang, X.; Guizani, M.; Gaanin, H.; Lin, N. Flying Social Networks: Architecture, Challenges and Open Issues. *IEEE Netw.* **2021**, *35*, 242–248. [CrossRef]
31. Ke, Z.; Vikalo, H. Real-Time Radio Technology and Modulation Classification via an LSTM Auto-Encoder. *IEEE Trans. Wirel. Commun.* **2022**, *21*, 370–382. [CrossRef]
32. Chang, S.; Huang, S.; Zhang, R.; Feng, Z.; Liu, L. Multitask-Learning-Based Deep Neural Network for Automatic Modulation Classification. *IEEE Internet Things J.* **2022**, *9*, 2192–2206. [CrossRef]
33. Srivastava, R.K.; Greff, K.; Schmidhuber, J. Training Very Deep Networks. In Proceedings of the Advances in Neural Information Processing Systems 28 (NIPS 2015) Conference, Montreal, QC, Canada, 7–12 December 2015; Cortes, C., Lawrence, N., Lee, D., Sugiyama, M., Garnett, R., Eds.; Curran Associates, Inc.: Red Hook, NY, USA, 2015; Volume 28, pp. 2377–2385.
34. Vaswani, A.; Shazeer, N.; Parmar, N.; Uszkoreit, J.; Jones, L.; Gomez, A.N.; Kaiser, L.; Polosukhin, I. Attention Is All You Need. In Proceedings of the Advances in Neural Information Processing Systems 30 (NIPS 2017) Conference, Long Beach, CA, USA, 4–9 December 2017.

Communication

Optimal Position and Target Rate for Covert Communication in UAV-Assisted Uplink RSMA Systems

Zhengxiang Duan [1], Xin Yang [1,*], Tao Zhang [2] and Ling Wang [1,*]

[1] School of Electronics and Information, Northwestern Polytechnical University, Xi'an 710072, China
[2] China Academy of Launch Vehicle Technology, Beijing 100076, China
*   Correspondence: xinyang@nwpu.edu.cn (X.Y.); lingwang@nwpu.edu.cn (L.W.)

Abstract: With the explosive increase in demand for wireless communication, the issue of wireless communication security has also become a growing concern. In this paper, we investigate a novel covert communication for unmanned aerial vehicle (UAV)-assisted uplink rate-splitting multiple access (RSMA) systems, where a UAV adopts the rate-splitting (RS) strategy to increase the total transmission rate while avoiding deteriorating the covert transmission of a ground user. In the proposed system, a ground user and a UAV adopt the RSMA scheme to simultaneously communicate with a base station surveilled by an evil monitor. The UAV acts as both the transmitter and the friendly jammer to cover the ground user's transmission with random power. To maximize the expected sum rate (ESR), we first study the RS strategy and obtain the optimal power allocation factor. Then, the closed-form of minimum detection error probability (DEP), ESR, and optimal target rate of the UAV are derived. Constrained by the minimum DEP and expected covert rate (ECR), we maximize the ESR by optimizing the position and target rate of the UAV. Numerical results show that the proposed scheme outperforms the traditional NOMA systems in terms of ESR with the same DEP and ECR.

Keywords: covert communication; RSMA; UAV; sum rate

Citation: Duan, Z.; Yang, X.; Zhang, T.; Wang, L. Optimal Position and Target Rate for Covert Communication in UAV-Assisted Uplink RSMA Systems. *Drones* **2023**, *7*, 237. https://doi.org/10.3390/drones7040237

Academic Editor: Maurizio Magarini

Received: 6 March 2023
Revised: 24 March 2023
Accepted: 27 March 2023
Published: 28 March 2023

1. Introduction

The rapid development of intelligent wireless terminals promotes a large amount of wireless private information, bringing broader attention to information security. The study of communication security at the physical layer has been segregated into two directions, namely physical layer security (PLS) [1,2] and covert communications [3,4]. The purpose of PLS is to ensure that the transmitted information is not intercepted by eavesdroppers, i.e., protecting the transmitted content. However, it's not always sufficient to focus solely on protecting information security, as the exposure of communication behavior can also pose potential risks and threats. For example, the exposure of the signal transmission would disclose the existence and position of a device to an adversary, ultimately resulting in an attack on the device. Different from PLS, covert communications focus on shielding the transmission behavior from potential watchful adversaries. Bash et al. initiated covert communication research and proposed a square root law as the fundamental limit in the additive white Gaussian noise (AWGN) channels [3]. In [4], the authors proved that it is possible to transmit $\mathcal{O}(n)$ bits covertly and reliably in n uses of AWGN channel when the monitor has uncertainty about the received power.

With the growing number of connected devices and increasing data traffic, there is a need for more efficient and effective ways to manage the available resources. Non-orthogonal multiple access (NOMA) is a promising technique that offers improvements over conventional orthogonal multiple access techniques in terms of spectral efficiency [5,6]. In [7], the PLS in NOMA systems assisted with a HAP and UAVs was studied. Rate-splitting multiple access (RSMA), which can further increase the sum rate, has recently emerged as a more general and robust transmission framework compared to NOMA [8–12]. In particular, the performances of uplink RSMA systems were studied in [10,11]. In [10], the authors investigated a rate-splitting (RS) strategy in uplink cognitive radio systems, where a secondary user splits its rate to guarantee the primary user's transmission. In [11], the optimal decoding order and maximum sum rate in uplink RSMA systems were studied. To protect the privacy information, the authors in [13] studied the security and energy efficiency of the cognitive RSMA-based satellite-terrestrial networks, where a beamforming scheme was proposed to prevent eavesdropping and increase energy efficiency.

Covert communications in NOMA systems have also been widely studied [14,15]. The author in [14] achieved covert communication in an uplink NOMA system via random power jamming generated by channel inversion power control. The study in [15] explored an intelligent reflecting surface (IRS)-assisted covert communication in both downlink and uplink NOMA systems. The randomness was brought about by the phase-shift uncertainty of IRS and the overlapping signal transmission. In addition, unmanned aerial vehicles (UAVs) have been used by virtue of their high mobility, which provides new degrees of freedom to enhance the covertness of communications [16,17]. In [16], the optimal transmit power and location for the UAV were studied to achieve covert communications. In [17], the authors used the geometric method to solve the trajectory problem. Most recently, covert communication in UAV-aided NOMA systems was investigated in [18].

To further increase the sum rate, we investigate covert communication in UAV-assisted uplink RSMA systems in this paper. In this system, a ground user and a UAV simultaneously communicate with a base station (BS), suffering the surveillance of an evil monitor. The UAV acts as both the transmitter and the friendly jammer, covering the ground user's transmission with random power. This work aims to maximize the expected sum rate (ESR) by designing the UAV's power allocation, position, and target rate while guaranteeing the ground user's covertness and throughput. The main contributions of this paper are given as follows.

- We investigate a novel application of RSMA systems, where a UAV splits its rate to avoid deteriorating the covert transmission of a ground user while increasing the ESR. To the best of the authors' knowledge, this is the first work that studied the covet communication in UAV-assisted uplink RSMA system.
- We derive the closed-form expressions of the ESR and obtain the optimal target rate of UAV which maximizes the ESR of the system. Subjected to minimum detection error probability (DEP) and expected covert rate (ECR) constraints, a joint position and target rate optimization problem is formulated for maximizing the ESR of uplink RSMA systems.
- The numerical results show that the proposed scheme outperforms NOMA systems in terms of ESR with the same DEP and ECR and illustrate the effect of constraints on the ESR.

2. System Model

2.1. Communication Scenario

We consider the covert communication in uplink RSMA system, which consists of a pair of RSMA users (U1 and U2), a BS, and a warden (Willie), as shown in Figure 1. U1 is a UAV deployed as both the communication node and the friendly jammer hovering at the constant altitude z_1. U2 wants to transmit covertly detected by Willie, who continuously senses whether U2 is transmitting by a radiometer. In order to increase the total transmission rate while protecting U2's covert communication, U1 adopts the RS strategy with random transmit power. Without loss of generality, we use a three-dimensional (3D) Cartesian coordinate system to describe locations. Each node is equipped with a single antenna.

Denote $\mathbf{q}_1 = [x_1, y_1, z_1]$, $\mathbf{q}_2 = [x_2, y_2, 0]$, $\mathbf{q}_b = [0, 0, z_b]$, and $\mathbf{q}_w = [x_w, y_w, 0]$ as the coordinate of U1, U2, BS and Willie, respectively. Considering an open area, the communication channel between BS and U2 is modeled as line-of-sight (LOS) links and AWGN channels. This assumption is based on the fact that in the urban macro, the probability of the LOS path is much higher than that of the non-LOS path when the horizontal distance between BS and terminals is less than 70 m according to 3GPP specification [19]. In addition, we assume that the channels between UAV and terrestrial nodes are mainly dominated by LOS components and the non-LOS path is negligible (as in e.g., [16,17,20]). U2 and Willie are ground users and the channel undergoes the block quasi-static fading, which means that the channel coefficients remain constant in one time slot, and change independently from one time slot to another. The large-scale fading coefficient from node i to node j is denoted as $L_{ij} = \beta_0 \|\mathbf{q}_i - \mathbf{q}_j\|^{-\beta}$, where $i \in \{1, 2\}$, $j \in \{b, w\}$, β_0 is the fading coefficient at the reference distance of 1 m, $\beta = 2$ is the free space path-loss factor, and $\|\cdot\|$ denotes the Euclidean norm. And the small-scale fading between U2 and Willie r_{2w} follows complex Gaussian distribution $\mathcal{CN}(0, 1)$. Therefore, the channel coefficient is denoted as

$$h_{ij} = \begin{cases} \sqrt{L_{ij}}, & ij \in \{1w, 1b, 2b\}, \\ \sqrt{L_{ij}}r_{2w}, & ij = 2w. \end{cases} \tag{1}$$

And the channel power gain is expressed by $g_{ij} = |h_{ij}|^2$. Suppose the location information is available for all nodes since Willie's location can be detected with a radar or camera by U1. This assumption has been also widely adopted in previous research on UAV-assisted covert communication [16,17,20]. In addition, we assume that full channel state information (CSI) is available for Willie, while legitimate users only possess statistical CSI between U2 and Willie.

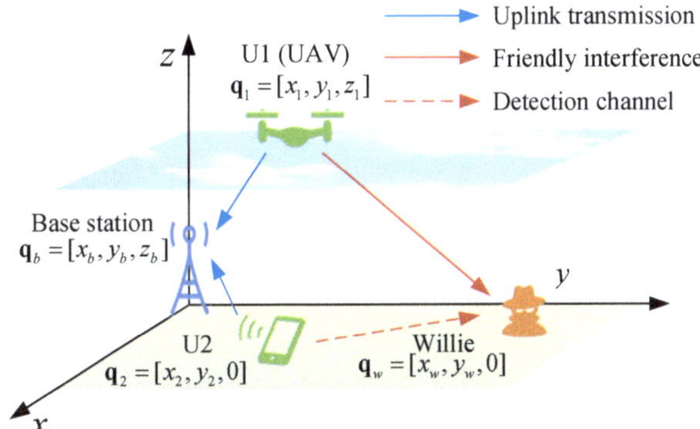

Figure 1. The uplink RSMA covert communication system model.

2.2. Proposed Transmission Scheme

U1 adopts the RS strategy with random transmit power to assist U2's covert and reliable transmission. U2 conveys secret messages at a fixed rate and probability of 0.5, while U1 transmits public information. To confuse Willie's detection, U1 adopts the random transmit power. Furthermore, U1 shares the same resource block with U2 while preventing U2's outage probability from deteriorating through an RS strategy.

Denoting P_i, $\mathbf{s}_i$, and $\hat{R}_i$ as the transmit power, messages, and target rate of Ui, respectively. For simplicity, we adopt μ_i to represent $2^{\hat{R}_i} - 1$. U2 transmits with a fixed power P_2 and rate $\hat{R}_2$. Then, the interference threshold of U1 is given by

$$\tau = \frac{g_{2b}P_2}{\mu_2} - \sigma_b^2, \tag{2}$$

where σ_b^2 represents the received noise power at BS. When the interference received by U2 is lower than τ, there is no outage. Conversely, U2's connection is always interrupted when the interference is large than τ. Note that to ensure $\tau \geq 0$, we have $\mu_2 \leq \frac{g_{2b}P_2}{\sigma_b^2}$.

To enhance the covertness of U2's transmission, P_1 changes from slot to slot, following a continuous uniform distribution within $[P_1^{\min}, P_1^{\max}]$. The probability density function of P_1 is given by

$$f_{P_1}(x) = \begin{cases} \frac{1}{P_1^{\max} - P_1^{\min}}, & \text{if } P_1^{\min} \leq x \leq P_1^{\max}, \\ 0, & \text{otherwise.} \end{cases} \tag{3}$$

In order to obfuscate Willie's detection and transmit more information, U1 continuously sends messages to BS. The random transmit power of U1 is designed to create ambiguity in Willie's received power. Consequently, it becomes challenging for Willie to determine whether the increase in received power is due to U2's transmission or simply a variation in U1's transmit power.

To increase the total throughput without causing interruptions to U2, U1 applies RS in each time slot. In uplink RSMA systems, U1 needs to split its messages $\mathbf{s}_1$ into two parts $\mathbf{s}_{11}$ and $\mathbf{s}_{12}$, as shown in Figure 2a. Note that Figure 2a depicts only one possible splitting scheme. There are also alternative approaches that can be considered. And the received signals at node j can be expressed as

$$\mathbf{y}_j[n] = \begin{cases} \sqrt{P_1}h_{1j}(\sqrt{\alpha}\mathbf{s}_{11}[n] + \sqrt{1-\alpha}\mathbf{s}_{12}[n]) + \mathbf{n}_j[n], & \mathcal{H}_0, \\ \sqrt{P_1}h_{1j}(\sqrt{\alpha}\mathbf{s}_{11}[n] + \sqrt{1-\alpha}\mathbf{s}_{12}[n]) + \sqrt{P_2}h_{2j}\mathbf{s}_2[n] + \mathbf{n}_j[n], & \mathcal{H}_1, \end{cases} \tag{4}$$

where $n = 1 \ldots N$ is the index of channel use, $\mathbf{n}_j[n]$ is the received AWGN at j with the variance of σ_j^2, and α is the power allocation factor satisfying $0 \leq \alpha \leq 1$. The hypotheses $\mathcal{H}_1$ and $\mathcal{H}_0$ represent the existence and non-existence of U2's secret transmission, respectively. It is assumed that $\mathbf{s}_k$, $k \in \{1,2,11,12\}$, is independently coded with the Gaussian codebook satisfying $\mathbb{E}\{\mathbf{s}_k[n]\mathbf{s}_k^*[n]\} = 1$, where $\mathbb{E}\{\cdot\}$ and $(\cdot)^*$ represent the expectation and conjugate transpose operators, respectively.

The decoding order for uplink RSMA is $\mathbf{s}_{11} \rightarrow \mathbf{s}_2 \rightarrow \mathbf{s}_{12}$ [11]. Thus, the signal-to-interference-plus-noise ratios (SINRs) for BS decoding $\mathbf{s}_{11}$, $\mathbf{s}_2$, and $\mathbf{s}_{12}$ are given by $\gamma_{11} = \frac{\alpha g_{1b}P_1}{g_{2b}P_2 + (1-\alpha)g_{1b}P_1 + \sigma_b^2}$, $\gamma_2 = \frac{g_{2b}P_2}{(1-\alpha)g_{1b}P_1 + \sigma_b^2}$, and $\gamma_{12} = \frac{(1-\alpha)g_{1b}P_1}{\sigma_b^2}$, respectively. Correspondingly, the achievable rates of $\mathbf{s}_{11}$, $\mathbf{s}_2$, and $\mathbf{s}_{12}$ are expressed as $R_{11} = \log_2(1 + \gamma_{11})$, $R_2 = \log_2(1 + \gamma_2)$ and $R_{12} = \log_2(1 + \gamma_{12})$, respectively.

To maximize the sum rate for the RS strategy, U1 needs to allocate the maximum possible power to $\mathbf{s}_{12}$. As per γ_{12}, $\mathbf{s}_{12}$ is free from interference, hence allocating power to $\mathbf{s}_{12}$ would be more efficient compared to $\mathbf{s}_{11}$. Considering that the interference received by U2 should be no large than τ to keep U2 uninterrupted, we have $(1-\alpha)g_{1b}P_1 \leq \tau$. Obviously, the allowed maximum power of $\mathbf{s}_{12}$ is τ/g_{1b} with $\alpha = 1 - \frac{\tau}{g_{1b}P_1}$. Meanwhile, U2's messages are not supposed to be decoded firstly for covertness, which results in $\alpha P_1 \geq 0$, i.e., $P_1 \geq \tau/g_{1b}$. The target rates of $\mathbf{s}_{12}$ and $\mathbf{s}_{11}$ are set as $\hat{R}_{12} = \log_2(1 + \frac{\tau}{\sigma_b^2})$ and $\hat{R}_{11} = \hat{R}_1 - \hat{R}_{12}$, respectively. The power allocation scheme and decoding order are shown in Figure 2b.

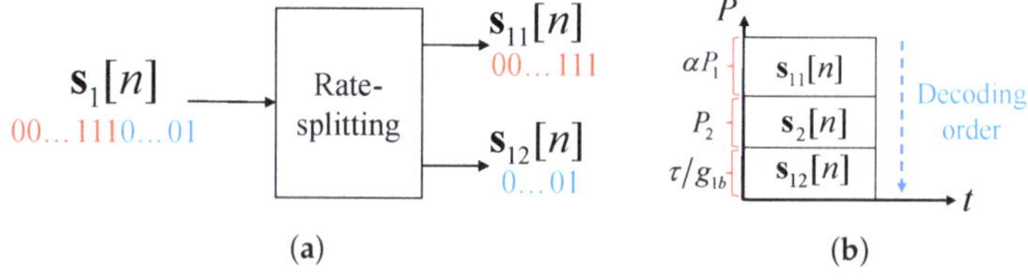

Figure 2. (**a**) A sample for splitting $\mathbf{s}_1$ into $\mathbf{s}_{11}$ and $\mathbf{s}_{12}$. (**b**) An illustration of power allocation and decoding order.

2.3. Detection Metrics at Willie

Willie tries to make a decision whether U2 is transmitting or not based on the received signals $\mathbf{y}_w[n]$. From the independent and identically distributed (i.i.d.) nature of Willie's received signals, the distribution of $\mathbf{y}_w[n]$ is expressed as

$$
\begin{cases}
\mathcal{CN}(0, g_{1w}P_1 + \sigma_w^2), & \mathcal{H}_0, \\
\mathcal{CN}(0, g_{1w}P_1 + g_{2w}P_2 + \sigma_w^2), & \mathcal{H}_1.
\end{cases}
\tag{5}
$$

According to the Neyman-Pearson criterion, the optimal decision rule at Willie is the likelihood ratio test (LRT) [14,15], which can be shown as a radiometer

$$
T_w \underset{\mathcal{D}_1}{\overset{\mathcal{D}_0}{\lessgtr}} \lambda,
\tag{6}
$$

where $T_w = \frac{1}{N}\sum_{n=1}^{N}|\mathbf{y}_w[n]|^2$ is the average power received at Willie in a time slot, λ is the detection threshold of Willie, $\mathcal{D}_1$ and $\mathcal{D}_0$ are the binary decisions for the hypotheses $\mathcal{H}_0$ and $\mathcal{H}_1$, respectively. Considering a long observation of Willie, i.e., $N \to \infty$, and employing the strong law of large numbers, i.e., $\mathcal{X}_{2N}^2/N \to 1$, T_w is given by

$$
\begin{aligned}
T_w &= \lim_{N\to\infty}
\begin{cases}
(g_{1w}P_1 + \sigma_w^2)\mathcal{X}_{2N}^2/N, & \mathcal{H}_0, \\
(g_{1w}P_1 + g_{2w}P_2 + \sigma_w^2)\mathcal{X}_{2N}^2/N, & \mathcal{H}_1,
\end{cases} \\
&=
\begin{cases}
g_{1w}P_1 + \sigma_w^2, & \mathcal{H}_0, \\
g_{1w}P_1 + g_{2w}P_2 + \sigma_w^2, & \mathcal{H}_1.
\end{cases}
\end{aligned}
\tag{7}
$$

The detection performance is measured by the DEP, which is denoted as

$$
\mathbb{P}_E \triangleq \mathbb{P}_{\mathrm{FA}} + \mathbb{P}_{\mathrm{MD}},
\tag{8}
$$

where $\mathbb{P}_{\mathrm{FA}} = \mathbb{P}\{T_w > \lambda | \mathcal{H}_0\}$ is the false alarm probability (FAP), $\mathbb{P}_{\mathrm{MD}} = \mathbb{P}\{T_w < \lambda | \mathcal{H}_1\}$ is the miss detection probability (MDP), $\mathbb{P}\{\cdot\}$ denotes probability operation and the prior probabilities of hypotheses $\mathcal{H}_0$ and $\mathcal{H}_1$ are assumed to be equal, i.e., $\mathbb{P}(\mathbb{H}_0) = \mathbb{P}(\mathbb{H}_1) = 1/2$.

3. Performance Analysis

In this section, we analyze the performances of the covertness and sum rate in the proposed system.

3.1. Covertness Analysis

Similar to the proof in [21], the FAP and DEP are given by

$$
\mathbb{P}_{\mathrm{FA}} =
\begin{cases}
1, & \frac{\lambda - \sigma_w^2}{g_{1w}} < P_1^{\min}, \\
\frac{g_{1w}P_1^{\max} + \sigma_w^2 - \lambda}{g_{1w}(P_1^{\max} - P_1^{\min})}, & P_1^{\min} \le \frac{\lambda - \sigma_w^2}{g_{1w}} \le P_1^{\max}, \\
0, & \frac{\lambda - \sigma_w^2}{g_{1w}} > P_1^{\max},
\end{cases}
\tag{9}
$$

$$
\mathbb{P}_{\mathrm{MD}} =
\begin{cases}
0, & \frac{\lambda - \sigma_w^2}{g_{1w}} < \rho_1, \\
\frac{\lambda - g_{1w}P_1^{\min} - g_{2w}P_2 - \sigma_w^2}{g_{1w}(P_1^{\max} - P_1^{\min})}, & \rho_1 \le \frac{\lambda - \sigma_w^2}{g_{1w}} \le \rho_2, \\
1, & \frac{\lambda - \sigma_w^2}{g_{1w}} > \rho_2,
\end{cases}
\tag{10}
$$

where $\rho_1 = P_1^{\min} + \frac{g_{2w}}{g_{1w}}P_2$, $\rho_2 = P_1^{\max} + \frac{g_{2w}}{g_{1w}}P_2$. Then, the DEP at Willie is given by

$$
\mathbb{P}_E = \begin{cases}
1, & \frac{\lambda-\sigma_w^2}{g_{1w}} < P_1^{\min}, \\
\frac{g_{1w}P_1^{\max}+\sigma_w^2-\lambda}{g_{1w}(P_1^{\max}-P_1^{\min})}, & P_1^{\min} \le \frac{\lambda-\sigma_w^2}{g_{1w}} < \rho_1, \\
1 - \frac{g_{2w}P_2}{g_{1w}(P_1^{\max}-P_1^{\min})}, & \rho_1 \le \frac{\lambda-\sigma_w^2}{g_{1w}} \le P_1^{\max}, \\
\frac{\lambda-g_{1w}P_1^{\min}-g_{2w}P_2-\sigma_w^2}{g_{1w}(P_1^{\max}-P_1^{\min})}, & P_1^{\max} < \frac{\lambda-\sigma_w^2}{g_{1w}} \le \rho_2, \\
1, & \frac{\lambda-\sigma_w^2}{g_{1w}} > \rho_2.
\end{cases}
\tag{11}
$$

Note that the condition $g_{1w}(P_1^{\max} - P_1^{\min}) \ge g_{2w}P_2$ needs to be satisfied; otherwise, Willie has zero probability of making detection errors.

According to the monotonicity of (11), the minimum DEP is given by

$$
\mathbb{P}_E^\dagger = 1 - \frac{g_{2w}P_2}{g_{1w}(P_1^{\max} - P_1^{\min})},
\tag{12}
$$

and the corresponding detection threshold satisfies $\rho_1 \le \frac{\lambda-\sigma_w^2}{g_{1w}} \le P_1^{\max}$. Since legitimate users don't have instantaneous CSI between Willie and U2, we consider the expected minimum DEP $\overline{\mathbb{P}_E^\dagger}$ over all possible realization of h_{2w} as the measurement of covertness from the perspective of legitimate users. $\overline{\mathbb{P}_E^\dagger}$ is given by

$$
\begin{aligned}
\overline{\mathbb{P}_E^\dagger} &= \int_0^{\frac{g_{1w}(P_1^{\max}-P_1^{\min})}{P_2}} \mathbb{P}_E^\dagger f_{g_{2w}}(x)dx \\
&= \int_0^{\frac{g_{1w}(P_1^{\max}-P_1^{\min})}{P_2}} \left[1 - \frac{xP_2}{g_{1w}(P_1^{\max}-P_1^{\min})}\right] \frac{1}{L_{2w}} e^{-\frac{x}{L_{2w}}} dx \\
&= 1 + \frac{L_{2w}P_2}{g_{1w}(P_1^{\max}-P_1^{\min})}\left[e^{-\frac{g_{1w}(P_1^{\max}-P_1^{\min})}{L_{2w}P_2}} - 1\right],
\end{aligned}
\tag{13}
$$

where $f_{g_{2w}}(x)$ is the probability density function of g_{2w}.

The results indicate that as the variation interval of P_1 increases, there is a corresponding rise in DEP, leading to a larger value of $\hat{R}_2$. In addition, U1 can modify its channel to Willie by repositioning to meet the covertness requirement.

3.2. Sum Rate Analysis

In the proposed scheme, the power allocation for s_{12} is designed to prevent connection outages during the decoding of both s_2 and s_{12}. However, an outage may still occur during the decoding of s_{11} due to the randomness of P_1. Therefore, the outage probability of the system is determined by that of decoding s_{11}. We respectively analyze the outage probabilities and ESR under $\mathcal{H}_0$ and $\mathcal{H}_1$ in the following.

1. Under $\mathcal{H}_1$

    The achievable rate under $\mathcal{H}_1$ of s_{11} is given by

$$
R_{11}^1 = \log_2\left(1 + \frac{g_{1b}P_1 - \tau}{g_{2b}P_2 + \tau + \sigma_b^2}\right).
\tag{14}
$$

Thus, the outage probability of $\mathbf{s}_{11}$ under $\mathcal{H}_1$ is expressed as

$$
\begin{aligned}
\mathbb{O}_{11}^1 &= \mathbb{P}\{R_{11}^1 < \hat{R}_{11}\} \\
&= \mathbb{P}\left\{P_1 < \frac{\mu_{11}(g_{2b}P_2 + \tau + \sigma_b^2) + \tau}{g_{1b}}\right\} \\
&= \max\left\{\frac{\mu_{11}(g_{2b}P_2 + \tau + \sigma_b^2) + \tau - g_{1b}P_1^{\min}}{g_{1b}(P_1^{\max} - P_1^{\min})}, 0\right\},
\end{aligned}
\tag{15}
$$

where $\mu_{11} = 2^{\hat{R}_{11}} - 1$. Eventually, the ESR under $\mathcal{H}_1$ is given by

$$
\overline{R}_{\text{sum}}^1 = (\hat{R}_{11} + \hat{R}_2 + \hat{R}_{12})(1 - \mathbb{O}_{11}^1).
\tag{16}
$$

2. Under $\mathcal{H}_0$

Similarly, the achievable rate under $\mathcal{H}_0$ of $\mathbf{s}_{11}$ is given by

$$
R_{11}^1 = \log_2\left(1 + \frac{g_{1b}P_1 - \tau}{\tau + \sigma_b^2}\right).
\tag{17}
$$

And the outage probability of $\mathbf{s}_{11}$ under $\mathcal{H}_0$ is expressed as

$$
\begin{aligned}
\mathbb{O}_{11}^0 &= \mathbb{P}\{R_{11}^0 < \hat{R}_{11}\} \\
&= \mathbb{P}\left\{P_1 < \frac{\mu_{11}(\tau + \sigma_b^2) + \tau}{g_{1b}}\right\} \\
&= \max\left\{\frac{\mu_{11}(\tau + \sigma_b^2) + \tau - g_{1b}P_1^{\min}}{g_{1b}(P_1^{\max} - P_1^{\min})}, 0\right\}.
\end{aligned}
\tag{18}
$$

Since fixed power is allocated to $\mathbf{s}_{12}$ to satisfy $\hat{R}_{12}$, $\mathbb{P}\{R_{11}^0 < \hat{R}_{11}\} = \mathbb{P}\{R_1^0 < \hat{R}_1\}$. The ESR under $\mathcal{H}_0$ is given by

$$
\overline{R}_{\text{sum}}^0 = (\hat{R}_{11} + \hat{R}_{12})(1 - \mathbb{O}_{11}^0).
\tag{19}
$$

Finally, the ESR of the system is expressed as

$$
\begin{aligned}
\overline{R}_{\text{sum}} &= \frac{1}{2}\left(\overline{R}_{\text{sum}}^0 + \overline{R}_{\text{sum}}^1\right) \\
&\triangleq f\left[\left(a - b2^{\hat{R}_{11}}\right)\hat{R}_{11} - c2^{\hat{R}_{11}} + d\right],
\end{aligned}
\tag{20}
$$

where $a = 2g_{1b}P_1^{\max} + g_{2b}P_2 + 2\sigma_b^2$, $b = g_{2b}P_2 + 2\tau + 2\sigma_b^2$, $c = (2\hat{R}_{12} + \hat{R}_2)(\tau + \sigma_b^2) + (\hat{R}_{12} + \hat{R}_2)g_{2b}P_2$, $d = (2\hat{R}_{12} + \hat{R}_2)(g_{1b}P_1^{\max} + \sigma_b^2) + (\hat{R}_{12} + \hat{R}_2)g_{2b}P_2$, and $f = \frac{1}{2g_{1b}(P_1^{\max} - P_1^{\min})}$. Component $1/2$ is due to $\mathbb{P}(\mathcal{H}_0) = \mathbb{P}(\mathcal{H}_1) = 1/2$. Equation (20) demonstrates that as $\hat{R}_{11}$ increases, the outage probabilities also increase, whereas the change in ESR is uncertain. Therefore, to maximize ESR, it is necessary to consider how to set $\hat{R}_{11}$.

Lemma 1. *The optimal $\hat{R}_{11}$ to maximize the ESR is given by*

$$
\hat{R}_{11}^\dagger = \max\left\{0, \log_2 e\left[W\left(\frac{ae2^{c/b}}{b}\right) - 1\right] - \frac{c}{b}\right\},
\tag{21}
$$

where e is Euler's number, and $W(\cdot)$ denotes Lambert W Function [22]. The corresponding $\overline{R}_{sum}$ is denoted as $\overline{R}_{sum}(\hat{R}_{11}^\dagger)$.

Proof. See Appendix A. $\square$

We notice that as g_{1w} increases, $\hat{R}_{11}^\dagger$ increases, resulting in a higher ESR. Therefore, ESR can be increased by placing U1 closer to the BS.

4. Optimization Problem

In this letter, we aim to maximize the ESR by optimizing the deployment of U1 and the target rate of s_{11}, subject to the covertness constraint and the ECR constraint. The optimization problem is formulated as

$$\max_{x_1, y_1, \hat{R}_{11}} \quad \overline{R}_{\text{sum}}, \tag{22a}$$

$$\text{s.t.} \quad \overline{\mathbb{P}_E^\dagger} \geq 1 - \delta, \tag{22b}$$

$$\overline{R}_2 = \hat{R}_2(1 - \mathbb{O}_{11}^1) \geq \epsilon, \tag{22c}$$

where (22b) is the covertness constraint, and (22c) is the ECR constraint. To solve the optimization problem (22), we decompose it into two subproblems, as shown in Figure 3. We first discuss the monotonicity of $\overline{R}_{\text{sum}}$, $\overline{\mathbb{P}_E}$, and $\overline{R}_2$ w.r.t. x_1 and y_1. We observe that as the distance between U1 and Bob, i.e., $\|\mathbf{q}_1 - \mathbf{q}_b\|$, decreases, both $\overline{R}_{\text{sum}}$ and $\overline{R}_2$ increases. On the other hand, $\overline{\mathbb{P}_E}$ decreases since U1 gets farther away from Willie. Then, the first subproblem is to optimize the placement of U1 under the covertness constraint (22b) to minimize the distance between U1 and Bob. The second subproblem is to optimize $\hat{R}_{11}$ under the ECR constraint (22c) and U1's optimal placement obtained from the first subproblem to maximize the ESR.

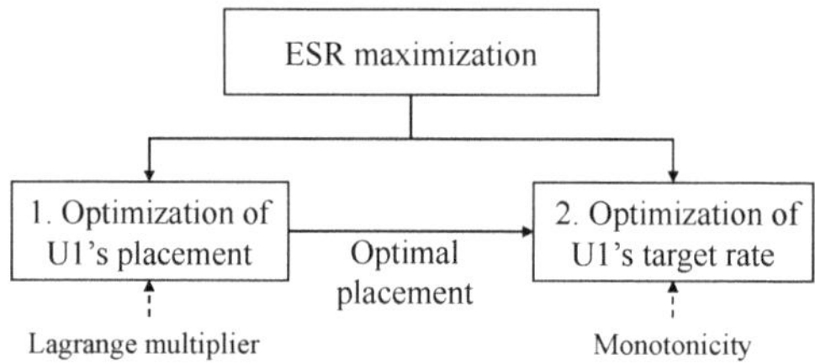

Figure 3. Procedure for solving optimization problem (22).

The first subproblem to optimize U1's placement is expressed as

$$\min_{x_1, y_1} \quad x_1^2 + y_1^2, \tag{23a}$$

$$\text{s.t.} \quad (x_1 - x_w)^2 + (y_1 - y_w)^2 \leq t, \tag{23b}$$

$$t = \frac{\delta \beta_0 (P_1^{\max} - P_1^{\min})}{g_{2w} P_2 \left[1 + W\left(-\frac{e^{-1/\delta}}{\delta}\right)\right]} - z_1^2. \tag{23c}$$

Lemma 2. *The optimal position of U1 is given by* $\mathbf{q}_1^\dagger = [x_1^\dagger, y_1^\dagger, z_1]$ *when* $0 < t < x_w^2 + y_w^2$, *where* $[x_1^\dagger, y_1^\dagger] = \left(1 - \sqrt{\frac{t}{x_w^2 + y_w^2}}\right)[x_w, y_w]$, *otherwise,* $x_1^\dagger = y_1^\dagger = 0$, *when* $t \geq x_w^2 + y_w^2$.

Proof. See Appendix B. $\square$

From Lemma 2, we notice that the optimal horizontal position of U1 lies on the line between BS and Willie.

The second subproblem to optimize $\hat{R}_{11}$ is expressed as

$$\max_{\hat{R}_{11}} \quad \overline{R}_{\text{sum}}(x_1^\dagger, y_1^\dagger), \tag{24a}$$

$$\text{s.t.} \quad \mathbb{O}_{11}(x_1^\dagger, y_1^\dagger) \leq 1 - \frac{\epsilon}{\hat{R}_2}, \tag{24b}$$

where $\overline{R}_{\text{sum}}(x_1^\dagger, y_1^\dagger)$ and $\mathbb{O}_{11}(x_1^\dagger, y_1^\dagger)$ represent substituting $(x_1^\dagger, y_1^\dagger)$ into $\overline{R}_{\text{sum}}$ and $\mathbb{O}_{11}$, respectively. It is shown in (18) that $\mathbb{O}_{11}$ is a monotonically increasing function w.r.t. $\hat{R}_{11}$. Thus, the upper limit of $\hat{R}_{11}$ is expressed as $\hat{R}_{11}^\epsilon$, where $\hat{R}_{11}^\epsilon$ is the solution of $\mathbb{O}_{11}(x_1^\dagger, y_1^\dagger) = 1 - \frac{\epsilon}{\hat{R}_2}$. Together with Lemma 1, the optimal choice of $\hat{R}_{11}$ is given by $\hat{R}_{11}^\dagger = \min\{\hat{R}_{11}^\epsilon, \hat{R}_{11}^\dagger\}$.

5. Numerical Results

In this section, we present numerical results to investigate the performance of the proposed covert communication scheme. Unless otherwise stated, we set $\beta_0 = -20$ dB, $\mathbf{q}_b = [0, 0, 10]$ m, $\mathbf{q}_w = [0, 100, 0]$ m, $\mathbf{q}_2 = [50, 50, 0]$ m, $z_1 = 25$ m, $P_1^{\max} = 10$ W, $P_1^{\min} = 1$ W, $P_2 = 2$ W, $\sigma_b^2 = \sigma_w^2 = -60$ dBm. In this section, we compared the proposed method with NOMA systems, which can be regarded as a special case of RSMA where all power is allocated to $\mathbf{s}_{11}$.

Figure 4 shows the maximum ESR versus the expected minimum DEP with different ECR constraints in RSMA and NOMA systems, where $\hat{R}_2 = 3$ bpcu. We observe that the curves remain stable initially, but decrease as the expected minimum DEP increases. When the covertness constraint is loose ($t \geq x_w^2 + t_w^2$), the placement of U1 remains unchanged at $x_1 = y_1 = 0$. As the covertness constraint increases, U1 moves closer to Willie while moving further from the BS, thereby leading to a decrement in ESR. Additionally, a higher ECR constraint results in a lower ESR, implying that $\mathbb{O}_{11}^1$ is not zero at the maximum ESR. Moreover, the proposed scheme has a higher ESR than NOMA with the same minimum DEP.

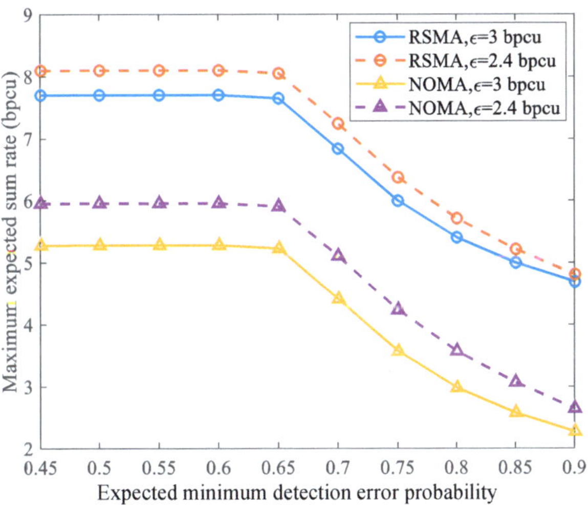

Figure 4. The maximum ESR versus the expected minimum DEP for different ECR constraints in RSMA and NOMA systems, where $\hat{R}_2 = 3$ bpcu.

Figure 5 depicts $\mathbb{O}_{11}^1$ of $\hat{R}_{11}^\dagger$ versus $\hat{R}_2$ for different covertness constraints in RSMA and NOMA systems. In this case, we do not consider the ESR constraint. We find that NOMA has a higher outage probability than RSMA to maximize the ESR. Moreover, as $\hat{R}_2$ increases, the $\mathbb{O}_{11}^1$ of RSMA increases, while that of NOMA decreases. When $\tau = 0$ (i.e., at maximum $\hat{R}_2$), the $\mathbb{O}_{11}^1$ of both systems are equal. In NOMA systems, fixed power is allocated to $\mathbf{s}_{11}$ ($\alpha = 1$). As $\hat{R}_2$ increases, decreasing $\mathbb{O}_{11}^1$ to increase $\overline{R}_2$ will result in a larger ESR. In RSMA systems, with the increment of $\hat{R}_2$, more power is allocated to $\mathbf{s}_{11}$, and more outages can be tolerated to increase the ESR.

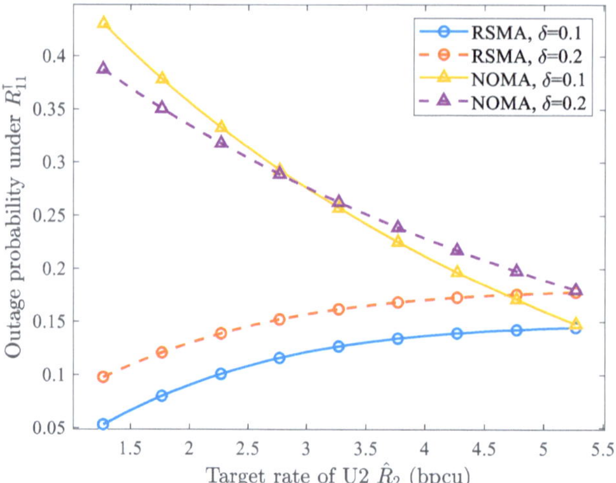

Figure 5. The outage probability of $\hat{R}_{11}^{\dagger}$ under $\mathcal{H}_1$ for different covertness constraints in RSMA and NOMA systems, where $\epsilon = 0.8\hat{R}_2$.

Figure 6 plots the maximum ESR versus $\hat{R}_2$ for different covertness constraints, where $\epsilon = 0.8\hat{R}_2$. Similar to Figure 5, we observe that the maximum ESR of RSMA increases while that of NOMA decreases as $\hat{R}_2$ increases. Specifically, when the covertness constraint is looser, RSMA achieves a higher $\mathbb{O}11^1$ and ESR compared to NOMA. This can be attributed to the fact that the channel gain between the BS and U1 is stronger. The stronger channel gain results in R_1 playing a more crucial role in determining ESR, thus leading to an increase in $\mathbb{O}_{11}^1$.

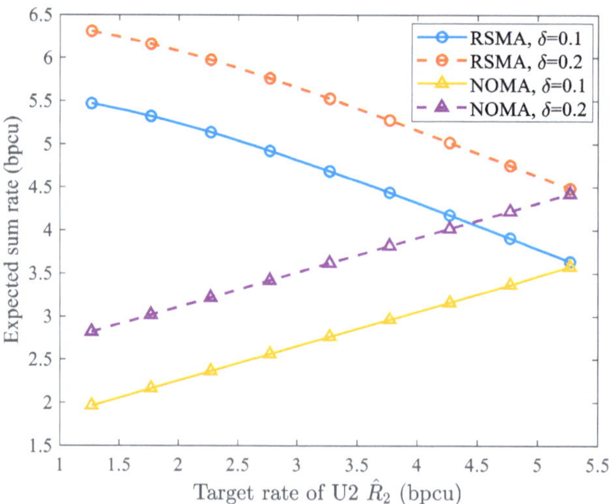

Figure 6. The maximum ESR versus the target rate of U2 $\hat{R}_2$ for different covertness constraints in RSMA and NOMA systems, where $\epsilon = 0.8\hat{R}_2$.

6. Conclusions

In this paper, we investigated the covert communication in UAV-assisted uplink RSMA system, where U1 adopts an RS strategy with random transmit power to guarantee U2's covert and reliable transmission. Specially, we studied the optimal power allocation factor for s_{11} and s_{12}. Then, we derived closed-form expressions of minimum DEP, ESR, and UAV's optimal target rate that maximizes ESR. Constrained by the minimum DEP and ECR, we maximized the ESR by optimizing the position and target rate of the UAV. Numerical results showed that the proposed scheme outperforms NOMA systems in terms of the ESR constraint by the same DEP and ECR.

Author Contributions: Conceptualization, Z.D. and X.Y.; methodology, Z.D.; software, Z.D.; validation, Z.D.; formal analysis, Z.D.; investigation, Z.D.; data curation, Z.D.; writing—original draft preparation, Z.D.; writing—review and editing, Z.D, X.Y. and L.W.; visualization, Z.D.; supervision, X.Y.; project administration, L.W. and T.Z.; funding acquisition, L.W., T.Z. and X.Y. All authors have read and agreed to the published version of the manuscript.

Funding: This work was supported in part by the National Natural Science Foundation of China under Grant No. 62271412, 62271399, and partly supported by Natural Science Basic Research Plan in Shaanxi Province of China under grant No. 2022KJXX-39.

Institutional Review Board Statement: Not applicable.

Informed Consent Statement: Not applicable.

Data Availability Statement: Not applicable.

Conflicts of Interest: The authors declare no conflict of interest.

Abbreviations

The following abbreviations are used in this manuscript:

PLS	Physical layer security
AWGN	Additive white Gaussian noise
UAV	Unmanned aerial vehicle
RSMA	Rate-splitting multiple access
NOMA	Non-orthogonal multiple access
RS	Rate-splitting
BS	Base station
ESR	Expected sum rate
DEP	Detection error probability
3D	Three-dimensional
CSI	Channel state information
LOS	Line-of-sight
SINR	Signal-to-interference-plus-noise ratio
LRT	Likelihood ratio test
FAP	False alarm probability
MDP	Miss detection probability
KKT	Karush–Kuhn–Tucker

Appendix A. Proof of Lemma 1

Taking the derivative of (20) w.r.t. $\hat{R}_{11}$ yields

$$\frac{\partial \overline{R}_{\text{sum}}}{\partial \hat{R}_{11}} = f\left\{ a - \left[b + (c + b\hat{R}_{11}) \log_e 2 \right] 2^{\hat{R}_{11}} \right\}. \tag{A1}$$

By setting $\frac{\partial \overline{R}_{\text{sum}}}{\partial \hat{R}_{11}} = 0$, we have

$$a - \left[b + (c + b\hat{R}_{11})\log_e 2\right]2^{\hat{R}_{11}} = 0$$

$$\left[1 + \left(\frac{c}{b} + \hat{R}_{11}\right)\log_e 2\right]e^{\left[1 + \left(\frac{c}{b} + \hat{R}_{11}\right)\log_e 2\right]} = \frac{ae2^{c/b}}{b}. \tag{A2}$$

According to the definition of Lambert W Function, i.e., $x = W(v)$ is the solution of $xe^x = v(v \geq 0)$, (A2) can be rephrased as

$$1 + \left(\frac{c}{b} + \hat{R}_{11}\right)\log_e 2 = W\left(\frac{ae2^{c/b}}{b}\right)$$

$$\hat{R}_{11} = \log_2 e\left[W\left(\frac{ae2^{c/b})}{b}\right) - 1\right] - \frac{c}{b}. \tag{A3}$$

It is obviously that $\overline{R}_{\text{sum}}$ is a concave function since (A1) is monotonically decreasing w.r.t $\hat{R}_{11}$. Therefore, the solution of (A3) is a maximum point. In addition, $\hat{R}_{11}$ should be no less than 0, which completes the proof.

Appendix B. Proof of Lemma 2

The Lagrangian of this problem is

$$L(x_1, y_1, \kappa) = x_1^2 + y_1^2 + \kappa[(x_1 - x_w)^2 + (y_1 - y_w)^2 - t]. \tag{A4}$$

Taking the derivative of L w.r.t to x_1 and y_1 obtains the Karush–Kuhn–Tucker (KKT) conditions

$$2x_1 + 2\kappa(x_1 - x_w) = 0, \tag{A5a}$$

$$2y_1 + 2\kappa(y_1 - y_w) = 0, \tag{A5b}$$

$$\kappa[(x_1 - x_w)^2 + (y_1 - y_w)^2 - t] = 0, \tag{A5c}$$

$$(x_1 - x_w)^2 + (y_1 - y_w)^2 \leq t \tag{A5d}$$

$$\kappa \geq 0. \tag{A5e}$$

When $\kappa = 0$, it's obvious that $x_1^\dagger = y_1^\dagger = 0$. Meanwhile, to satisfy (A5d), one obtains $t \geq x_w^2 + y_w^2$.

When $\kappa > 0$, after some manipulations, we have $x_1^\dagger = \frac{\kappa x_w}{1+\kappa}$, $y_1^\dagger = \frac{\kappa y_w}{1+\kappa}$, $\kappa = -1 + \sqrt{\frac{x_w^2 + y_w^2}{t}}$, and $0 < t < x_w^2 + y_w^2$. The proof is completed.

References

1. Wang, D.; He, T.; Zhou, F.; Cheng, J.; Zhang, R.; Wu, Q. Outage-driven link selection for secure buffer-aided networks. *Sci. China Inf. Sci.* **2022**, *65*, 1–16. [CrossRef]
2. Lin, Z.; Lin, M.; Champagne, B.; Zhu, W.P.; Al-Dhahir, N. Secrecy-Energy Efficient Hybrid Beamforming for Satellite-Terrestrial Integrated Networks. *IEEE Trans. Commun.* **2021**, *69*, 6345–6360. [CrossRef]
3. Bash, B.A.; Goeckel, D.; Towsley, D. Limits of Reliable Communication with Low Probability of Detection on AWGN Channels. *IEEE J. Sel. Top. Signal Process.* **2013**, *31*, 1921–1930. [CrossRef]
4. Sobers, T.V.; Bash, B.A.; Guha, S.; Towsley, D.; Goeckel, D. Covert Communication in the Presence of an Uninformed Jammer. *IEEE Trans. Wirel. Commun.* **2017**, *16*, 6193–6206. [CrossRef]
5. Wang, D.; Zhou, F.; Lin, W.; Ding, Z.; Al-Dhahir, N. Cooperative Hybrid Non-orthogonal Multiple Access-Based Mobile-Edge Computing in Cognitive Radio Networks. *IEEE Trans. Cogn. Commun. Netw.* **2022**, *8*, 1104–1117. [CrossRef]
6. Lin, Z.; Lin, M.; Wang, J.B.; de Cola, T.; Wang, J. Joint Beamforming and Power Allocation for Satellite-Terrestrial Integrated Networks With Non-Orthogonal Multiple Access. *IEEE J. Sel. Top. Signal Process.* **2019**, *13*, 657–670. [CrossRef]
7. Wang, D.; Wu, M.; He, Y.; Pang, L.; Xu, Q.; Zhang, R. An HAP and UAVs Collaboration Framework for Uplink Secure Rate Maximization in NOMA-Enabled IoT Networks. *Remote Sens.* **2022**, *14*, 4501. [CrossRef]

8. Zhou, G.; Mao, Y.; Clerckx, B. Rate-Splitting Multiple Access for Multi-Antenna Downlink Communication Systems: Spectral and Energy Efficiency Tradeoff. *IEEE Trans. Wirel. Commun.* **2022**, *21*, 4816–4828. [CrossRef]
9. Li, X.; Fan, Y.; Yao, R.; Wang, P.; Qi, N.; Miridakis, N.I.; Tsiftsis, T.A. Rate-Splitting Multiple Access-Enabled Security Analysis in Cognitive Satellite Terrestrial Networks. *IEEE Trans. Veh. Technol.* **2022**, *71*, 11756–11771. [CrossRef]
10. Liu, H.; Bai, Z.; Lei, H.; Pan, G.; Kim, K.J.; Tsiftsis, T.A. A New Rate Splitting Strategy for Uplink CR-NOMA Systems. *IEEE Trans. Veh. Technol.* **2022**, *71*, 7947–7951. [CrossRef]
11. Yang, Z.; Chen, M.; Saad, W.; Xu, W.; Shikh-Bahaei, M. Sum-Rate Maximization of Uplink Rate Splitting Multiple Access (RSMA) Communication. *IEEE. Trans. Mob. Comput.* **2022**, *21*, 2596–2609. [CrossRef]
12. Lin, Z.; Lin, M.; de Cola, T.; Wang, J.B.; Zhu, W.P.; Cheng, J. Supporting IoT With Rate-Splitting Multiple Access in Satellite and Aerial-Integrated Networks. *IEEE Internet Things J.* **2021**, *8*, 11123–11134. [CrossRef]
13. Lin, Z.; Lin, M.; Champagne, B.; Zhu, W.P.; Al-Dhahir, N. Secure and Energy Efficient Transmission for RSMA-Based Cognitive Satellite-Terrestrial Networks. *IEEE Wirel. Commun. Lett.* **2021**, *10*, 251–255. [CrossRef]
14. Wang, M.; Yang, W.; Lu, X.; Hu, C.; Liu, B.; Lv, X. Channel Inversion Power Control Aided Covert Communications in Uplink NOMA Systems. *IEEE Wirel. Commun. Lett.* **2022**, *11*, 871–875. [CrossRef]
15. Lv, L.; Wu, Q.; Li, Z.; Ding, Z.; Al-Dhahir, N.; Chen, J. Covert Communication in Intelligent Reflecting Surface-Assisted NOMA Systems: Design, Analysis, and Optimization. *IEEE Trans. Wirel. Commun.* **2022**, *21*, 1735–1750. [CrossRef]
16. Yan, S.; Hanly, S.V.; Collings, I.B. Optimal Transmit Power and Flying Location for UAV Covert Wireless Communications. *IEEE J. Sel. Areas Commun.* **2021**, *39*, 3321–3333. [CrossRef]
17. Rao, H.; Xiao, S.; Yan, S.; Wang, J.; Tang, W. Optimal Geometric Solutions to UAV-Enabled Covert Communications in Line-of-Sight Scenarios. *IEEE Trans. Wirel. Commun.* **2022**, *21*, 10633–10647. [CrossRef]
18. Su, Y.; Fu, S.; Si, J.; Xiang, C.; Zhang, N.; Li, X. Optimal Hovering Height and Power Allocation for UAV-aided NOMA Covert Communication System. *IEEE Wirel. Commun. Lett.* **2023**, 1–5. [CrossRef]
19. 3GPP. *Technical Specification Group Radio Access Network*; Study on Channel Model for Frequencies from 0.5 to 100 GHz (Release 14); Technical Specification (TS) 38.901; Version 14.3.0; 3rd Generation Partnership Project (3GPP). 2018. Available online: drones-2295032pleaseconfirmthis.Ifthereisnolocation,pleaseaddthewebsiteandaccesseddatehttps://www.etsi.org/deliver/etsi_tr/138900_138999/138901/14.00.00_60/tr_138901v140000p.pdf (accessed on 6 March 2023).
20. Zhou, X.; Yan, S.; Shu, F.; Chen, R.; Li, J. UAV-Enabled Covert Wireless Data Collection. *IEEE J. Sel. Areas Commun.* **2021**, *39*, 3348–3362. [CrossRef]
21. Shahzad, K.; Zhou, X.; Yan, S.; Hu, J.; Shu, F.; Li, J. Achieving Covert Wireless Communications Using a Fulsl-Duplex Receiver. *IEEE Trans. Wirel. Commun.* **2018**, *17*, 8517–8530. [CrossRef]
22. Corless, R.M.; Gonnet, G.H.; Hare, D.E.; Jeffrey, D.J.; Knuth, D.E. On the Lambert W function. *Adv. Comput. Math.* **1996**, *5*, 329–359. [CrossRef]

Article

Multi-UAV Trajectory Planning during Cooperative Tracking Based on a Fusion Algorithm Integrating MPC and Standoff

Bo Li [1,*], Chao Song [1], Shuangxia Bai [1], Jingyi Huang [1], Rui Ma [2], Kaifang Wan [1] and Evgeny Neretin [3]

1   School of Electronics and Information, Northwestern Polytechnical University, Xi'an 710072, China
2   Xi'an Electronic Engineering Research Institute, Xi'an 710100, China
3   School of Robotic and Intelligent Systems, Moscow Aviation Institute, 125993 Moscow, Russia
*   Correspondence: libo803@nwpu.edu.cn; Tel.: +86-133-5921-2759

Abstract: In this paper, an intelligent algorithm integrating model predictive control and Standoff algorithm is proposed to solve trajectory planning that UAVs may face while tracking a moving target cooperatively in a complex three-dimensional environment. A fusion model using model predictive control and Standoff algorithm is thus constructed to ensure trajectory planning and formation maintenance, maximizing UAV sensors' detection range while minimizing target loss probability. Meanwhile, with this model, a fully connected communication topology is used to complete the UAV communication, multi-UAV formation can be reconfigured and planned at the minimum cost, keeping off deficiency in avoiding real-time obstacles facing the Standoff algorithm. Simulation validation suggests that the fusion algorithm proves to be more capable of maintaining UAVs in stable formation and detecting the target, compared with the model predictive control algorithm alone, in the process of tracking the moving target in a complex 3D environment.

Keywords: UAV trajectory planning; model predictive control; standoff algorithm; formation tracking control; intelligent computing

Citation: Li, B.; Song, C.; Bai, S.; Huang, J.; Ma, R.; Wan, K.; Neretin, E. Multi-UAV Trajectory Planning during Cooperative Tracking Based on a Fusion Algorithm Integrating MPC and Standoff. *Drones* **2023**, *7*, 196. https://doi.org/10.3390/drones7030196

Academic Editor: Francesco Nex

Received: 10 February 2023
Revised: 5 March 2023
Accepted: 9 March 2023
Published: 14 March 2023

1. Introduction

The increasingly complex mission environment in recent years has given UAVs their favored market, seeing them widely used for reconnaissance and monitoring missions due to their low cost, high autonomy and reusability [1,2]. Tracking a moving target, whether for single or cooperative tacking, is a significant sub-problem for UAVs performing monitoring tasks. Yet, a single UAV can hardly meet its actual task requirements as it works on its own [3,4], because its sensor's range of view may be easily blocked and therefore its ability to accomplish tasks limited. Cooperation of several UAVs, however, helps make target tracking and monitoring easier. Cooperative efforts made by UAVs can reduce the risk of target loss [5,6], and ensure the accomplishment of a task with multi-sensor data fusion, which means multi-UAV collaboration used in trajectory planning for moving target tracking purposes.

At present, multi-UAV collaborative planning mainly involves artificial potential field method [7,8], bionic algorithm and control algorithm. When the artificial potential field method is applied to the collaborative planning process, it is easy to fall into local optimality and difficult to establish a complete mathematical model. Bionic algorithms, which mainly include ant colony algorithms [9], and particle swarm algorithms [10], also prove to be challenging to meet the real-time demand due to their limited processing efficiency. Control algorithms mainly cover PID control [11], optimal control [12], H-infinity robust control [13], sliding mode control [14], and model predictive control [15,16], etc. Most of these algorithms, such as PID control, optimal control, H-infinity robust control and sliding mode control, are not suitable for complex variable control problems such as cooperative planning of multiple UAVs given their limited control variables and application scenarios that appear quite poor, while the model predictive control algorithm, as the only control method that can explicitly handle constraints at present, has leveled itself up to the acknowledged standard for handling complex constrained variable control problems. It adopts a form of rolling optimization and feedback correction, i.e., the predicted trajectory will be corrected online at each sampling cycle. With strong anti-interference ability and strong robustness, it has attracted widespread attention from scholars at home and abroad. Animesh Sahu [17] and others conducted a study on multi-UAV tracking of multiple moving targets in two dimensions based on the model predictive control algorithm and developed a data-driven Gaussian process (GP) based model that relates the hyperparameters used in model predictive control to mission efficiency. Marc Ille [18] and others carried out research on multi-UAV formation collision avoidance in two-dimensional environments based on the model predictive control algorithm, optimized model predictive control cost functions using penalty term methods, and controlled UAVs' track planning as they tracked a moving target based on formation avoidance constraints. However, relevant research on [17,18] UAV formation control is rare. Tagir Z. Muslimov [19] and others proposed a method based on the Lyapunov vector field for multi-UAV cooperative tracking of the moving target in a two-dimensional environment. The method is grounded around dispersed guided Lyapunov vector fields for path planning. Based on the two-dimensional environment, Q. Guo [20] and others proposed a performance guaranteed $5\frac{1}{3}$-approximation algorithm for the UAV scheduling problem when ignoring the limited flying time of each UAV, such that the maximum spent time of UAVs in their flyingtours is minimized. A fusion algorithm for adaptive multi-model traceless Kalman particle filter was adopted by Niu Yifeng [21] and others to carry out a study on coordinated tracking of ground multi-target trajectory for UAV swarms in complex two-dimensional environments. A pioneering exploration is Zhang Yi [22] and others who solved the problems regarding non-convergence of initial heading and long phase coordination time among UAVs in the process of cooperatively tracking a moving target based on Standoff method, following which Zhu Qian [23] and his team also studied two aircrafts' cooperative tracking of a moving target by means of angle measurement.

A comprehensive analysis of the above research found that most of the current research on multi-UAV trajectory planning through cooperative formation stays in two-dimensional space, still challenged by problems such as large model calculation and insufficient real-time. At the same time, the current research faces great difficulty in establishing a complete non-linear UAV 3D motion model, and thus fails to meet actual mission requirements [24]. As for traditional multi-UAV sensors, their limited detection coverage as well as weak formation and retention capabilities [25] prevent them from being the hot spot in this field, leaving UAV trajectory planning that integrates collision avoidance and obstacle avoidance not fully explored.

Against such a background, this paper proposes a fusion algorithm that combines the model predictive control algorithm [26] and the Standoff algorithm. The model predictive control algorithm solves the problem of large-scale real-time optimal control in limited time [27] and uses the preview capability to achieve optimal maneuver control in a constrained, non-linear, model-uncertain and unpredictable environment to generate smooth flyable paths suitable for the actual flight of the formation [28]. The Standoff algorithm [29], one of the main algorithms for formation control, maximizes sensor detection range and reduces the probability of target loss with safe distances as grounds [30]. Compared with the traditional multi-UAV cooperative trajectory planning method, the fusion algorithm simplifies the mathematical modelling of UAVs' three-dimensional motion [31], reduces the computational complexity which is caused by strong non-linearity as defined in the dynamics [32], and enhances real-time performance that an algorithm can show compared with the two papers [33,34]. It integrates the maximization of the sensor's observation coverage to establish UAV sensors' monitoring model, and more importantly, reduces the probability that UAVs lose their moving target compared with the sensor detection model proposed by the thesis [35]; Inspired by the minimum long-term operational cost suggested by the paper [36], the present study designs the reconfiguration planning of UAV formation at the minimum cost. As the distributed learning principle reported in the research [37] indicates, it constructs a multi-UAV track planning model using a distributed model predictive control algorithm to transform the challenge of centralized UAV formation mentioned in the paper [38] into that of a distributed flight control optimization, verifying the effectiveness of the fusion algorithm by means of unexpected artificially implanted obstacles.

The remainder of this paper is as follows: Section 2 introduces the trajectory planning model that UAVs take while they cooperatively track the moving target in a complex three-dimensional environment, followed by how it is configured and designed based on the fusion algorithm in Section 3, in addition to the cooperative formation reconfiguration and planning when an unexpected situation occurs to the vehicles. Simulation validation is carried out in Section 4 to demonstrate the effectiveness of the fusion algorithm applied to multi-UAV collaborative tracking of moving target trajectory planning. The following section conducts a study on the effectiveness and monitoring capability of multiple UAVs in coordinated formation to track moving targets, and illustrates that the fusion algorithm has better tracking effectiveness and monitoring capability in the test, while the last section offers a conclusion.

2. UAV Model and Environment Model

2.1. UAV Motion Model

Different from most of the previous literature that used the two-dimensional plane to establish the motion model of the UAV, this paper regards the UAV as a mass point and builds a three-dimensional motion model based on the inertial reference system without considering the influence of external disturbances, noise and air resistance on the UAV dynamics, and carries out discretization processing on it. Assuming that the sampling time is Δt, the UAV motion model is expressed as Equation (1).

$$\begin{cases} x(k+1) = x(k) + v(k)\cos\theta(k)\sin\varphi(k)\Delta t \\ y(k+1) = y(k) + v(k)\cos\theta(k)\cos\varphi(k)\Delta t \\ z(k+1) = z(k) + v(k)\sin\theta(k)\Delta t \\ v(k+1) = v(k) + a(k+1)\Delta t \\ \varphi(k+1) = \varphi(k) + \dot{\varphi}(k+1)\Delta t \\ \theta(k+1) = \theta(k) + \dot{\theta}(k+1)\Delta t \\ s(k) = [x(k), y(k), z(k)]' \in S \\ u(k) = [v(k), \varphi(k), \theta(k)]' \in U \end{cases} \tag{1}$$

where $s(k)$ denotes the UAV state sampling at time k; S denotes the feasible state set; $u(k)$ denotes the control input of the UAV at time k; U denotes the feasible input set; $(x(k), y(k), z(k))$ is the real time position of the UAV; $v(k)$, $\varphi(k)$ and $\theta(k)$ denote the real time speed, heading angle and pitch angle of the UAV respectively, and a dotes the acceleration of the UAV.

2.2. UAV Collision Avoidance Model

Since UAVs need to fly as ultra-low as possible in order to avoid radar detection, the complex ground environment and its obstacles become the primary threat to UAV trajectory planning. This paper creates a map model based on undulating terrain topography to fulfill the actual task requirements, as shown in Figure 1. To improve the robustness of the method, a safety buffer zone is established around the UAV, and the obstacles are divided into static obstacle modelling and emergent obstacles. The static obstacle model is approximated by a cylinder whose co-ordinate center is set to P_o, whose co-ordinates are $[P_{ox}, P_{oy}]$, and whose radius and height are denoted by P_{or} and P_{oz}. A collision zone (denoted by L_{od} and ΔH_{od}) and a threat zone (denoted by L_{OD} and ΔH_{OD}) are established around it. L_{od} is the minimum proximity safety distance while ΔH_{od} is the minimum height proximity distance, and if the distance between the UAV and the static obstacle is less than L_{od} and ΔH_{od}, the UAV will collide. L_{OD} and ΔH_{OD} are the maximum threat distance of the static obstacle, and if the distance between the UAV and the static obstacle is less than L_{OD} and ΔH_{OD}, the UAV may have the risk of collision. The sudden obstacle model is approximated by a sphere, the centre of which is set to P_t, with specific coordinates $[P_{tx}, P_{ty}, P_{tz}]$ and a radius of R_t. The collision zone (represented by a sphere with a radius of R_p) and the threat zone (represented by a sphere with a radius of R_w) are also set up, and the specific UAV collision avoidance and collision avoidance model is shown in Equation (2).

$$
\begin{cases}
z_i(k) - z_{all}(k) > \Delta H_d \\[4pt]
\left| \begin{matrix} x_i(k), y_i(k), z_i(k) \\ x_j(k), y_j(k), z_j(k) \end{matrix} \right| \geq 2R_a \\[8pt]
\sqrt{(x_i(k) - P_{tx})^2 + (y_i(k) - P_{ty})^2 + (z_i(k) - P_{tz})^2} \geq R_w \\[8pt]
\sqrt{(x_i(k) - P_{ox})^2 + (y_i(k) - P_{oy})^2} \geq L_{OD} \quad or \quad z_i(k) - P_{oz} \geq \Delta H_{OD}
\end{cases}
\tag{2}
$$

where $(x_i(k), y_i(k), z_i(k))$ denotes the current UAV position coordinates; R_a denotes the UAV minimum collision avoidance safety distance; $(x_j(k), y_j(k), z_j(k))$ denotes the adjacent UAV position coordinates; $z_{all}(k)$ denotes the height of the ground coordinates $(x_i(k), y_i(k))$; and ΔH_d denotes the UAV near-ground minimum safety distance.

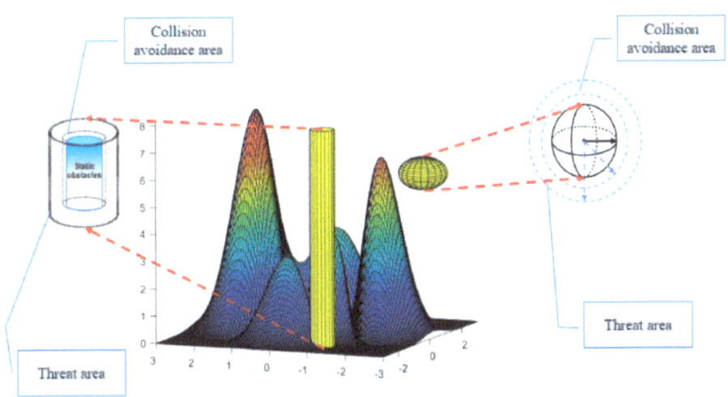

Figure 1. Schematic diagram of modeling of 3D environment and obstacles.

2.3. Moving Target Model

The establishment of a rationalized moving target motion model is the prerequisite for the successful track planning of UAVs when they cooperatively track the moving target. This paper defines the target motion model as Equation (3). To simplify the operation, the moving target's trajectory is compressed from the three-dimensional space to the two-dimensional yoz plane, i.e., the x coordinate of the moving target is set to a constant value.

$$
\begin{bmatrix} \dot{x}_u(k) \\ \dot{y}_u(k) \\ \dot{z}_u(k) \\ \dot{v}_u(k) \\ \dot{\theta}_u(k) \\ \dot{\varphi}_u(k) \end{bmatrix}
=
\begin{bmatrix} v_u(k)\cos\theta_u(k)\sin\varphi_u(k) \\ v_u(k)\cos\theta_u(k)\cos\varphi_u(k) \\ v_u(k)\sin\theta_u(k) \\ -g\sin\theta_u(k) \\ \dfrac{-g\cos\theta_u(k)}{v_u(k)} \\ 0 \end{bmatrix}
+
\begin{bmatrix} 0 & 0 & 0 \\ 0 & 0 & 0 \\ 0 & 0 & 0 \\ 1 & 0 & 0 \\ 0 & \dfrac{1}{v_u(k)} & 0 \\ 0 & 0 & \dfrac{1}{v_u(k)\cos\theta_u(k)} \end{bmatrix}
\begin{bmatrix} a_1 \\ a_2 \\ a_3 \end{bmatrix}
\tag{3}
$$

where $v_u(k)$ denotes the velocity of the moving target at k, $\theta_u(k)$ denotes the pitch angle of the moving target, $\varphi_u(k)$ denotes the heading angle of the moving target $(\varphi_u(k) = 0)$, g is the acceleration of gravity, a_1 denotes the horizontal acceleration of the moving target, a_2 denotes the vertical acceleration of the moving target and a_3 denotes the angular acceleration of the moving target, and the motion constraint of the target can be completed by adjusting according to parameter $a = [a_1, a_2, a_3]$.

2.4. Target Observation Coverage Modelling

The modeling of target observation coverage is based on the UAV sensors. In this paper, the mathematical modeling of target observation coverage is based on four factors: P_f, P_w, L_{max} and L_{min}. P_f indicates the probability that the sensor detects the target effectively, P_w indicates the probability that the sensor detects the target incorrectly, and $P_f, P_w \in (0.1]$. L_{max} indicates the maximum detection distance of the sensor, and L_{min} indicates the effective distance that the sensor detects completely. When the distance between the sensor and the target is less than L_{min}, $P_f = 1$. Given the influence of multiple obstacles encountered during UAV trajectory planning, it does not meet the actual needs to only use the maximum detection distance of the sensor as the measurement standard. Based on this, this paper defines that the UAV is only likely to detect a target when the target enters an area where it can be seen by the vehicle, and the sensor is only capable of detecting the target when the target is within its coverage. The intersection of the area where the target is visible and the sensor's coverage area is defined as the target observation coverage, which circumvents the obstruction of the UAV's line of sight by environmental obstacles and ensures effective monitoring of the moving target by multiple UAVs in formation, as shown in Figure 2, whose discretization modelling is expressed as Equation (4). Define the effective detection range of UAV sensors and the radius of the target coverage area to be the same, both of which are L_{min} = 40 m . The UAV can monitor the moving target when it is within the sensor's detection range.

$$p(L_t) = \begin{cases} 1 & L_t < L_{min} \\ p_f - \dfrac{(p_f - p_w)(L_t - L_{min})}{L_{max} - L_{min}} & L_{min} < L_t < L_{max} \\ p_w & L_t > L_{max} \end{cases} \tag{4}$$

where $P(L_t)$ denotes the probability of the sensor effectively monitoring the target and L_t denotes the real-time distance between the UAV sensor and the target.

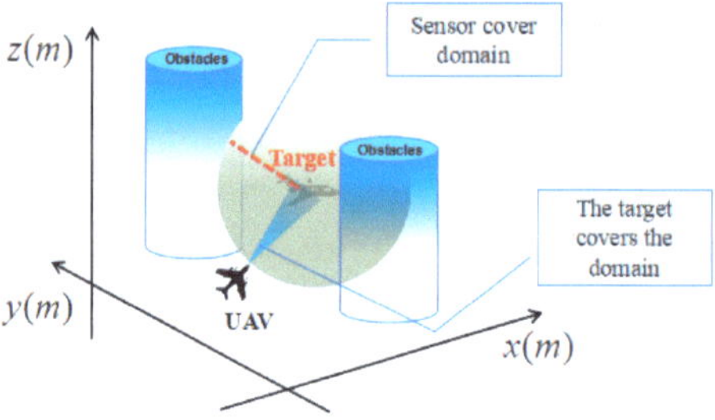

Figure 2. Schematic diagram of observation coverage.

3. Designing a Multi-UAV Cooperative Tracking System Based on the Fusion Algorithm

3.1. System Design

After each UAV receives the tracking task, it initializes the system model according to the prior obstacle information, target movement information and its own motion information. In view of constraints such as obstacle avoidance and collision avoidance, the model predictive control algorithm is used to predict the trajectory of multiple UAVs at the minimum planning cost. In terms of formation and maintenance of multi-UAV formation, the Standoff algorithm is used to complete the multi-UAV formation control, so that the UAV swarm is evenly distributed around the target, and then multi-UAV sensors can maximize the monitoring of the moving target. The specific system framework is shown in Figure 3. The cooperative collision avoidance control module is mainly responsible for obstacle collision avoidance and inter-UAV collision avoidance, taking into account the UAV motion state, obstacle information, map boundaries and other factors to plan a safe and collision-free flight path. The model prediction control module is responsible for predicting the UAV trajectory at the minimum flight cost, and the distributed cooperative controller plans and coordinates the global trajectory. Standoff control module is mainly responsible for UAV formation maintenance, real-time acquisition of multi-UAV phase distribution, and maximizing UAV sensors' coverage. The formation reconfiguration module means that during the flight of multiple UAVs in accordance with the established formation, the formation needs to carry out reconstruction planning due to unexpected situations.

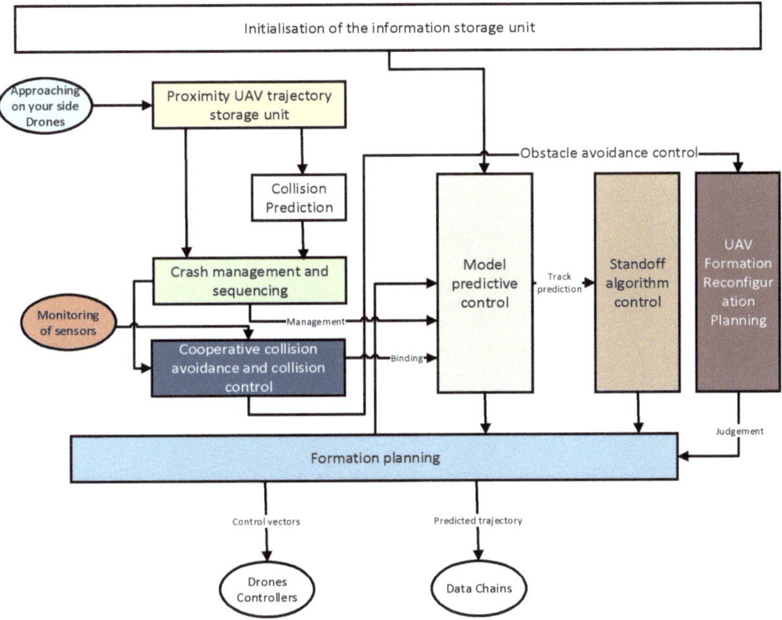

Figure 3. Framework diagram of how UAVs make track planning as they track the moving target through cooperative formation.

3.2. Multi-UAV Cooperative Trajectory Planning Based on the Fusion Algorithm

In this paper, a fusion of model predictive control algorithm and Standoff algorithm is used to promote UAVs' trajectory planning as they reach cooperative formation when tracking the moving target, as illustrated by Figure 4.

3.2.1. Multi-UAV Formation Control Based on the Standoff Algorithm

Multi-UAV formation control research using the Standoff algorithm is carried out in the following steps: introduce UAV-target relative desired distance and UAV sensors' observation coverage information; use Lyapunov vector field guidance algorithm to guide the UAVs' trajectory planning during moving target tracking to ensure that the moving target is within UAV sensors' detection range to the maximum extent possible; control the UAV trajectory rotation characteristics to make it more flexible when optimizing the trajectory, and then better approach the desired position to reduce the probability of target loss. Figure 5 shows a schematic diagram of the UAV swarm model for tracking the moving target based on the Standoff algorithm.

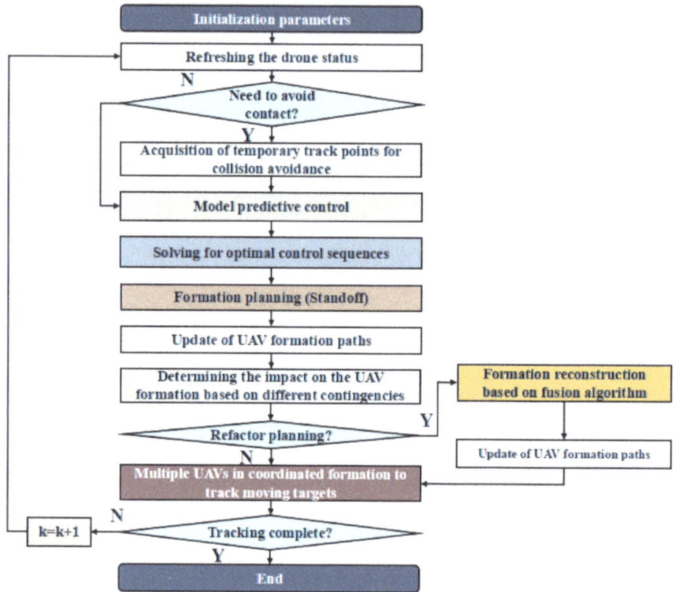

Figure 4. Framework diagram of multi-UAV trajectory planning based on the fusion algorithm.

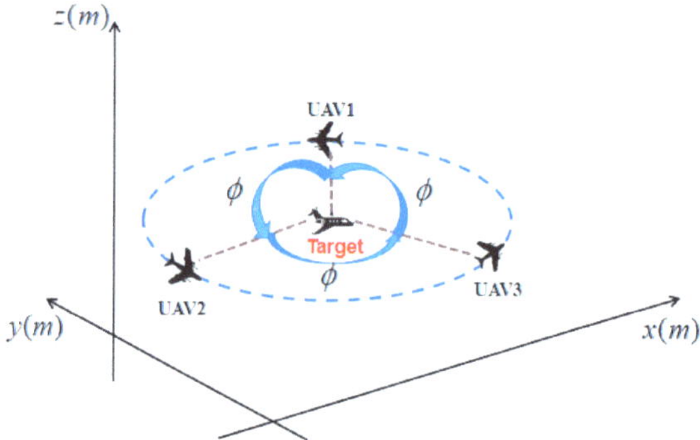

Figure 5. Schematic of formation control using the Standoff algorithm.

Set the target motion state is known, the multi-UAV cooperative formation moves around the target circular motion through the Lyapunov function, and the multi-UAV speed adjustment is assisted by the feedback-correction mechanism, so as to maintain the ideal tracking of the multi-UAV formation and the moving target. In this paper, the radius of circular distribution is set as D_r, and the corresponding Lyapunov energy function is the distance function, as shown in Equation (5).

$$\begin{cases} L_d(x,y,z) = (r^2 - D_r^2)^2 \\ |r - D_r| \leq \xi \end{cases} \tag{5}$$

where r is the radial distance between UAV position (x_r, y_r, z_r) and moving target position (x_d, y_d, z_d), $r = \sqrt{(x_r - x_d)^2 + (y_r - y_d)^2 + (z_r - z_d)^2}$, and ξ denotes the formation coordination error.

Assuming that three UAVs are performing a moving target tracking task at the same time, the positioning process requires any two UAVs to be positioned in comparison to each other to maintain the relative balance of the three UAVs' positions. In order to simplify the operation, this paper sets three UAVs distributed in the same plane, so only the influence of phase angle positioning needs to be considered. Assuming that the phase angles of any two UAVs are ϕ_i and ϕ_j respectively, and the expected relative phase angle is ϕ_z, the phase distribution function of multi-UAV cooperative formation is calculated based on the Lyapunov stability theory as shown in Equation (6).

$$\begin{cases} \Phi_p = (\phi_i - \phi_j - \phi_z)^2 \\ \phi_z = \frac{2\pi}{N}, N \geq 2 \end{cases} \tag{6}$$

where N denotes the number of drones and $N = 3$.

The speed calculation of any two UAVs is shown in Equation (7).

$$\begin{aligned} v_i &= v \\ v_j &= k \cdot (\phi_i - \phi_j - \phi_z) \cdot D_r + v \end{aligned} \tag{7}$$

where v represents the real-time velocity of the moving target.

The phase angular velocity of any two UAVs is calculated as Equation (8).

$$\begin{aligned} \dot{\phi}_i &= v_i / D_r \\ \dot{\phi}_j &= k \cdot (\phi_i - \phi_j - \phi_z) + v_j / D_r \end{aligned} \tag{8}$$

where k is the function coefficient.

Assuming that the moving target position and velocity are known, the optimal desired velocity of the UAV formation can be calculated by combining the multi-UAV predicted velocity with the moving target velocity correction term, which is calculated as Equation (9).

$$\begin{bmatrix} \dot{x}_t \\ \dot{y}_t \\ \dot{z}_t \end{bmatrix} = \begin{bmatrix} \dot{x}_i - \dot{x} \\ \dot{y}_i - \dot{y} \\ \dot{z}_i - \dot{z} \end{bmatrix} \tag{9}$$

where $(\dot{x}_i, \dot{y}_i, \dot{z}_i)$ is the predicted velocity value of the UAV and $(\dot{x}, \dot{y}, \dot{z})$ is the target velocity correction value.

The predicted speed v_t, heading angle φ_t and pitch angle θ_t of the multi-UAV formation can be calculated according to Equation (10).

$$\begin{aligned} v_t &= \sqrt{\dot{x}_t^2 + \dot{y}_t^2 + \dot{z}_t^2} \\ \varphi_t &= \arctan(\dot{y}_t / \dot{x}_t) \\ \theta_t &= \arctan(\dot{z}_t / \sqrt{\dot{x}_t^2 + \dot{y}_t^2}) \end{aligned} \tag{10}$$

3.2.2. Track Planning UAVs Take during Cooperative Tracking of the Moving Target Based on the Fusion Algorithm

Inspired by the fact that the model predictive control algorithm can predict UAV trajectories in real time, and the applicability of the Standoff algorithm to UAV formation control, this paper reports on the trajectory planning UAVs take during cooperative tracking of the moving target based on the fusion of the two algorithms. Taking the i-th UAV as an example, given constraints such as multi-UAV collision avoidance and collision avoidance, the predicted motion state of the UAV in the finite time domain is constructed based on the model predictive control framework, the UAV cooperative trajectory planning model is constructed based on minimizing the UAV trajectory planning cost, while the fusion Standoff algorithm is used to carry out formation control, based on a "feedback-correction" mechanism using a moving target speed correction term to correct the optimal desired speed of the UAV in real time. With the scaling factor of UAV speed and angular speed added, the predicted velocity v_t and predicted angular velocity ω_t of the multi-UAV formation are calculated in real time, as shown in Equation (11). Each UAV is solved at each sampling moment using the quadratic programming method to obtain its own optimal control sequence and local predicted trajectory, and the information at the current sampling moment is calculated on the basis of control sequence. The specific algorithm flow is displayed in Algorithm 1.

$$
\left\{
\begin{array}{l}
\min(-f_1^i, f_2^i, f_3^i) \\
s.t. \\
\begin{bmatrix} x_i(k+p+1|k) \\ y_i(k+p+1|k) \\ z_i(k+p+1|k) \end{bmatrix} = \begin{bmatrix} x_i(k+p|k) \\ y_i(k+p|k) \\ z_i(k+p|k) \end{bmatrix} + \begin{bmatrix} v_i(k+p|k)\cos\theta_i(k+p|k)\sin\varphi_i(k+p|k) \\ v_i(k+p|k)\cos\theta_i(k+p|k)\cos\varphi_i(k+p|k) \\ v_i(k+p|k)\sin\theta_i(k+p|k) \end{bmatrix}\Delta t \\
\begin{bmatrix} \dot{x}_t(k+p+1|k) \\ \dot{y}_t(k+p+1|k) \\ \dot{z}_t(k+p+1|k) \end{bmatrix} = \begin{bmatrix} \dot{x}_i(k+p+1|k) - \dot{x}(k+p+1|k) \\ \dot{y}_i(k+p+1|k) - \dot{y}(k+p+1|k) \\ \dot{z}_i(k+p+1|k) - \dot{z}(k+p+1|k) \end{bmatrix} \\
v_t(k+p+1|k) = v_t(k+p|k) + (u_i^v(k+p|k) - v_t(k+p|k))/\tau_v \\
\omega_t(k+p+1|k) = \omega_t(k+p|k) + (u_i^\omega(k+p|k) - \omega_t(k+p|k))/\tau_\omega
\end{array}
\right.
\tag{11}
$$

where $u_i^v(k+p|k)$ and $u_i^\omega(k+p|k)$ are the velocity and angular velocity control inputs of the i-th UAV in the predicted time domain; $v_i(k|k)$ is the UAV velocity; $\omega_i(k|k)$ is the UAV angular velocity; $(x_i(k+p+1|k), y_i(k+p+1|k), z_i(k+p+1|k))$ is the three-dimensional position coordinates of this UAV in the predicted time domain; τ_v and τ_ω are the UAV velocity and angular velocity scaling factors respectively; f_1^i is the UAV monitoring target coverage, f_2^i is the control input cost, and f_3^i is the formation planning cost, consisting of two parts: regular planning and reconfiguration planning. Set the formation planning cost in the interval $[0, j)$ for predicted trajectory flight, in the interval $[j, J)$, reconfiguration planning is required based on the unexpected situation multi-UAV formation, in the interval $[J, k)$, the UAV completes the formation planning and continues to fly in accordance with the established formation, as shown in Equation (12).

$$
\begin{aligned}
f_3^i =\ & \sum_{i=0}^{j} (w_1 f_L^i + w_2 f_H^i + w_3 f_T^i) + \\
& \sum_{i=j}^{J-1} (\|x_i(k+j|k) - x_g\|_{A_i}^2 + \|u_i(k+j|k)\|_{B_i}^2) + \\
& \sum_{i=J}^{J+k} (w_1 f_L^i + w_2 f_H^i + w_3 f_T^i)
\end{aligned}
\tag{12}
$$

where $x_i(k+j|k)$ denotes the UAV $J-1$ step state; x_g denotes the terminal target state; $u_i(k+j|k)$ denotes the UAV $J-1$ step control input; A and B are symmetric positive definite weight matrices; $w = (w_1, w_2, w_3)^T$ is the weight vector; f_T^i denotes the environmental threat cost, calculated by Equation (13); f_L^i denotes the energy consumption cost, calculated by Equation (14); and f_H^i denotes the UAV altitude cost, which is calculated by Equation (15).

$$
f_T^i(x_i, y_i, z_i) = \begin{cases} \infty & \text{No fly zones} \\ 1 & \text{Safety zones} \end{cases}
\tag{13}
$$

where (x_i, y_i, z_i) denotes the coordinates of the current UAV track point.

$$
f_L^i = \sqrt{(x_i - x_l)^2 + (y_i - y_l)^2 + (z_i - z_l)^2}
\tag{14}
$$

where (x_l, y_l, z_l) denotes the coordinates of the current moving target.

$$f_H^i = \begin{cases} z_1 & z_i < \Delta H_d \\ z_i - \Delta H_d & \Delta H_d \leq z_i \leq \Delta H_{\max} \\ z_2 & z_i > \Delta H_{\max} \end{cases} \tag{15}$$

where z_i denotes the current track point altitude; ΔH_{max} denotes the maximum flight altitude; and z_1 and z_2 denote the altitude penalty values.

Assuming a fully connected communication topology between UAVs, where each real UAV can obtain information sent by others in real time and without delay within a sampling period, the inter-aircraft communication distance constraint needs to be considered, and the specific fusion algorithm constraint is shown in Equation (16).

$$\begin{cases} u_i^{v\min} \leq u_i^v(k+P|k) \leq u_i^{v\max} \\ |u_i^\omega(k+P|k)| \leq u_i^{\omega\max} \\ z_i(k+P+1|k) - z_{all} > \Delta H_d \\ \left| \begin{matrix} x_i(k+P+1|k), y_i(k+P+1|k), z_i(k+P+1|k) \\ x_j(k+P+1|k), y_j(k+P+1|k), z_j(k+P+1|k) \end{matrix} \right| \geq 2R_a \\ \left| \begin{matrix} x_i(k+P+1|k), y_i(k+P+1|k), z_i(k+P+1|k) \\ x_j(k+P+1|k), y_j(k+P+1|k), z_j(k+P+1|k) \end{matrix} \right| < R_T \\ x_{ix} = x_i(k+P+1|k) - P_{tx} \\ y_{iy} = y_i(k+P+1|k) - P_{ty} \\ z_{iz} = z_i(k+P+1|k) - P_{tz} \\ \sqrt{(x_{ix})^2 + (y_{iy})^2 + (z_{iz})^2} \geq R_w \\ \sqrt{(x_i(k+P+1|k) - P_{ox})^2 + (y_i(k+P+1|k) - P_{oy})^2} \geq L_{OD} \\ i \in \{1, \dots, N_v\} \\ j \in \{1, \dots, N_v\} \\ x_i(k|k) = x_i(k), y_i(k|k) = y_i(k), z_i(k|k) = z_i(k) \\ v_i(k|k) = v_i(k), \theta_i(k|k) = \theta_i(k), \varphi_i(k|k) = \varphi_i(k) \end{cases} \tag{16}$$

where $(x_j(k+p+1|k), y_j(k+p+1|k), z_j(k+p+1|k))$ is the three-dimensional position coordinates of formation j-th UAVs in the predicted time domain; $(x_i(k+p+1|k), y_i(k+p+1|k), z_i(k+p+1|k))$ is the three-dimensional position coordinates of formation i-th UAVs in the predicted time domain; R_T is the maximum communication radius of the formation UAVs; $u_i^{v\max}$ and $u_i^{v\min}$ are the maximum and minimum velocity constraints of the UAVs and $u_i^{\omega\max}$ is the maximum angular velocity constraint of the UAVs. At moment k, the optimization problem above is solved and the first term $u_i(k|k)$ of the control sequence is applied to the UAV system, and the process above is repeated at moment $k+1$.

3.3. Application Steps of Multi-UAV Cooperative Tracking of the Moving Target Based on the Fusion Algorithm

The following steps are taken to plan the coordinated tracking of the moving target by multiple UAVs.

Step 1: Consider the UAV's own constraints, collision avoidance constraints and other conditions, and determine the number of participating tracking UAVs and UAV formation according to the type of the moving target and tracking needs.

Step 2: The Standoff algorithm and the model predictive control algorithm are fused to complement each other and form a fusion algorithm with more optimized performance. The specific fusion algorithm is as follows: given the basic information of prediction time domain, sampling period, UAV control input $u_i(k-1|k)$ and UAV state quantity $[x_i(k|k), y_i(k|k), z_i(k|k)]$ at the current k moments, build the planning model that UAVs follow when tracking the moving target, carry out UAV finite time domain prediction trajectory based on collision avoidance constraint, at the same time use the Standoff algorithm to calculate UAV formation phase distribution value, and then build the multi-UAV formation model to reach cooperative tracking of the moving target.

Step 3: In the process of multi-UAV formation movement, determine in real time whether the UAV formation encounters an unexpected situation. If yes, go to step 4; if no, continue to track the moving target.

Step 4: When the UAV formation encounters an unexpected situation during the tracking process, UAVs need to use the fusion algorithm to carry out real-time trajectory planning, and the 'feedback-correction' mechanism to correct the trajectory until they resume the formation after the unexpected situation is resolved to continue tracking the moving target.

Algorithm 1: Fusion Algorithm Based on MPC and Standoff.

---

1. Initialize map environment information
2. Initialize fusion algorithm information
3. Initialize multi-UAV movement information
4. **For** step = $1, 2, \dots, N$:
5.   Obtain the initial state of UAVs in environments $(x_r, y_r, z_r), v$ and ϕ_z
6.   **For** k = $1, \dots, J$:
7.   **if** multi-UAV formations encounter no surprises:
8.   Comprehensive consideration of UAV trajectory planning constraints: $u^{v\max}, u^{v\min}, u^{w\max}$
9.   Input prediction of velocity and angular velocity control in the time domain $u^v(k + p|k)$, $u^\omega(k + p|k)$
10.   "red" UAV in the environment executing the previous control input of the drone $u_1(k + j|k)$ and correcting speed variables $(\dot{x}_1(k + p + 1|k), \dot{y}_1(k + p + 1|k), \dot{z}_1(k + p + 1|k))$ based on the Standoff algorithm, and obtains the next state $u_1(k + j + 1|k + j)$
11.   "yellow" UAV in the environment executing the previous control input of the drone $u_2(k + j|k)$ and correcting speed variables $(\dot{x}_2(k + p + 1|k), \dot{y}_2(k + p + 1|k), \dot{z}_2(k + p + 1|k))$ based on the Standoff algorithm, and obtains the next state $u_2(k + j + 1|k + j)$
12.   "green" UAV in the environment executing the previous control input of the drone $u_3(k + j|k)$ and correcting speed variables $(\dot{x}_3(k + p + 1|k), \dot{y}_3(k + p + 1|k), \dot{z}_3(k + p + 1|k))$ based on the Standoff algorithm, and obtains the next state $u_3(k + j + 1|k + j)$
13.   Store the above track planning information in the model predictive control module
14.   **if** multi-UAV formations encounters an unexpected obstacle:
15.   UAV reconfiguration planning based on Computational (12)
16.   Update drone location information (x_i, y_i, z_i) based on minimum generation value
17.   **end if**
18.   else: break
19.   **end if**
20.   **end for**
21.   step = step + 1
22. **end for**

---

4. Simulation Verification

With parameters of UAVs and the moving target initialized according to the known information, simulation results have verified that UAVs are able to make trajectory planning through coordinated formation to track the moving target, under the premise that each UAV's own constraints as well as constraints related to collision avoidance and obstacle avoidance are all considered. Under this verification, an ideal distance and angle between the UAV formation and the moving target is maintained, which makes the UAVs' monitoring possible and effective. Simulation verification on reconfiguration of multi-UAV formation and trajectory replanning is also carried out, in which different contingencies are handled at the minimized formation planning cost so that UAV trajectory planning can be less dependent on priori information. Initialization information is shown in Table 1.

Table 1. Initialization of system parameters.

Serial Number	Parameters Name	Parameter Value
1	UAV1 starting position	(200 m, 5 m, 115 m)
2	UAV2 starting position	(160 m, 5 m, 75 m)
3	UAV3 starting position	(240 m, 5 m, 75 m)
4	Target starting position	(200 m, 5 m, 95 m)
5	UAV initial speed	25 m/s
6	UAV speed range	[20 m/s, 40 m/s]
7	Maximum yaw angle of UAV	$\pi/4$ rad
8	Maximum pitch angle of UAV	$\pi/4$ rad
9	Minimum turning radius for UAV	10 m
10	Number of UAVs N	3
11	Maximum speed constraint for UAVs $u_i^{v_{max}}$	40 m/s
12	Minimum speed constraint for UAVs $u_i^{v_{min}}$	10 m/s
13	Maximum angular velocity constraint for UAVs $u_i^{w_{max}}$	0.25 rad/s

In this paper, the simulation environment is based on MATLAB R2020b software. The map modelling is based on the undulating terrain of the mountainous landscape, the terrain obstacle composition is mainly derived from the original terrain and the threat of mountain peaks, and the mathematical model of the terrain is artificially formulated. To further approximate the real flight scenario, a safety buffer zone is set up around the UAV and the obstacles are divided into static obstacle modelling and emergent obstacles, with the static obstacle model being approximated by a cylinder and the emergent obstacle model by a sphere. In addition, to further enhance the accuracy of the simulation, the rasterised map environment, i.e., taking into account terrain obstacles, no-fly zones, threat zones, etc., rasterises the map, with each grid called a cell, converts the 3D mathematical model of the map into vector structure data and then into a raster structure, giving each raster cell unique attributes to represent entities. In this paper, the rasterized map unit length is determined to be 5 m with an accuracy of 0.1 m, and its 3D height information is formulated by human.

In this paper, the moving target is set as a low altitude slow speed target, the UAV collision avoidance safety distance is defined as 15 m, the maximum communication radius between UAVs is 90 m, and the UAV detection coverage range is 40 m. For specific sudden obstacle model information, see Table 2. The simulation system randomly selects the established sudden obstacle model for testing the fusion algorithm applied to UAV trajectory planning and its formation reconfiguration capability. When the simulation system selects the sudden obstacle 1, UAVs in formation follow the way as planned by conventional trajectory in their flight, taking into account constraints such as collision avoidance and obstacle avoidance, and maximizing multi-UAV sensors' monitoring coverage. For simulation details, see Figure 6. When the simulation system selects the sudden obstacle 2, it needs to use the fusion algorithm to quickly develop a reconfiguration plan for UAV cooperative formation. Simulation results are shown in Figure 7. To verify the effectiveness of the fusion algorithm, this paper uses the model predictive control algorithm to carry out comparative simulations of same-state trajectory planning, as shown in Figures 8 and 9.

Table 2. Sudden obstacle information.

Serial Number	Coordinate Position	Radius Size of Obstacle
1	(100 m, 270 m, 250 m)	50 m
2	(200 m, 300 m, 250 m)	50 m

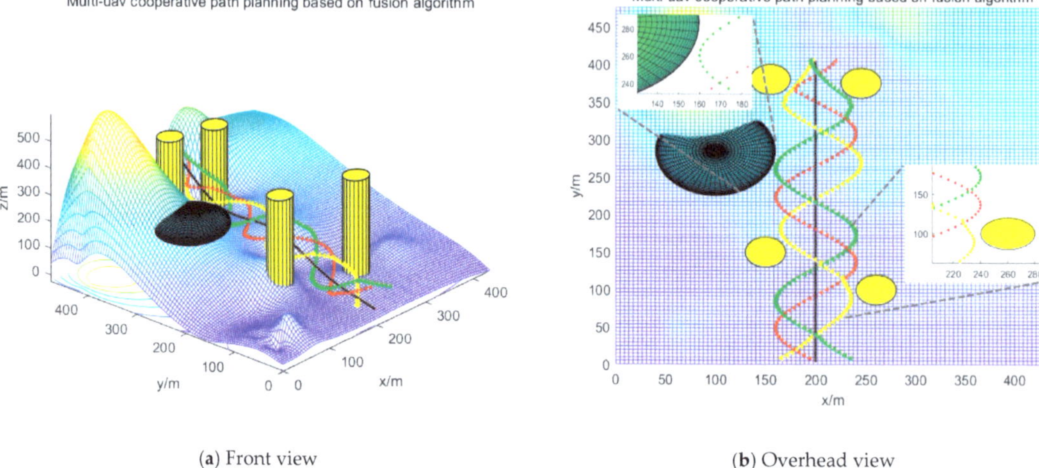

(**a**) Front view

(**b**) Overhead view

Figure 6. Scene 1-simulation of multiple UAVs using the fusion algorithm for coordinated formation tracking.

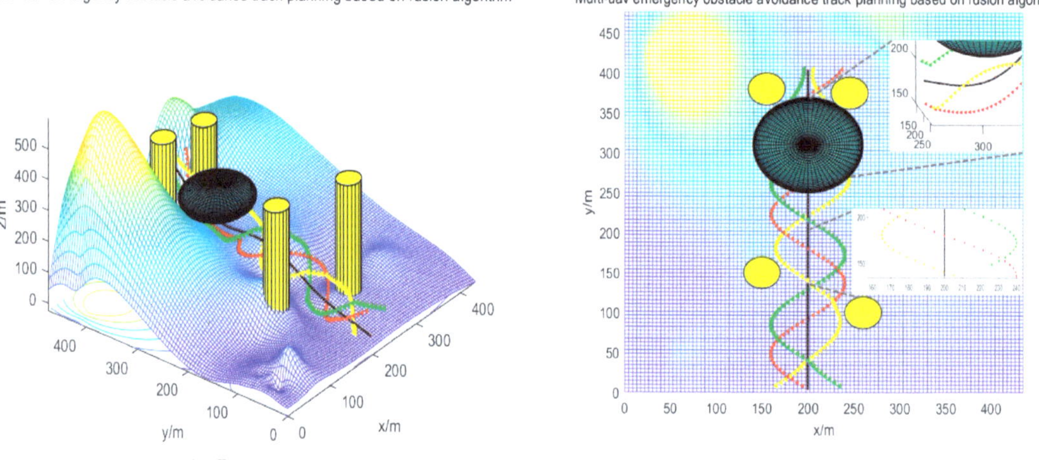

(**a**) Front view

(**b**) Overhead view

Figure 7. Scene 1-simulation of multiple UAVs using the fusion algorithm for reconfiguration of cooperative formation.

The black trajectory in Figures 6–9 is the trajectory of the moving target, and the red, yellow, and green trajectories respectively represent the trajectory planning results of UAV1, UAV2, and UAV3 tracking the moving target. According to the figures, it can be seen that the three UAVs can satisfy several conditions such as their own flight constraints, constraints related to collision avoidance and obstacle avoidance, and carry out real-time stable formation tracking of the moving target. As illustrated by Figures 6 and 7, the simulation of the fusion algorithm makes it possible for UAVs to stably track the target that moves along the established trajectory. Four static obstacles, together with some sudden obstacles, are avoided, which justifies advantages and effectiveness of the fusion algorithm. Unlike the Standoff algorithm that proves to be poor in real-time obstacle avoidance, the fusion algorithm works well in this regard: the three UAVs are distributed around the moving target to maximize the detection coverage of UAV sensors. Thus, the formation reconfiguration task is effectively completed and the unexpected obstacle is successfully bypassed. Figures 8 and 9 only use a single model predictive control algorithm to carry out track planning. Although the vehicles can continue tracking the moving target, their formation is unstable, and the detection coverage for the moving target is insufficient, as shown in Figures 10 and 11.

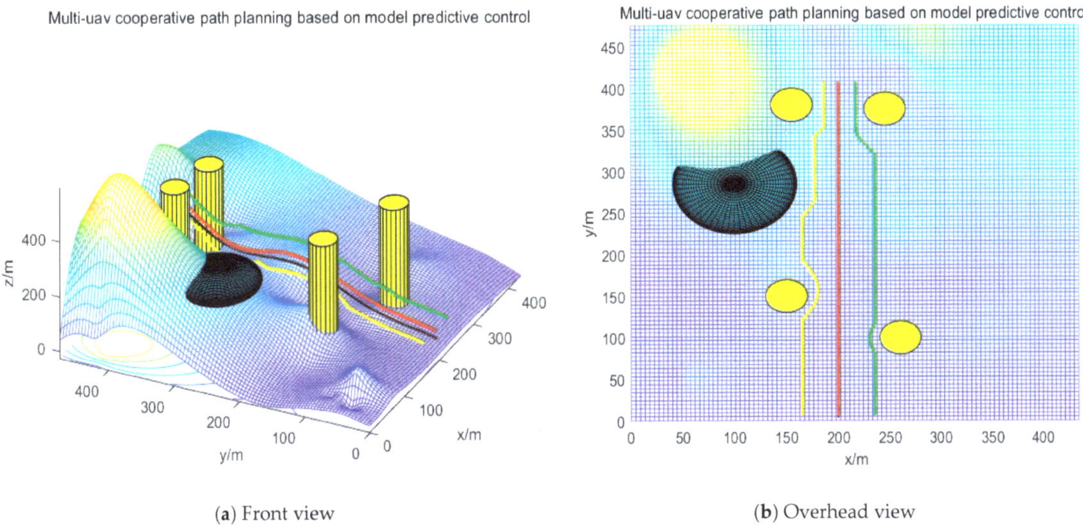

(a) Front view   (b) Overhead view

Figure 8. Scene 1-simulation of multi-UAV coordinated formation tracking using the model predictive control algorithm.

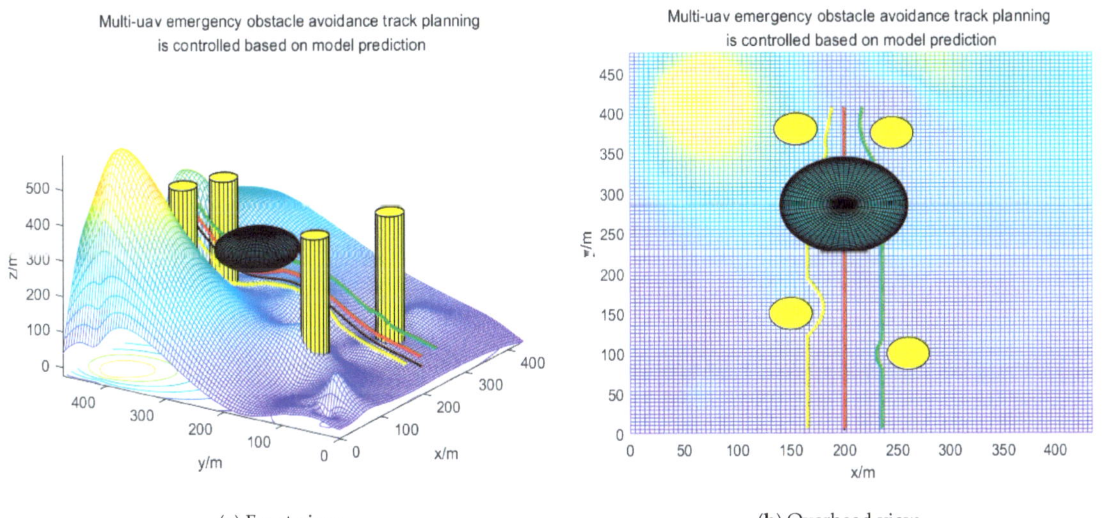

(a) Front view   (b) Overhead view

Figure 9. Scene 1-simulation of multiple UAVs for reconfiguration of cooperative formation using the model predictive control algorithm.

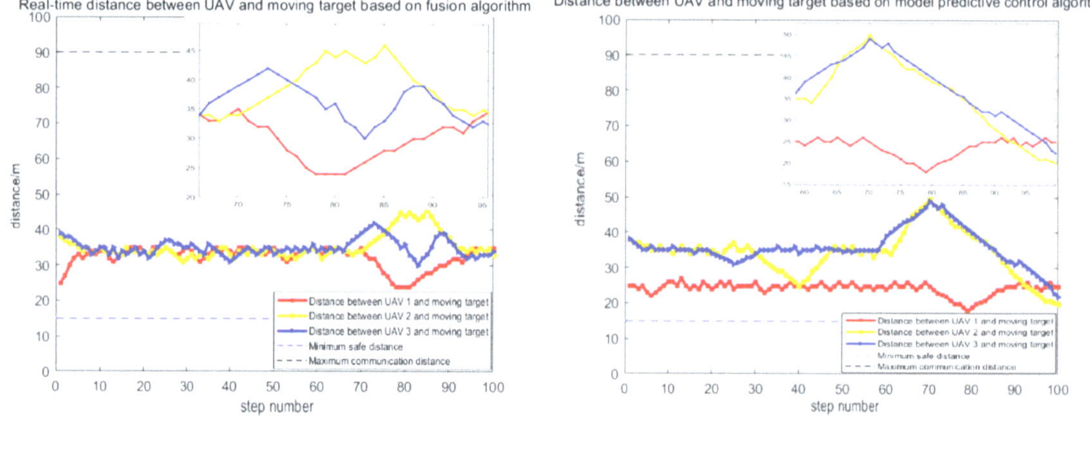

(**a**) Fusion algorithm

(**b**) Model predictive control algorithm

Figure 10. Scene 1-simulation of real-time distance data with the moving target during formation reconstruction of multiple UAVs tracking the moving target.

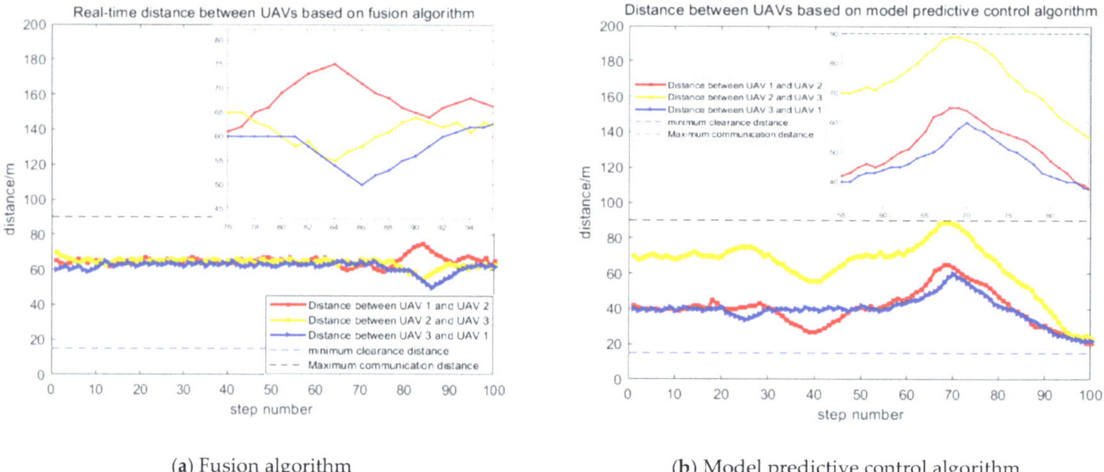

(**a**) Fusion algorithm

(**b**) Model predictive control algorithm

Figure 11. Scene 1-simulation of real-time distance data between multiple UAVs during formation reconfiguration.

A comparison of the simulated data in Figures 10 and 11 verifies that the fusion algorithm is effective in avoiding unexpected obstacles when applied to the trajectory planning process UAVs take through cooperative formation when tracking the moving target. Compared with the model predictive control algorithm alone, the fusion algorithm shows its advantage in formation control with the help of the Standoff algorithm, allowing multiple UAVs to move in a circular motion around the target, maximizing UAV sensors' monitoring range and enabling cooperative formation to track the target. As can be seen in Figure 10, the fusion-based algorithm results in a smaller distance between the UAV and the moving target in real time, and a tighter formation which can be maintained after emergency obstacle avoidance. In Figure 11, the fusion-based UAV spacing remains more stable and less volatile regarding the distance each UAV keeps from the other.

In order to further verify the effectiveness of the fusion algorithm applied to UAVs' tracking of a moving target, and to verify the real-time obstacle avoidance capability of the fusion algorithm, the number of static obstacles is increased to six in this paper, and the specific system simulation results are shown in Figures 12 and 13. At the same time, the same state comparison simulation experiments are carried out using the model predictive control algorithm, as shown in Figures 14 and 15.

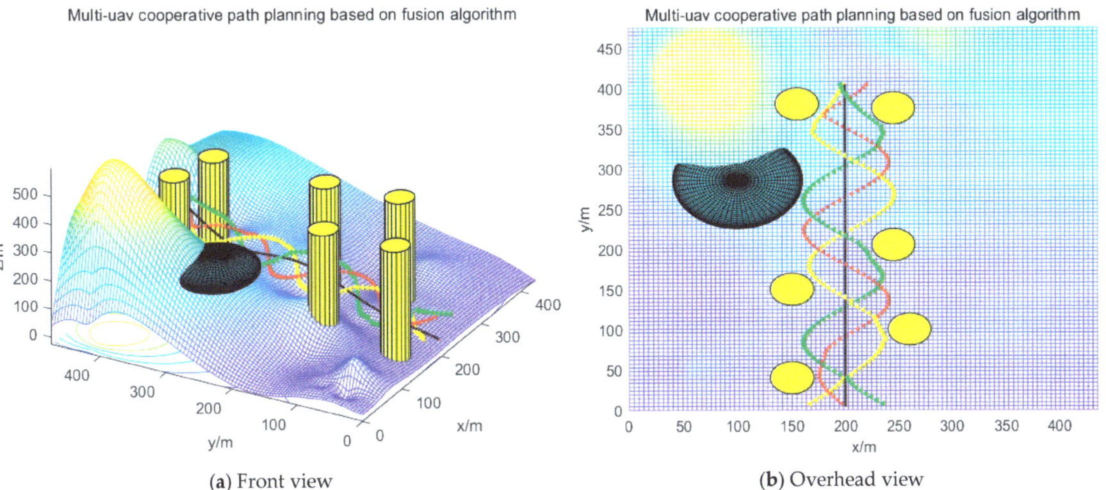

(**a**) Front view

(**b**) Overhead view

Figure 12. Scene 2-simulation of multiple UAVs using the fusion algorithm for coordinated formation tracking.

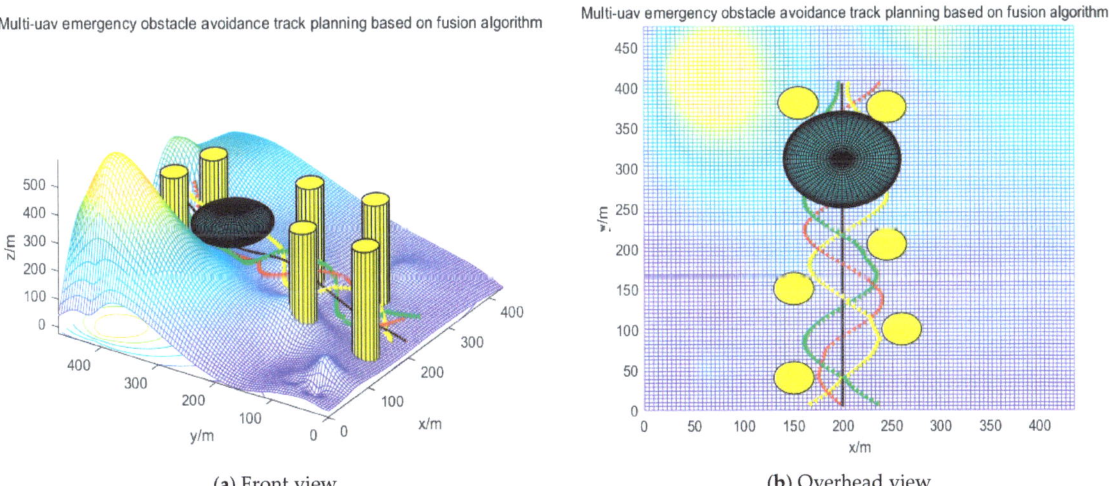

(**a**) Front view

(**b**) Overhead view

Figure 13. Scene 2-simulation of multiple UAVs using the fusion algorithm for reconfiguration of cooperative formation.

As can be seen from the figure above, by increasing the number of static obstacles to six in the scenario, the three UAVs can still satisfy multiple conditions such as their own flight constraints and obstacle avoidance constraints, and be distributed around the moving target in a class circle to maximize the UAV sensor's detection coverage, and effectively complete the task of formation reconstruction and real-time stable formation tracking of the moving target on the basis of collaborative formation trajectory planning in complex environments. A comparison between Figures 13 and 15 shows that the single model predictive control algorithm for track planning, although also capable of continuously tracking moving targets, has an unstable formation and thus insufficient detection coverage for moving targets. Specific tracking accuracy parameters are shown in Figures 16 and 17.

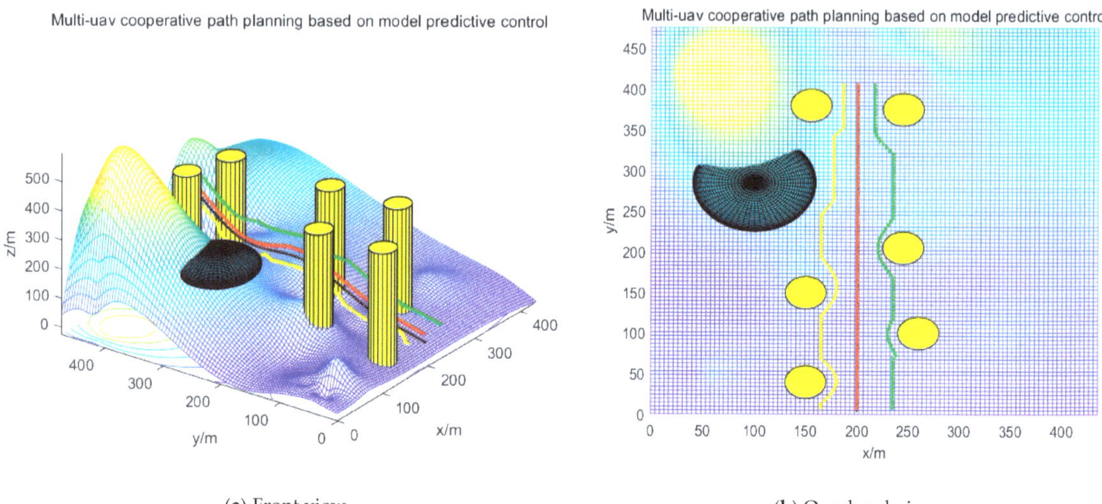

(**a**) Front view　　　　(**b**) Overhead view

Figure 14. Scene 2-simulation of multi-UAV coordinated formation tracking using the model predictive control algorithm.

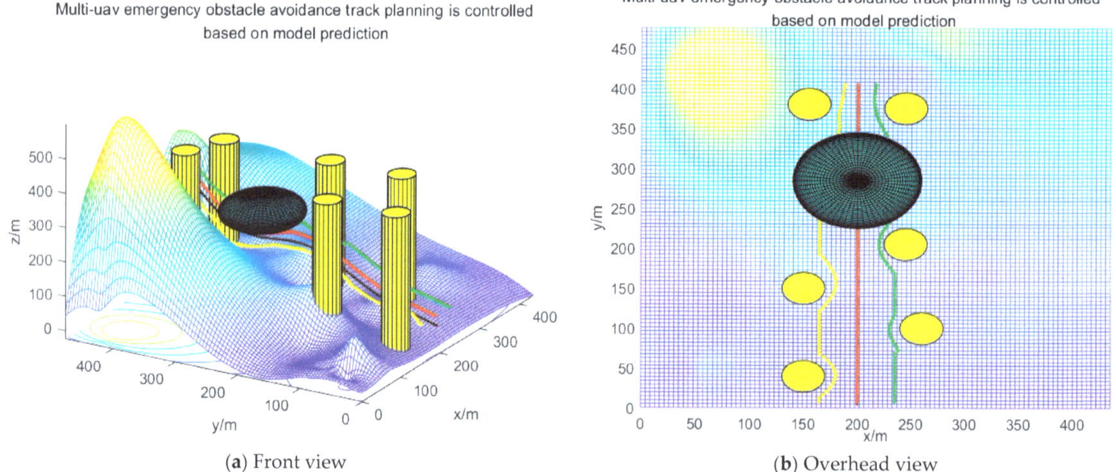

(**a**) Front view　　　　(**b**) Overhead view

Figure 15. Scene 2-simulation of multiple UAVs for reconfiguration of cooperative formation using the model predictive control algorithm.

In order to test the effectiveness of the fusion optimization algorithm applied to UAV cooperative formation tracking moving target trajectory planning for different trajectory targets, this paper changes the established motion trajectory of the moving target, increases the degrees of freedom of the moving target, expands the 2-dimensional motion of the moving target to 3-dimensional motion, and at the same time adjusts the complex 3-dimensional environment model and changes the dynamic obstacle position, the specific simulation results are shown in Figures 18 and 19. Using the model predictive control algorithm to carry out the same state comparison simulation experiments, as shown in Figures 20 and 21.

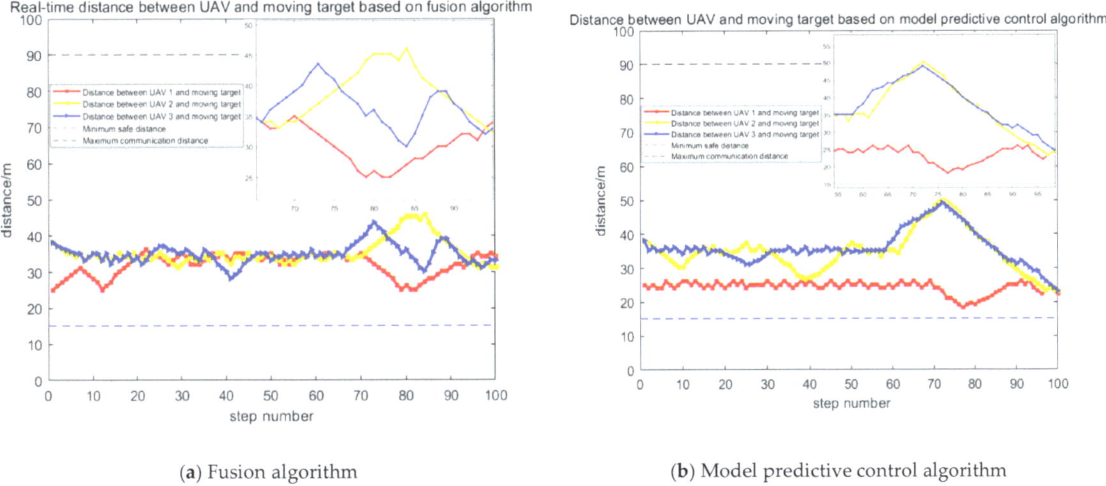

(**a**) Fusion algorithm

(**b**) Model predictive control algorithm

Figure 16. Scene 2-simulation of real-time distance data with the moving target during formation reconstruction of multiple UAVs tracking the moving target.

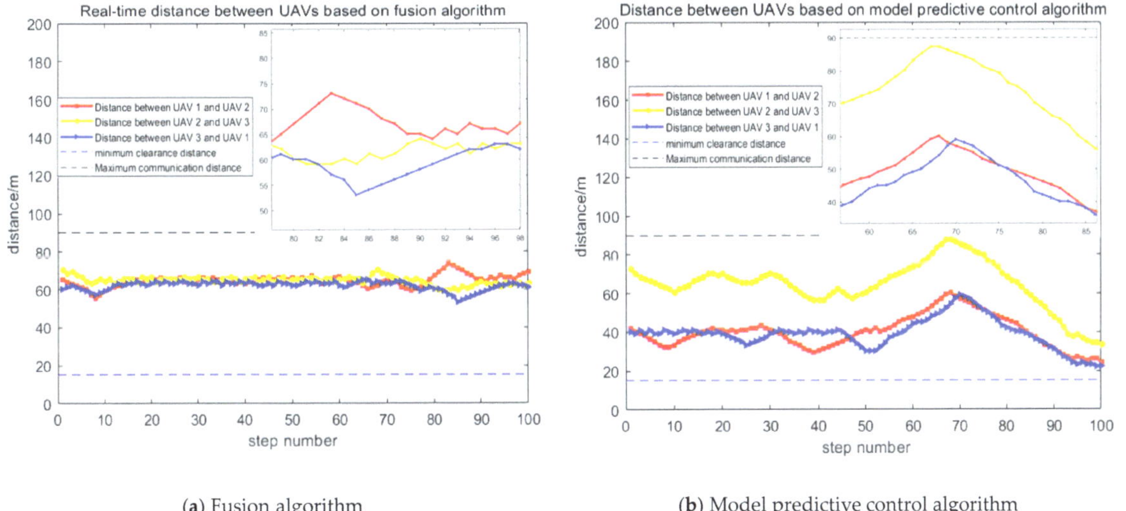

(**a**) Fusion algorithm

(**b**) Model predictive control algorithm

Figure 17. Scene 2-simulation of real-time distance data between multiple UAVs during formation reconfiguration.

According to the figure above, in the context of changing the complex map environment and changing the trajectory of the moving target, the UAVs can still satisfy multiple conditions such as their own flight constraints, collision avoidance and obstacle avoidance constraints, etc., and distribute around the moving target in a class circle to maximize the detection coverage of the UAV sensors, and effectively complete the task of formation reconstruction based on the realization of trajectory planning of multi-UAVs in cooperative formation in a complex environment, and carry out real-time stable formation tracking of the moving target. This demonstrates the effectiveness of the fusion algorithm for tracking moving targets in a complex and variable environment. Specific tracking accuracy parameters are shown in Figures 22 and 23.

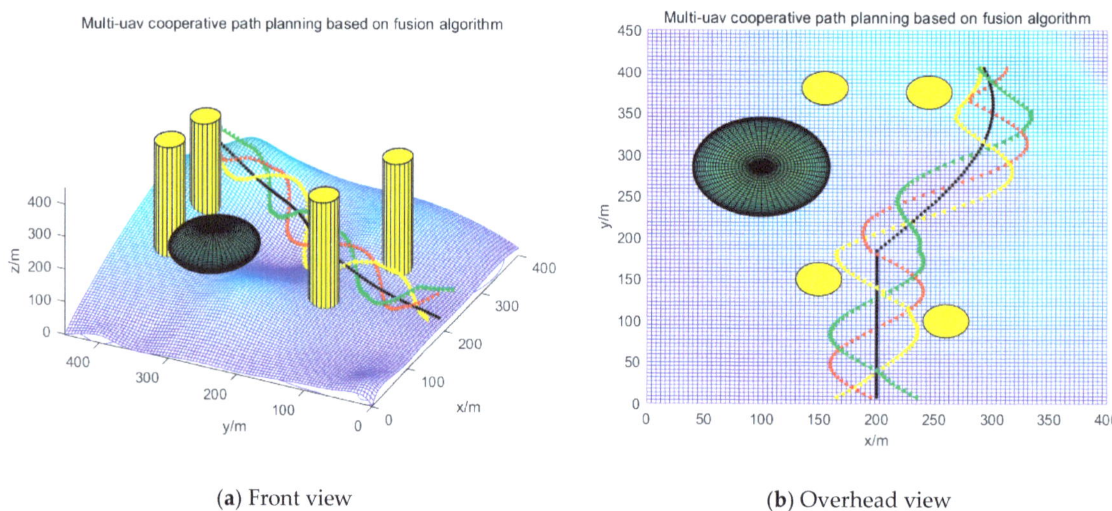

(**a**) Front view (**b**) Overhead view

Figure 18. Scene 3-simulation of multiple UAVs using the fusion algorithm for coordinated formation tracking.

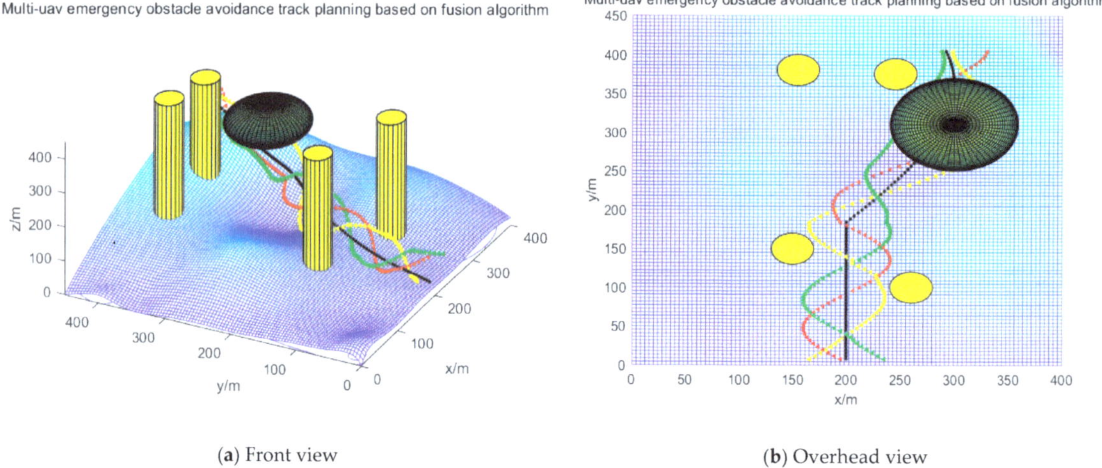

(**a**) Front view (**b**) Overhead view

Figure 19. Scene 3-simulation of multiple UAVs using the fusion algorithm for reconfiguration of cooperative formation.

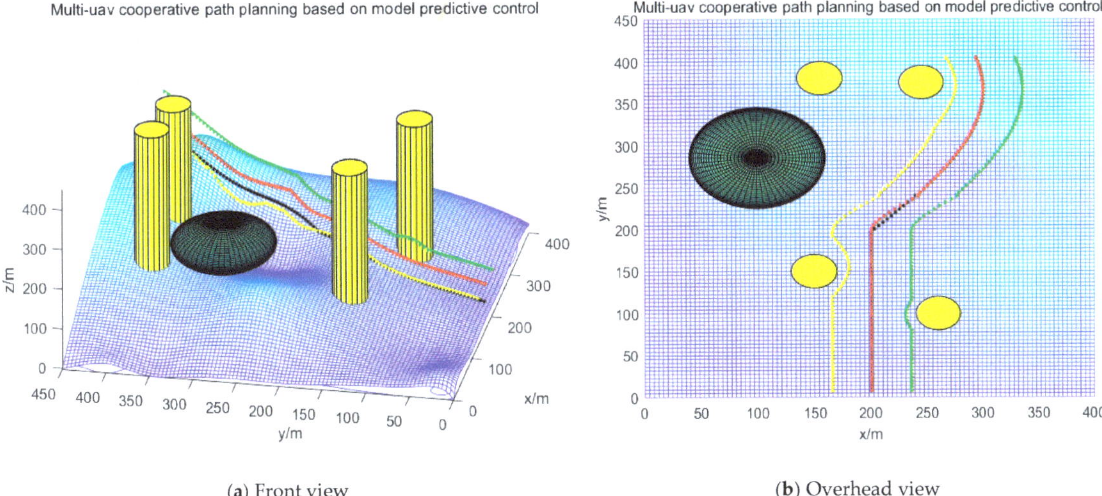

(**a**) Front view

(**b**) Overhead view

Figure 20. Scene 3-simulation of multi-UAV coordinated formation tracking using the model predictive control algorithm.

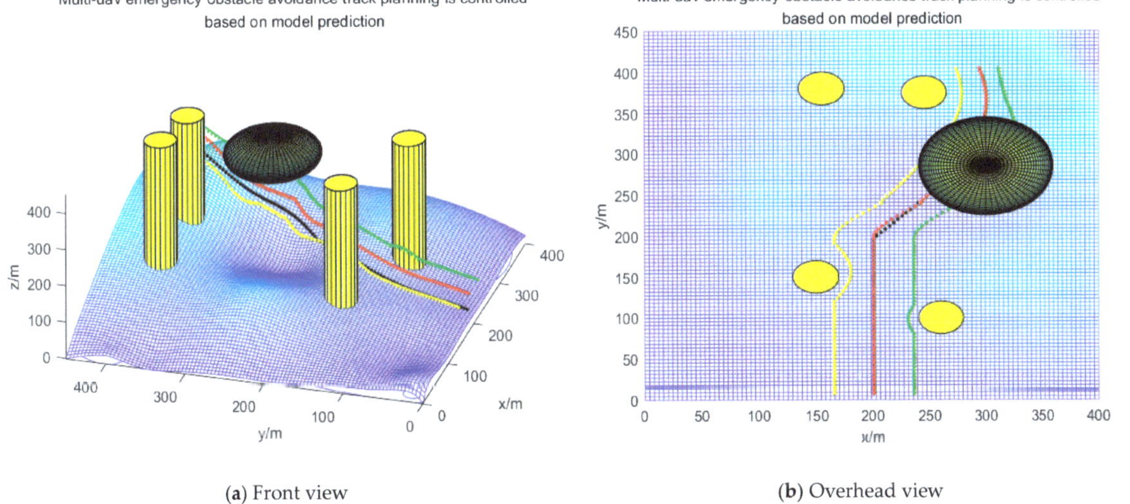

(**a**) Front view

(**b**) Overhead view

Figure 21. Scene 3-simulation of multiple UAVs for reconfiguration of cooperative formation using the model predictive control algorithm.

As can be seen in Figure 22, the fusion-based algorithm has a smaller distance between the UAV and the moving target in real time and maintains a tighter formation, which can be maintained even after emergency obstacle avoidance. In Figure 23, the fusion-based UAV spacing remains stable and less volatile when comparing distances between UAVs.

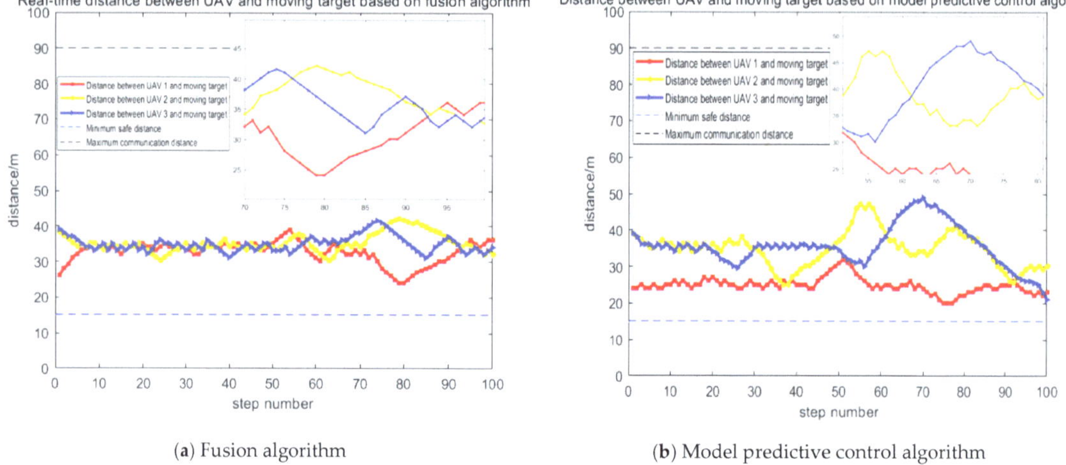

(**a**) Fusion algorithm

(**b**) Model predictive control algorithm

Figure 22. Scene 3-simulation of real-time distance data with the moving target during formation reconstruction of multiple UAVs tracking the moving target.

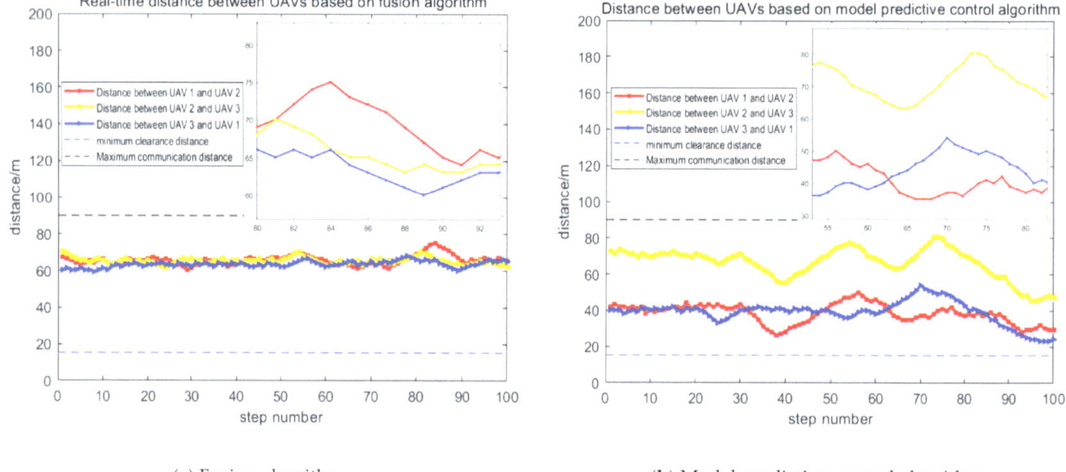

(**a**) Fusion algorithm

(**b**) Model predictive control algorithm

Figure 23. Scene 3-simulation of real-time distance data between multiple UAVs during formation reconfiguration.

5. Discussion

In order to evaluate the proposed fusion algorithm, this paper makes a judgment about the sensors' detection coverage during multi-UAV tracking of a moving target in coordinated formation, while maximizing their detection range and minimizing the probability of target loss in UAV formation, and compares it with the use of a single model predictive control algorithm to verify that the fusion algorithm helps to improve UAV target monitoring capabilities.

For the target tracking effect and monitoring capability, this paper compares the fusion algorithm and the single model predictive control algorithm in the same environment, guiding multiple UAVs to cooperate in formation as they track the moving target, counting the frequency of UAV sensors to effectively monitor the moving target. Experimental results are shown in Table 3, according to which, the three UAVs effectively monitored target coverage using the fusion algorithm a total of 286 times in Scene 1, compared with 268 effective monitoring times using the single model predictive control algorithm, resulting in a 6.72% increase in combined monitoring coverage; in Scene 2 the three UAVs effectively monitored target coverage a total of 283 times with the help of the fusion algorithm, compared with 264 effective monitoring times using the single model predictive control algorithm, resulting in a 7.20% increase in combined monitoring coverage; and in Scene 3 three UAVs effectively monitored target coverage a total of 287 times with the fusion algorithm, while the effective number of monitoring using a single model predictive control algorithm was 269, with a 6.69% increase in comprehensive monitoring coverage, which in turn can be derived from the advantages of the fusion algorithm in terms of tracking and monitoring effectiveness. The improvement of monitoring ability comes from the effective integration of the model predictive control algorithm and the Standoff algorithm in Sections 3.2.1 and 3.2.2. The former uses a 'feedback-correction' mechanism to correct UAV trajectories, ensuring real-time tracking of moving target trajectory planning, while enabling reconfiguration and planning of multiple UAVs in formation reaching the least-cost goal. The latter ensures cooperative formation control of multiple UAVs, builds UAV sensor monitoring models, maximizes sensors' monitoring range and reduces the probability of UAVs losing the moving target. Clearly, the fusion algorithm displays a better tracking effect and monitoring capability in the test.

Table 3. Count of effective detection UAV sensors.

UAV Category	Usage	Effective Number of Detected Steps (Scene 1-Total: 100)	Effective Number of Detected Steps (Scene 2-Total: 100)	Effective Number of Detected Steps (Scene 3-Total: 100)
UAV1	Fusion algorithm	100	100	100
	Model predictive control algorithm	100	100	100
UAV2	Fusion algorithm	89	88	92
	Model predictive control algorithm	86	84	88
UAV3	Fusion algorithm	97	95	95
	Model predictive control algorithm	82	80	81

The fusion algorithm promotes the construction of a multi-UAV track planning model, which obtains a more adaptive tracking strategy and effectively solves the problem of multi-UAV formation reconfiguration and obstacle avoidance in emergency situations. From the experimental results, it can be seen that the algorithm has great advantages in terms of tracking effectiveness and monitoring capability, and can support UAV target tracking in uncertain environments. Although some work has been done in this paper on UAV tracking effectiveness and monitoring capability, there are still some challenges in deploying the algorithm to real UAVs. In practice, external interference, noise and air resistance have a dynamic effect on UAV trajectory planning, making it difficult to keep the UAV maneuvering at all times, and time delays in communication between multiple UAVs may occur. No matter how good the UAV's trajectory planning is in the simulation environment, it is still far from real application. However, we can keep increasing the realism of the scenarios and models in the simulation environment, and thus get closer to the real environment. For future research, we will consider implementing more detailed UAV control, including controlling the UAV with motor speed, acquiring target information through the UAV's vision sensors and acquiring range information through LIDAR as status information, thus achieving target tracking in a more realistic 3D scene.

6. Conclusions

In this paper, a fusion and optimization method is proposed for trajectory planning UAVs make through cooperative formation when tracking the moving target, a framework for the multi-UAV tracking system is designed, and research on stable tracking is carried out to maximize UAV sensors' coverage as they monitor the moving target, which in turn reduces the probability of target loss in the tracking process. Against a complex three-dimensional environment in which priori information is insufficient, the fusion algorithm promotes the reconfiguration and planning of multi-UAV formation at the minimum cost, and thus ensures the existence and maintenance of the multi-UAV formation. The simulation verifies the effectiveness of the fusion algorithm applied to multi-UAV cooperative formation, keeping off deficiency in avoiding real-time obstacles facing the Standoff algorithm.

Some future work includes implementing more detailed UAV control for 3D spatial and target tracking in more complex environments, setting up more realistic scenarios (different flight scenarios with different numbers of tracked targets) for extensive simulation validation, and adding on-board sensors to obtain more data as status information, allowing multiple UAVs to carry out collaborative tracking of a moving target closer to realistic scenarios, so that fusion optimization algorithms can find their market in actual UAV trajectory planning in the future.

Author Contributions: Resources, B.L. and R.M.; methodology, C.S.; validation, C.S., S.B. and J.H.; writing—original draft preparation, B.L. and C.S.; writing—review and editing, K.W. and E.N.; funding acquisition, K.W. and B.L. All authors have read and agreed to the published version of the manuscript.

Funding: This research was partially funded by Project supported by the National Nature Science Foundation of China under grant no. 62003267, the Fundamental Research Funds for the Central Universities under grant no. G2022KY0602, the Technology on Electromagnetic Space Operations and Applications Laboratory under grant no. 2022ZX0090, the Key Research and Development Program of Shaanxi Province under grant no. 2023-GHZD-33, and the key core technology research plan of Xi'an under grant no. 21RGZN0016.

Data Availability Statement: Data sharing not applicable.

Conflicts of Interest: The authors declare no conflict of interest.

References

1. Huang, G.; Hu, M.; Yang, X.; Lin, P. Multi-UAV Cooperative Trajectory Planning Based on FDS-ADEA in Complex Environments. *Drones* **2023**, *7*, 55. [CrossRef]
2. Li, B.; Gan, Z.; Chen, D.; Aleksandrovich, S. UAV Maneuvering Target Tracking in Uncertain Environments Based on Deep Reinforcement Learning and Meta-Learning. *Remote Sens.* **2020**, *12*, 3789. [CrossRef]
3. Zhang, J.; Yan, J.; Zhang, P.; Kong, X. Design and Information Architectures for an Unmanned Aerial Vehicle Cooperative Formation Tracking Controller. *IEEE Access* **2018**, *6*, 45821–45833. [CrossRef]
4. Li, B.; Yang, Z.P.; Chen, D.Q.; Liang, S.Y.; Ma, H. Maneuvering target tracking of UAV based on MN-DDPG and transfer learning. *Def. Technol.* **2021**, *17*, 10. [CrossRef]
5. Bian, L.; Sun, W.; Sun, T. Trajectory Following and Improved Differential Evolution Solution for Rapid Forming of UAV Formation. *IEEE Access* **2019**, *7*, 169599–169613. [CrossRef]
6. Liu, W.; Zheng, X.; Luo, Y. Cooperative search planning in wide area via multi-UAV formations based on distance probability. In Proceedings of the 2020 3rd International Conference on Unmanned Systems (ICUS), Harbin, China, 27–28 November 2020; pp. 1072–1077. [CrossRef]
7. Li, Y.; Tian, B.; Yang, Y.; Li, C. Path planning of robot based on artificial potential field method. In Proceedings of the 2022 IEEE 6th Information Technology and Mechatronics Engineering Conference (ITOEC), Chongqing, China, 4–6 March 2022; pp. 91–94. [CrossRef]
8. Liang, Q.; Zhou, H.; Xiong, W.; Zhou, L. Improved artificial potential field method for UAV path planning. In Proceedings of the 2022 14th International Conference on Measuring Technology and Mechatronics Automation (ICMTMA), Changsha, China, 15–16 January 2022; pp. 657–660. [CrossRef]

9. Zong, C.; Yao, X.; Fu, X. Path Planning of Mobile Robot based on Improved Ant Colony Algorithm. In Proceedings of the 2022 IEEE 10th Joint International Information Technology and Artificial Intelligence Conference (ITAIC), Chongqing China, 17–19 June 2022; pp. 1106–1110. [CrossRef]
10. Gao, Y. An Improved Hybrid Group Intelligent Algorithm Based on Artificial Bee Colony and Particle Swarm Optimization. In Proceedings of the 2018 International Conference on Virtual Reality and Intelligent Systems (ICVRIS), Hunan, China, 10–11 August 2018; pp. 160–163. [CrossRef]
11. Ma, F.; Lu, J.; Liu, L.; He, Y. Application of Improved Single Neuron Adaptive PID Control Method in the Angle Predefined Loop of Active Radar Seeker for Anti-radiation Missile. In Proceedings of the 2021 IEEE 4th Advanced Information Management, Communicates, Electronic and Automation Control Conference (IMCEC), Chongqing, China, 18–20 June 2021; pp. 2160–2164. [CrossRef]
12. Xingke, L.; Xuesong, C.; Shuting, C. Smoothing Method for Nonlinear Optimal Control Problems with Inequality Path Constraints. In Proceedings of the 2019 Chinese Control and Decision Conference (CCDC), Nanchang, China, 3–5 June 2019; pp. 5350–5353. [CrossRef]
13. Anastasiou, D.; Nanos, K.; Papadopoulos, E. Robust Model-based H∞ control for Free-floating Space Manipulator Cartesian Motions. In Proceedings of the 2022 30th Mediterranean Conference on Control and Automation (MED), Vouliagmeni, Greece, 28 June–1 July 2022; pp. 598–603. [CrossRef]
14. Yu, L.; He, G.; Wang, X.; Zhao, S. Robust Fixed-Time Sliding Mode Attitude Control of Tilt Trirotor UAV in Helicopter Mode. *IEEE Trans. Ind. Electron.* **2022**, *69*, 10322–10332. [CrossRef]
15. Vazquez, S.; Rodriguez, J.; Rivera, M.; Franquelo, L.G.; Norambuena, M. Model Predictive Control for Power Converters and Drives: Advances and Trends. *IEEE Trans. Ind. Electron.* **2017**, *64*, 935–947. [CrossRef]
16. Rodriguez, J.; Kazmierkowski, M.P.; Espinoza, J.R.; Zanchetta, P.; Abu-Rub, H.; Young, H.A.; Rojas, C.A. State of the Art of Finite Control Set Model Predictive Control in Power Electronics. *IEEE Trans. Ind. Inform.* **2013**, *9*, 1003–1016. [CrossRef]
17. Sahu, A.; Kandath, H.; Krishna, K.M. Model predictive control based algorithm for multi-target tracking using a swarm of fixed wing UAVs. In Proceedings of the 2021 IEEE 17th International Conference on Automation Science and Engineering (CASE), Lyon, France, 23–27 August 2021; pp. 1255–1260.
18. Ille, M.; Namerikawa, T. Collision avoidance between multi-UAV systems considering formation control using MPC. In Proceedings of the 2017 IEEE International Conference on Advanced Intelligent Mechatronics (AIM), Munich, Germany, 3–7 July 2017; pp. 651–656. [CrossRef]
19. Muslimov, T.Z.; Munasypov, R.A. Coordinated UAV Standoff Tracking of Moving Target Based on Lyapunov Vector Fields. In Proceedings of the 2020 International Conference Nonlinearity, Information and Robotics (NIR), Innopolis, Russia, 3–6 December 2020; pp. 1–5. [CrossRef]
20. Guo, Q.; Peng, J.; Xu, W.; Liang, W.; Jia, X.; Xu, Z.; Yang, Y.; Wang, M. Minimizing the Longest Tour Time Among a Fleet of UAVs for Disaster Area Surveillance. *IEEE Trans. Mob. Comput.* **2022**, *21*, 2451–2465. [CrossRef]
21. Niu, Y.; Liu, J.; Xiong, J.; Li, J.; Shen, L. Research on cooperative ground multi-target guidance method for UAV swarm tracking. *China Sci. Technol. Sci.* **2020**, *50*, 403–422.
22. Zhang, Y.; Fang, G.-W.; Yang, X.-X. Cooperative tracking of multiple UAVs under command decision. *Flight Mech.* **2020**, *38*, 28–33. [CrossRef]
23. Zhu, Q.; Zhou, R.; Dong, Z.-N.; Li, H. Two-machine cooperative standoff target tracking under angular measurement. *J. Beijing Univ. Aeronaut. Astronaut.* **2015**, *41*, 2116–2123. [CrossRef]
24. Wang, D.; Wu, M.; He, Y.; Pang, L.; Xu, Q.; Zhang, R. An HAP and UAVs Collaboration Framework for Uplink Secure Rate Maximization in NOMA-Enabled IoT Networks. *Remote Sens.* **2022**, *14*, 4501. [CrossRef]
25. Wang, D.; He, T.; Zhou, F.; Cheng, J.; Zhang, R.; Wu, Q. Outage-driven link selection for secure buffer-aided networks. *Sci. China Inf. Sci.* **2022**, *65*, 182303. [CrossRef]
26. Parisio, A.; Rikos, E.; Glielmo, L. A Model Predictive Control Approach to Microgrid Operation Optimization. *IEEE Trans. Control. Syst. Technol.* **2014**, *22*, 1813–1827. [CrossRef]
27. Dantec, E.; Taix, M.; Mansard, N. First Order Approximation of Model Predictive Control Solutions for High Frequency Feedback. *IEEE Robot. Autom. Lett.* **2022**, *7*, 4448–4455. [CrossRef]
28. Harinarayana, T.; Hota, S. Coordinated Standoff Target Tracking by Multiple UAVs in Obstacle-filled Environments. In Proceedings of the 2021 Seventh Indian Control Conference (ICC), Mumbai, India, 20–22 December 2021; pp. 111–116. [CrossRef]
29. Song, R.; Long, T.; Wang, Z.; Cao, Y.; Xu, G. Multi-UAV Cooperative Target Tracking Method using sparse A search and Standoff tracking algorithms. In Proceedings of the 2018 IEEE CSAA Guidance, Navigation and Control Conference (CGNCC), Xiamen, China, 10–12 August 2018; pp. 1–6. [CrossRef]
30. Abedini, A.; Bataleblu, A.A.; Roshanian, J. Robust Backstepping Control of Position and Attitude for a Bi-copter Drone. In Proceedings of the 2021 9th RSI International Conference on Robotics and Mechatronics (ICRoM), Tehran, Iran, 17–19 November 2021; pp. 425–432. [CrossRef]
31. Cheng, Z.; Zhao, L.; Shi, Z. Decentralized Multi-UAV Path Planning Based on Two-Layer Coordinative Framework for Formation Rendezvous. *IEEE Access* **2022**, *10*, 45695–45708. [CrossRef]
32. Wang, D.; Zhou, F.; Lin, W.; Ding, Z.; Al-Dhahir, N. Cooperative Hybrid Non-Orthogonal Multiple Access Based Mobile-Edge Computing in Cognitive Radio Networks. *IEEE Trans. Cogn. Commun. Netw.* **2022**, *8*, 1104–1117. [CrossRef]

33. Shaowu, D.; Chaolun, Z.; Fei, L.; Xu, H.; Guorong, Z. A model predictive control algorithm for UAV formations under multiple constraints. *Control. Decis. Mak.* **2023**, *38*, 706–714. [CrossRef]
34. Fuchun, L.; Huanli, G. Research on model predictive control algorithms for small unmanned helicopters. *Control. Theory Appl.* **2018**, *35*, 1538–1545.
35. Haiou, L.; Yuxuan, H.; Qingxiao, L.; Shihao, L.; Huiyan, C.; Li, C. Research on the search strategy of different detection distance sensors. *J. Beijing Univ. Technol.* **2023**, *43*, 151–160. [CrossRef]
36. Fan, G.; Zhao, Y.; Guo, Z.; Jin, H.; Gan, X.; Wang, X. Towards Fine-Grained Spatio-Temporal Coverage for Vehicular Urban Sensing Systems. In Proceedings of the IEEE INFOCOM 2021—IEEE Conference on Computer Communications, Vancouver, BC, Canada, 10–13 May 2021; pp. 1–10. [CrossRef]
37. Wang, H.; Liu, C.H.; Dai, Z.; Tang, J.; Wang, G. Energy-Efficient 3D Vehicular Crowdsourcing for Disaster Response by Distributed Deep Reinforcement Learning. In Proceedings of the 27th ACM SIGKDD Conference on Knowledge Discovery & Data Mining (KDD '21), Washington, DC, USA, 14–18 August 2021; Association for Computing Machinery: New York, NY, USA, 2021; pp. 3679–3687. [CrossRef]
38. Cao, Y.; Cheng, X.; Mu, J. Concentrated Coverage Path Planning Algorithm of UAV Formation for Aerial Photography. *IEEE Sens. J.* **2022**, *22*, 11098–11111. [CrossRef]

drones

MDPI

Article

Files Cooperative Caching Strategy Based on Physical Layer Security for Air-to-Ground Integrated IoV

Weiguang Wang [1,2], Hui Li [1,*], Yang Liu [1], Wei Cheng [1] and Rui Liang [1]

[1] School of Electronics and Information, Northwestern Polytechnical University, Xi'an 710129, China
[2] Department of Computer Science, University of Victoria, Victoria, BC V8P 5C2, Canada
* Correspondence: lh@nwpu.edu.cn

Abstract: Mobile edge cache (MEC)-enabled air-to-ground integrated Internet of Vehicles (IoV) technology can solve wireless network backhaul congestion and high latency, but security problems such as eavesdropping are often ignored when designing cache strategies. In this paper, we propose a joint design of cache strategy and physical layer transmission to improve the security offloading ratio of MEC-enabled air-to-ground IoV. By using the random geometry theory and Laplace transform, we derive the closed-form expression of the network security offloading ratio, which is defined as the probability that the request vehicle (RV) successfully finds the required file around it and obtains the file with a data rate larger than a given threshold. During the file acquisition process, we collectively consider the impact of the successful connection and secure transmission in the vehicle wireless communication. Then, we establish an optimization problem for maximizing the network security offloading ratio, in which the cache strategy and the secure transmission rate are jointly optimized. Furthermore, we propose an alternating optimization algorithm to solve the joint optimization problem. Simulation experiments verify the correctness of our theoretical derivation, and prove that the proposed cache strategy is superior to other existing cache strategies.

Keywords: air-ground collaborative IoV; vehicle-to-vehicle (V2V); caching strategy; the secure transmission rate; physical layer security (PLC)

Citation: Wang, W.; Li, H.; Liu, Y.; Cheng, W.; Liang, R. Files Cooperative Caching Strategy Based on Physical Layer Security for Air-to-Ground Integrated IoV. *Drones* **2023**, *7*, 163. https://doi.org/10.3390/drones7030163

Academic Editor: Vishal Sharma

Received: 9 January 2023
Revised: 23 February 2023
Accepted: 23 February 2023
Published: 26 February 2023

1. Introduction

According to Cisco's 2018–2023 Internet Annual Report, global devices and connected devices will grow at a compound annual growth rate (CAGR) of 10%, and IoV-based applications will grow at a CAGR of 17% [1,2]. The explosive growth of mobile data will bring a heavy burden to the core network. Currently, if a proper solution is not found to address the explosive growth of data traffic, it may degrade the quality experience of user vehicle (UV) and even cause congestion on backhaul links in the future. However, the present technologies can not meet all requirements in fifth generation communication network (5G), and there will be higher requirements for wireless network latency, coverage, spectrum and energy efficiency [3–7] in sixth generation communication network (6G). In order to satisfy the above requirements, 6G will need a paradigm shift to provide intelligent services for mobile devices. MEC-enabled air-to-ground integrated IoV technology has been proven to be a key technology in vehicle wireless networks, which can fully utilize the cache space of edge UV to improve network resiliency, reduce latency, and backhaul traffic [8–11]. Vehicle-to-Vehicle (V2V) communication can allow UV to directly share files to other UV around without going through the core network. The previous theoretical and practical research on mobile edge cache (MEC) based on V2V communication shows that V2V communication technology can effectively improve the throughput of IoV [12–14]. Therefore, MEC-enabled air-to-ground integrated IoV technology will have better application prospects in the 6G wireless communication network.

By utilizing the V2V communication technology, the adjacent UVs with MEC capability can communicate with each other directly without relying on the data forwarding of the air base station (ABS), which will further improve the quality of service (QoS) of UVs and network performance. Since the cache capacity of the UV is also limited, it is necessary to design an optimal caching strategy to reasonably cache content to maximize network utility. Most of the existing caching strategy research use hit ratio, energy efficiency (EE), and network delay as optimization indicators. The significant breakthroughs have been made in these research directions [15–22]. Dai et al. [15] proposed a cooperative caching-multicast strategy based on V2V communication technology to improve the timeliness and spatial coverage of content services. They analyzed and verified the effectiveness of the strategy through the main technical indicators of average transmission coverage and average transmission delay. In order to obtain a more accurate caching strategy, Ning et al. [16] considered the effects of self-offloading, V2V offloading, general user preference, individual UV preference, and the peak time change of content preference on the design of the caching strategy. Therefore, it was also confirmed that the network data offloading performance could be effectively improved by rationally designing the caching strategy between UVs and making full use of the self-caching capability. Anjum et al. [17] proposed a cache method based on two-tiered segment, which divided the storage capacity of each mobile device into two areas. Their research could effectively reduce the startup and playback delay of video in the network. Ma et al. [18] investigated the application of the cluster center caching strategy in data sharing, and analyzed the effectiveness of the cluster center caching strategy by using network coverage probability, average completion ratio, and cache hit ratio. Lee et al. [19] studied the optimal caching strategy and cooperative distance design of V2V caching network from the aspects of network throughput and EE. Cai et al. [20] proposed a social-aware mobile edge caching strategy based on network coding, considering the impact of location proximity and UV social relations on caching strategies. S. Sinem Kafıloğlu et al. [21] proposed two cooperative cache replacement algorithms based on distance and priority classification to optimize network energy consumption. Because the battery capacity of each UV was limited, Li et al. [22] studied the design of caching strategy for V2V-assisted wireless networks from the perspective of network offloading gain and energy consumption. The above researches have made important contributions from various perspectives based on V2V wireless caching networks, but they all ignore the communication security issues in MEC-enabled air-to-ground integrated IoV.

In the era of advanced network technology, the use of various applications generates a large amount of unknown personal privacy data. Once the personal privacy data is leaked, it can seriously affect the privacy of UVs and even the safety of UVs' property and life. Therefore, people's privacy and security problems in MEC-enabled air-to-ground integrated IoV must be paid great attention to. Physical layer security [23,24] and wireless caching can be easily integrated in a low complexity and high flexibility manner, mainly including two reasons: (1) Physical layer security achieves wireless secrecy by using eavesdropping channel coding, which is different from source encryption. This encryption method can enable the cached files to be reused, thereby improving the reuse probability of the content in the edge cache. (2) Physical layer security can exploit the inherent randomness of wireless channels without necessarily relying on keys. The security problem in wireless networks is gradually attracting researchers' attention. Refs. [25–28] have done some research on the security problem of random wireless networks, but the research on using physical layer security to ensure that file transmissions in edge cache are not eavesdropped is still rare. There even lacks a basic theoretical security performance analysis framework and optimization from the perspective of random geometry. Wang and Zheng [25] investigated the physical layer security of random cellular networks, which laid the foundation for the study of wireless network security. Liu et al. [26] derived the exact expression of outage probability of large-scale access to wireless networks through physical layer security. Zheng et al. [27] studied the joint design of small cell network-based cache placement and physical layer transmission in the presence of randomly distributed eavesdroppers to improve the secure content delivery probability of small cell networks. Ren et al. [28]

proposed a mobile-aware cooperative coding caching strategy for the high-speed mobility of users and the secure transmission of content. Inspired by the above researches, this paper focuses on the research of security problems in the MEC-enabled air-to-ground integrated IoV to prevent the important data of UVs from being forged or tampered by attackers and provide a strong guarantee for UVs' privacy.

In this paper, we mainly investigate the cache strategy design and physical layer security in the MEC-enabled air-to-ground integrated IoV to improve the data offloading performance and the anti-eavesdropping capability. The main work and achievements are described as follows

- We propose a novel mobile edge cache strategy based on physical layer security, which enhances the adjacent discovery capability of files and improves the probability of secure transmission. Based on random geometry theory, we calculate the precise expression of the MEC-enabled air-to-ground integrated IoV security offloading ratio. Taking the security offloading ratio as the objective function, we build a joint optimization problem about the cache strategy and the secure transmission rate;
- Since the cache strategy and the secure transmission rate are tightly coupled in the objective function, it is difficult to directly obtain the joint optimal solution. Therefore, we propose an alternating optimization algorithm, which can obtain the joint optimal solution of the cache strategies and the secure transmission rate to maximize the network security offloading ratio;
- Through a numerical simulation of the key technical parameters, the results show that the network security offloading performance of the proposed caching strategy is superior to the existing caching strategies.

Other sections of the paper are arranged as follows. Section 2 is the air-to-ground integrated IoV system model, and Section 3 shows the problem formulation and analysis. The cache strategy optimization problem is presented in Section 4. In Section 5, theoretical analysis and numerical simulation results are described. Finally, the conclusions of this paper are drawn in Section 6.

2. System Model

This section mainly introduces the MEC-enabled air-to-ground integrated IoV network model and file access model considered in our research content.

2.1. Network Model

In this paper, we consider a MEC-enabled air-to-ground integrated IoV model, in which cooperative vehicles (CVs), RVs and eavesdropping vehicles (EVs) are modeled as a homogeneous Poisson point process (HPPP) [29] with density λ_p, λ_r and λ_e, respectively, as shown in Figure 1. This is a small cell network, such as a single road or an intersection of two roads. In this case, we can model network nodes into Poisson point process and use random geometric theory analysis. The PPP model is usually more accurate for the vehicle network formed on this sparse road layout [30–33]. Each UV has a single antenna with transmission power P_t. For large-scale fading we consider a standard fading model $r^{-\alpha}$. r represents the communication distance between UVs, α is the fading factor and $\alpha > 2$ is the pre-condition. For small-scale fading, we consider Rayleigh fading, in which the channel gain follows the exponential distribution with unit mean independently, i.e., $G \sim \exp(1)$ [34,35]. The reason is that this work focuses on Highrise Urban scenarios, consisting of many ground obstructions. Therefore, traditional air-to-ground channels (e.g., Nakagami fading channels [36,37]) may not be suitable for our considered scenarios. This assumption has been widely used for vehicular communications [38,39]. If the UV caches the required files in their own storage space, the UV can get it directly without consuming other resources. Otherwise, the required files will be obtained through cooperation between UVs. We assume that the ABS caches all files and knows all UV information to coordinate V2V cooperative communication.

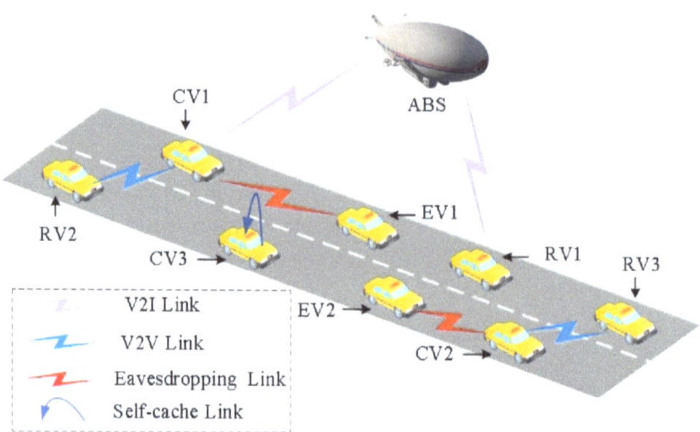

Figure 1. MEC-enabled air-to-ground integrated IoV model.

We consider a limited file library $\mathcal{F} = \{1, 2, \cdots, F\}$, where 1-st is the most popular file. Each CV has a limited cache capacity S ($S << F$). Due to the size and cost constraints assumptions in Refs. [22,40], we assume that each CV can cache one file, which easily generalizes to the multiple file case. We assume that each UV can obtain the file independently, which obeys the Zipf distribution [41]. The probability that the $f - th$ file is requested by the UE can be expressed as $p_f = f^{-\varepsilon} / \sum_{i=1}^{F} i^{-\varepsilon}$, where ε is the popularity factor. Each CV can cache files according to optimized cache strategy $\mathbf{q} = \left\{ q_1, \cdots, q_f, \cdots, q_F \right\}$, where q_f represents the probability that the $f - th$ file is cached. Based on the thinning property [42], the location of the CV that cached file f follows a HPPP distribution with density $q_f \lambda_p$.

2.2. File Access Model

We consider that the ABS caches all files and knows the RVs' request information and the CVs' cache status. The ABS will schedule the content according to the known information. If the requester sends the file request information, there are two ways to obtain the files, namely self-cache and V2V cache. When self-cache and V2V caching fail, the ABS provides the required files to the RV.

- **Case 1: self-cache:** If the RV caches the desired file in the local cache, and the RV will get the desired file directly from the local cache without associating other CVs. This is a special case consideration in edge caching, which is often ignored in existing related researches [43–45];
- **Case 2: V2V cache:** If the RV does not cache the required file in the self-cache, then the RV will obtain the required file from surrounding CVs through V2V communication. This process involves the successful establishment of V2V communication and the secure transmission of files. The research will be discussed in later chapters.

3. Problem Formulation and Analysis

In this section, we mainly analyze the data offloading performance of the MEC-enabled air-to-ground integrated IoV, and take the network security offloading ratio as the main quantitative indicator. The network security offloading ratio is defined as the probability that a file is successfully found around the RV and transmitted confidentially at a given data rate threshold. Therefore, the total network security offloading ratio of the MEC-enabled air-to-ground integrated IoV H_{total} is defined as

$$H_{total} = \sum_{f=1}^{F} p_f (P_s + P_{V2V} D_s D_c),$$ (1)

where P_s and P_{V2V} are the self-cache data offload ratio and the probability that V2V successfully finds the file, respectively. D_s and D_c are the probability of successful V2V connection and the probability of secure transmission, respectively.

3.1. Self-Cache Offloading Ratio

We calculate the probability that the RV finds the required file in the local cache library as the self-cache offloading ratio. Therefore, the self-cache data offloading ratio can be calculated as

$$P_s = \sum_{f=1}^{F} p_f q_f. \tag{2}$$

3.2. V2V Cache Offloading Ratio

If the RV cannot obtain the required files through self-caching, then the V2V cache needs to be enabled to obtain the required files from neighboring CVs. We assume that CV_1 caches the file f, because the location of CV_1 follows a HPPP distribution, so the probability density function of the RV and CV_1 with association distance r is calculated as $f = 2\pi q_f \lambda_p r e^{-\pi q_f \lambda_p r^2}$ [46]. Therefore, the probability that the RV successfully sensing the file f within the communication range z can be calculated as

$$P_{V2V,f} = \left(1 - q_f\right)\left(1 - e^{-\pi q_f \lambda_p z^2}\right). \tag{3}$$

From Equation (3), we can calculate the probability that the RV successfully sensing all files in the file library $\mathcal{F} = \{1, 2, \cdots, F\}$ as

$$P_{V2V} = \sum_{f=1}^{F} p_f \left(1 - q_f\right)\left(1 - e^{-\pi q_f \lambda_p z^2}\right). \tag{4}$$

Based on Shannon's theorem, when the transmission capacity C_b between UVs is $C_b \geq R_s + R_v$, the V2V communication can be successfully established. When EV's channel capacity C_e is $C_e \leq R_v$, the communication can be transmitted confidentially, where R_s and R_v represent the original transmission rate and redundant transmission rate, respectively. Therefore, the successful connection probability and security transmission probability of the cache-enabled V2V communication can be calculated as $D_s = \mathbb{P}\{\log_2(1 + SINR_b) \geq R_s + R_v\}$ and $D_c = \mathbb{P}\{\log_2(1 + SINR_e) \leq R_v\}$. Specifically, $SINR_b = \frac{P_t g_{b,0} r_{b,0}^{-\alpha}}{I_b + N_0 w_1}$ is the signal-to-interference-and-noise ratio (SINR) at RV, where I_b represents the interference generated by surrounding the RVs, whose location obeys a HPPP Φ_1 with density $\lambda_r P_{D2D}$. $N_0 w_1$ is the noise power at the receiver, $g_{b,0}$ is the channel gain, and $r_{b,0}$ represents the cooperation distance between UVs. $SINR_e = \frac{P_t g_{e,0} r_{e,0}^{-\alpha}}{I_e + N_0 w_2}$ is the SINR at EV, where $r_{e,0}$ is the distance between the CV and the EV, $g_{e,0}$ is the channel gain. The random variable $I_e = \sum_{i \in \Phi_2 \backslash CP_0} P_t g_i r_i^{-\alpha}$ refers to the interference of EVs around, and $N_0 w_2$ represents the received noise power. We do further calculations of probabilities D_s and D_c, the expression can be rewritten as

$$D_s = \mathbb{P}\left\{\log_2\left(1 + \frac{P_t g_{b,0} r_{b,0}^{-\alpha}}{I_b + N_0 w_1}\right) \geq R_s + R_v\right\}$$

$$\overset{(1)}{\approx} \mathbb{P}\left\{g_{b,0} \geq I_b P_t^{-1} r_{b,0}^{\alpha} 2^{R_s + R_v}\right\} \overset{(2)}{=} \mathbb{E}_{I_b}\left(e^{-I_b P_t^{-1} r_{b,0}^{\alpha} 2^{R_s + R_v}}\right), \tag{5}$$

$$\overset{(2)}{=} \mathcal{L}_{I_b}\left(P_t^{-1} r_{b,0}^{\alpha} 2^{R_s + R_v}\right)$$

where step (1) considers the interference restriction between UVs, step (2) considers small-scale fading that follows an exponential distribution $g_{b,0} \sim \exp(1)$ [34,35], and step (3) is the Laplace transform of the random variable I_b [47,48], where $I_b = \sum_{i \in \Phi_1 \backslash CP_0} P_t g_i r_i^{-\alpha}$. The Laplace transform of the random variable I_b can be calculated as

$$
\begin{aligned}
\mathcal{L}_{I_b}(s) &= \mathbb{E}_{\Phi_{1,g_i}} \left(e^{-s \sum_{i \in \Phi_1 \backslash CP_0} P_t g_i r_i^{-\alpha}} \right) \\
&= \mathbb{E}_{\Phi_{1,g_i}} \left[\prod_{i \in \Phi_1 \backslash CP_0} \left(1 + s P_t r_i^{-\alpha} \right)^{-1} \right] \\
&= \exp \left[-2\pi \lambda_r P_{V2V} \int_0^\infty \left(1 - \frac{1}{1 + s P_t x^{-\alpha}} \right) x \, dx \right] \\
&= \exp \left\{ -2\pi \lambda_r P_{V2V} \int_0^\infty \left[1 - \frac{x}{1 + (s P_t)^{-1} x^\alpha} \right] dx \right\}' \\
&= \exp \left(-2\pi \lambda_r P_{V2V} \int_0^\infty \frac{y^{\frac{2}{\alpha}-1}}{1 + (s P_t)^{-1} y} \frac{1}{\alpha} dy \right) \\
&\overset{(1)}{=} \exp \left[-2\pi^2 \lambda_r P_{V2V} (s P_t)^{\frac{2}{\alpha}} \csc\left(\frac{2}{\alpha} \right) \alpha^{-1} \right]
\end{aligned}
\tag{6}
$$

where step (1) is obtained by (Ref. [49] Equation (3.194.4)) when the path loss factor satisfies $\alpha > 2$. Then, by substituting $s = P_t^{-1} r_{b,0}^\alpha 2^{R_s + R_v}$ into Equation (6) we can rewrite Equation (5) as

$$
\begin{aligned}
D_s &= \mathcal{L}_{I_b} \left(P_t^{-1} r_{b,0}^\alpha 2^{R_s + R_v} \right) \\
&= \exp \left[-2\pi^2 \lambda_r P_{V2V} 2^{2(R_s + R_v)/\alpha} \csc\left(2\pi\alpha^{-1} \right) \alpha^{-1} r_{b,0}^2 \right].
\end{aligned}
\tag{7}
$$

Furthermore, we calculate the secure transmission probability of the MEC-enabled air-to-ground integrated IoV D_c. The SINR of EV can be expressed as $SINR_e = \frac{P_t g_{e,0} r_{e,0}^{-\alpha}}{I_e + N_0 w_2}$. The position at EVs follow a HPPP distribution Φ_2 with density λ_e. Therefore, the secure transmission probability D_c can be recalculated as

$$
\begin{aligned}
D_c &= \mathbb{P}\{\log_2(1 + SINR_e) \leq R_v\} \\
&= \mathbb{P}\left\{ \log_2\left(1 + \frac{P_t g_{e,0} r_{e,0}^{-\alpha}}{I_e + N_0 w_2} \right) \leq R_v \right\} \\
&= \mathbb{P}\left\{ g_{e,0} \leq I_e P_t^{-1} r_{e,0}^\alpha 2^{R_v} \right\} = 1 - \mathbb{P}\left\{ g_{e,0} > I_e P_t^{-1} r_{e,0}^\alpha 2^{R_v} \right\} \\
&= 1 - \mathbb{E}_{I_e} \left(e^{-I_e P_t^{-1} r_{e,0}^\alpha 2^{R_v}} \right) = 1 - \mathcal{L}_{I_e} \left(P_t^{-1} r_{e,0}^\alpha 2^{R_v} \right)
\end{aligned}
\tag{8}
$$

Then, we do the Laplace transform of the random variable I_e in Equation (8). The calculation process is as follows

$$
\begin{aligned}
\mathcal{L}_{I_e} \left(P_t^{-1} r_{e,0}^\alpha 2^{R_v} \right) &= \mathbb{E}_{\Phi_{2,g_i}} \left(e^{-s \sum_{i \in \Phi_2 \backslash CP_0} P_t g_i r_i^{-\alpha}} \right) \\
&= \mathbb{E}_{\Phi_{2,g_i}} \left[\prod_{i \in \Phi_2 \backslash CP_0} \left(1 + r_{e,0}^\alpha 2^{R_v} r_i^{-\alpha} \right)^{-1} \right] \\
&= \exp \left[-2\pi \lambda_e \int_0^\infty \left(1 - \frac{1}{1 + r_{e,0}^\alpha 2^{R_v} x^{-\alpha}} \right) x \, dx \right]. \\
&= \exp \left(-2\pi \lambda_e \int_0^\infty \frac{y^{\frac{2}{\alpha}-1}}{1 + \left(r_{e,0}^\alpha 2^{R_v} \right)^{-1} y} \frac{1}{\alpha} dy \right) \\
&= \exp \left[-2\pi^2 \lambda_e 2^{2R_v/\alpha} \csc\left(2\pi\alpha^{-1} \right) \alpha^{-1} r_{e,0}^2 \right]
\end{aligned}
\tag{9}
$$

Therefore, through the calculation of Equations (8) and (9), we can get the expression of the secure transmission probability D_c as

$$D_c = 1 - \exp\left[-2\pi^2 \lambda_e 2^{2R_v/\alpha} \csc\left(2\pi\alpha^{-1}\right)\alpha^{-1} r_{e,0}^2\right].$$ (10)

Finally, by substituting Equations (2), (3), (7) and (10) into Equation (1), we can get a closed-form expression for the security offloading ratio of the MEC-enabled air-to-ground integrated IoV as

$$H_{total} = \sum_{f=1}^{F} p_f \left\{ q_f + \left(1 - q_f\right)(1 - e^{-\pi q_f \lambda_p z^2}) D_s D_c \right\}.$$ (11)

4. The Cache Strategy and Secure Transmission Rate Optimization Problem

In this section, we investigate the joint effect of caching strategy and the secure transmission rate on the security offloading ratio of MEC-enabled air-to-ground integrated IoV. The redundancy rate and content cache probability are jointly optimized to maximize the network security offload probability. We study the optimal trade-off between file sharing and privacy security. According to the theoretical derivation results of the Section 3, we can construct the joint optimization as

$$\mathbf{P1} : \max_{\mathbf{q}, R_v} H_{total}$$ (12a)

$$s.t. \sum_{f=1}^{F} q_f \leq S$$ (12b)

$$0 \leq q_f \leq 1$$ (12c)

$$R_v \geq 0$$ (12d)

where the objective function (12a) represents the probability that the requester finds the desired file and obtains it successfully in the MEC-enabled air-to-ground integrated IoV. Constraint (12b) indicates that the cache capacity of each CV is limited. Constraint (12c) is the cache probability of each file. Constraint (12d) ensures that the secure transmission rate of the file is positive.

Due to the complexity brought by the exponential term in the objective function H_{total}, the joint optimization problem **P1** is an NP-hard problem [50]. It is difficult for us to directly obtain the joint optimal solution. From the objective function H_{total} we can observe that if the security transmission rate is increased, the probability of the successful V2V connection will decrease. Conversely, if the security transmission rate is too small, the security transmission probability will be reduced. Therefore, there may be an optimal secure transmission rate to maximize the security offloading ratio of the MEC-enabled air-to-ground integrated IoV. Furthermore, the caching strategy and the secure transmission rate are tightly coupled, so each caching strategy may correspond to an optimal secure transmission rate. Therefore, we propose an alternating joint optimization method. First, we transform the original problem **P1** into two sub-problems (**P1 − a** and **P1 − b**) for independent optimization, and then propose a joint optimization algorithm, which can finally solve the optimal solution of the joint cache strategy and the secure transmission rate.

4.1. Optimal Secure Transmission Rate for a Given Cache Strategy

In this subsection, our work focuses on optimizing the secure transmission rate of the MEC-enabled air-to-ground integrated IoV under a given caching strategy. Therefore, the sub-optimization problem is defined as

$$\mathbf{P1} - \mathbf{a} : \max_{R_v} H_{total}$$

$$s.t. \ R_v \geq 0$$ (13)

In order to get the optimal solution of the secure transmission rate R_v, we must first judge the Hessian matrix of of the objective function H_{total}. The first derivative of H_{total} can be solved as

$$\frac{\partial H_{total}}{\partial R_v} = \sum_{f=1}^{F} p_f \left[q_f + \left(1 - q_f\right) P_{V2V} \frac{\partial D_s D_c}{\partial R_v}\right]$$
$$= \sum_{f=1}^{F} p_f \left[q_f + \left(1 - q_f\right) P_{V2V} \left(\frac{\partial D_s}{\partial R_v} \cdot D_c + \frac{\partial D_c}{\partial R_v} \cdot D_s\right)\right] \tag{14}$$

For simplicity, we set $\varphi_b = 2\pi^2 \lambda_r \csc\left(2\pi\alpha^{-1}\right)\alpha^{-1} r_{b,0}^2 2^{2R_s/\alpha}$ and $\varphi_e = 2\pi^2 \lambda_e \csc\left(2\pi\alpha^{-1}\right)\alpha^{-1} r_{e,0}^2$. So D_s and D_c are rewritten as $D_s = \exp\left(-k_b P_{V2V} 2^{2R_v/\alpha}\right)$, and $D_c = 1 - \exp\left(-k_e 2^{2R_v/\alpha}\right)$. The first derivative of D_s and D_c with respect to R_v can be calculated as

$$\frac{\partial D_s}{\partial R_v} = -\frac{2}{\alpha} k_b P_{V2V} \exp\left(-k_b P_{V2V} 2^{2R_v/\alpha}\right) 2^{2R_v/\alpha} \ln 2. \tag{15}$$

$$\frac{\partial D_c}{\partial R_v} = \frac{2}{\alpha} k_e \exp\left(-k_e 2^{2R_v/\alpha}\right) 2^{2R_v/\alpha} \ln 2. \tag{16}$$

By substituting Equations (15) and (16) into Equation (14) , we can further rewrite $\frac{\partial H_{total}}{\partial R_v}$ as

$$\frac{\partial H_{total}}{\partial R_v} = \sum_{f=1}^{F} p_f \left[q_f + \left(1 - q_f\right) P_{V2V} \frac{\partial D_s D_c}{\partial R_v}\right]$$
$$= \sum_{f=1}^{F} p_f \left\{ \begin{array}{l} q_f + \left(1 - q_f\right) P_{V2V} \frac{2}{\alpha} 2^{2R_v/\alpha} \ln 2 \exp\left(-k_b P_{V2V} 2^{2R_v/\alpha}\right) \\ \times \left[-k_b P_{V2V} + (k_b P_{V2V} + k_e) \exp\left(-k_e 2^{2R_v/\alpha}\right)\right] \end{array} \right\} \tag{17}$$

By analyzing Equation (17), it can be seen that $\left(1 - q_f\right) \exp\left(-k_b P_{V2V} 2^{2R_v/\alpha}\right) \ln 2 \times P_{V2V} \frac{2}{\alpha} 2^{2R_v/\alpha}$ is a positive term, so the positive or negative of $\frac{\partial H_{total}}{\partial R_v}$ is determined by $\vartheta\left(q_f\right) = -k_b P_{V2V} + (k_b P_{V2V} + k_e) \exp\left(-k_e 2^{2R_v/\alpha}\right)$. Obviously $\vartheta\left(q_f\right)$ belongs to the exponential function. So according to the properties of the exponential function, $\vartheta\left(q_f\right)$ is a monotonically decreasing function. Therefore, we set $\vartheta\left(q_f\right)$ equal to 0 to get the extreme point R_v^* of the function H_{total}, which is calculated as

$$R_v^* = \frac{\alpha}{2} \log_2 \left\{ -\frac{\ln\left[\varphi_b P_{V2V} / \left(\varphi_b P_{V2V} + \varphi_e\right)\right]}{\varphi_e} \right\}. \tag{18}$$

This means that $\frac{\partial H_{total}}{\partial R_v}$ is positive within the interval of $0 < R_v < R_v^*$ and negative within $R_v > R_v^*$. Thus, it can be determined that H_{total} is a concave function within the interval $R_v > 0$, and the maximum point is R_v^*. We can optimize R_v^* by a fixed $\mathbf{q}$.

4.2. Optimal Cache Strategy for a Given Secure Transmission Rate

In this subsection, we optimize the cache strategy based on the given secure transmission rate. The sub-problem with respect to cache strategy $\mathbf{q} = \left\{q_1, \cdots, q_f, \cdots, q_F\right\}$ is formulated as

$$\mathbf{P1} - \mathbf{b} : \max_{\mathbf{q}} H_{total}$$

$$s.t. \sum_{f=1}^{F} q_f \leq C \tag{19}$$

$$0 \leq q_f \leq 1$$

Proposition 1. *The proposed optimization problem* $\mathbf{P1} - \mathbf{b}$ *is a convex optimization problem with regard to* $0 \leq q_f \leq 1$.

Proof of Proposition 1. See Appendix A. $\square$

Through Proposition 1, we know that the optimization problem $\mathbf{P1} - \mathbf{b}$ about the caching strategy is a convex programming problem. Generally, the optimization problem $\mathbf{P1} - \mathbf{b}$ can obtain the closed expression of the optimal cache strategy $\mathbf{q}^*$ through the analytical method, but the complex structure introduced by the exponential term in the objective function H_{total} and the existence of inequality constraints make it difficult for the optimization problem $\mathbf{P1} - \mathbf{b}$ to obtain the closed expression of the optimal cache strategy $\mathbf{q}^*$. Therefore, we use the *fmincon* module of MATLAB to solve the optimization problem $\mathbf{P1} - \mathbf{b}$ [19,51,52]. It can ensure that the constrained optimization problem $\mathbf{P1} - \mathbf{b}$ converges to the global optimal solution.

4.3. Iterative Algorithm for Joint Optimization

In this section, we jointly optimize the caching strategy $\mathbf{q}$ and the secure transmission rate R_v to maximize the security offloading rate of the MEC-enabled air-to-ground integrated IoV. From the previous theoretical analysis, it can be seen that the cache strategy $\mathbf{q}$ and the secure transmission rate R_v are the product relationship in the expression of the network security offloading ratio H_{total}, which makes the joint optimization more complicated. Therefore, we propose an alternating optimization algorithm to obtain the joint optimal solution of the caching strategy and the secure transmission rate. The details of the joint optimization algorithm are shown in Algorithm 1. In Algorithm 1, we first obtain the optimal secure transmission rate through a given caching strategy, and then solve the optimal caching strategy by obtaining the secure transmission rate, and alternately optimized each until the network security offloading ratio converges. Finally, a set of joint optimal solutions of the cache strategy $\mathbf{q}^*$ and the secure transmission rate R_v^* are output.

Algorithm 1 Joint optimization algorithm.

1: Initialize the cache strategy $\mathbf{q}$ to a feasible value.
2: **Repeat** Loop:
3:    (a) Calculate the security transmission rate R_v by Equation (18).
4:    (b) Update the file cache strategy $\mathbf{q}$ by solving the convex optimization problem $\mathbf{P1} - \mathbf{b}$ for fixed R_v.
5:    (c) Update the secure transmission rate R_v in Equation (18) using the cache strategy solved in step (b).
6: **Until** the network security offloading ratio H_{total}, the optimal secure transmission rate R_v^* and the optimal cache strategy $\mathbf{q}^*$
7: Output H_{total}, R_v and $\mathbf{q}$

---

5. Simulation and Numerical Results

In this section, we use key technical parameters to verify the performance of the proposed caching strategy and the correctness of the theoretical analysis. To verify the security offload performance and cache efficiency performance of files, we compare the proposed caching strategy with the PAEH caching strategy [29] and the Uniform-baseline caching strategy [53]. The PAEH caching strategy considered the impact of the self-caching and the successful transmission probability, which were the focus of current research in cooperative caching. By comparing the proposed cache strategy with the PAEH caching strategy, we can verify the security offloading ability of the proposed cache strategy. The uniform-baseline caching strategy is a cache strategy that does not consider the change of content popularity. This strategy, as the baseline cache strategy, appears in many related studies to prove the cache efficiency improvement ability of the proposed cache strategy. All the caching strategies consider the effect of self-caching. Unless otherwise specified, the simulation environment parameter settings in this paper are shown in Table 1.

Table 1. Simulation parameters.

Parameters	Value
Intensity of CVs λ_p	$4 \times 10^{-3}/\mathrm{m}^2$
Intensity of EVs λ_e	$2 \times 10^{-3}/\mathrm{m}^2$
V2V bandwidth W	20 MHz
Path loss exponent α	3.68
Noise power σ^2	-174 dBm/Hz
The number of files F	10 files
Each CV's cache capacity S	1 file
Zip parameter ε	0.6, 1

In Figure 2, we introduce the distribution probability of files under the proposed caching strategy, the PAEH caching strategy, and the uniform-baseline caching strategy. From Figure 2, we can easily see that the uniform-baseline caching strategy caches all files with the same probability. The proposed caching strategy and the PAEH caching strategy only cache a small number of high-ranked files. When the Zipf factor $\varepsilon = 1$, the proposed caching strategy and the PAEH caching strategy only need to cache the top 3 files to maximize the network security offloading rate, because an increase in the Zipf factor ε means that the probability of the file being requested becomes more concentrated. In order to increase the network security offload ratio, the proposed caching strategy and the PAEH caching strategy increase the caching probability of the top files. This is consistent with the high demand for a certain file in a certain period of time in the actual network. Lower-ranked files may not need to be cached due to the CV's limited cache capacity.

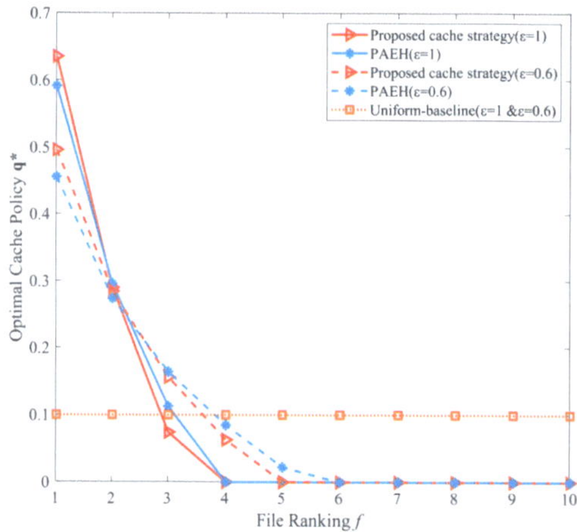

Figure 2. Distribution of caching strategies with different Zipf factors.

Figure 3 corresponds to optimization problem **P1** $-$ **a**, which illustrates the optimization of the secure transmission rate for a given caching strategy. The specific values of the caching strategy adopted in Figure 3 are given in Table 2. As can be seen from Figure 3, with the secure transmission rate increasing, the network security offloading ratio curve first increases to the extreme point and then drops rapidly. This phenomenon also verifies that our optimization scheme has an optimal solution. In addition, we can also observe that there is still the network security offloading ratio even when the secure transmission rate is zero, because the effect of self-caching is considered in our proposed caching strategy. We comprehensively take into account the factors of successful connection and security transmission of V2V communication in the proposed caching strategy. Therefore, under the condition of a very low secure transmission rate, the main way to obtain files by the proposed caching strategy may still be through self-caching. This may lead to the phenomenon that the curve starts. When the secure transmission rate is very low, the network security offloading ratio of the $\mathbf{q_1}$ cache strategy is lower than that of $\mathbf{q_3}$.

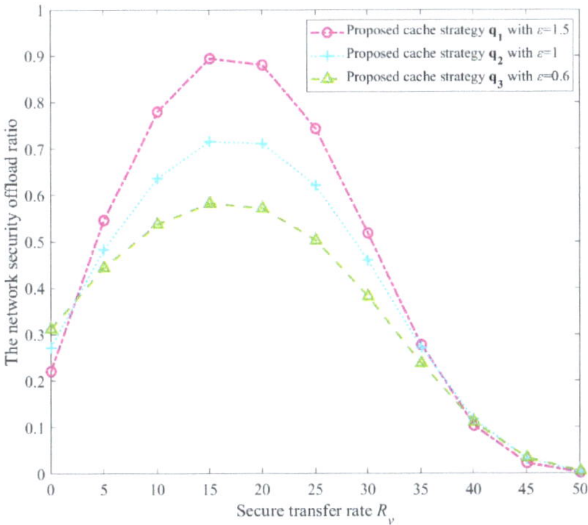

Figure 3. The curve of the network security offloading ratio versus the secure transmission rate.

Table 2. Three caching strategies adopted for Figure 3.

Zipf Parameters	Cache Probability of Files
$\mathbf{q_1}(\varepsilon = 1.5)$	0.7550 0.2450 0 0 0 0 0 0 0 0
$\mathbf{q_2}(\varepsilon = 1)$	0.6351 0.2904 0.0745 0 0 0 0 0 0 0
$\mathbf{q_3}(\varepsilon = 0.6)$	0.4960 0.2848 0.1562 0.0629 0 0 0 0 0 0

Figure 4 compares the proposed caching strategy, PAEH caching strategy and uniform-baseline caching strategy, with the increasing Zipf parameters. From Figure 4, we can easily see that the network security offloading ratio brought by the proposed caching strategy and PAEH caching strategy will increase rapidly with the increase of Zipf parameters. However, the network security offloading ratio of the uniform-baseline caching strategy is fixed on a horizontal line and does not change with the increase of the Zipf parameters. This result is the same as we expected, because the uniform-baseline caching strategy does not take into account the popularity of files, but caches all files with equal probability. Of course, the network security offloading ratio of the Uniform-baseline caching strategy is also the worst. Furthermore, we can also see that when the Zipf parameter is small, the proposed caching strategy is significantly better than the PAEH caching strategy, but the gap gradually decreases as the Zipf parameter increases. This is because the caching probability of our proposed scheme is strongly correlated with the probability of requesting files. The increase of popularity factor ε means that the probability of being requested for the most popular file increases. This will lead to the cache probability of the most popular file approaching 1, and the cache probability of other files approaching 0. When the popularity factor ε gradually increases, both the proposed caching strategy and the PAEH caching strategy cache the top files, so the network security offloading ratio is gradually approaching.

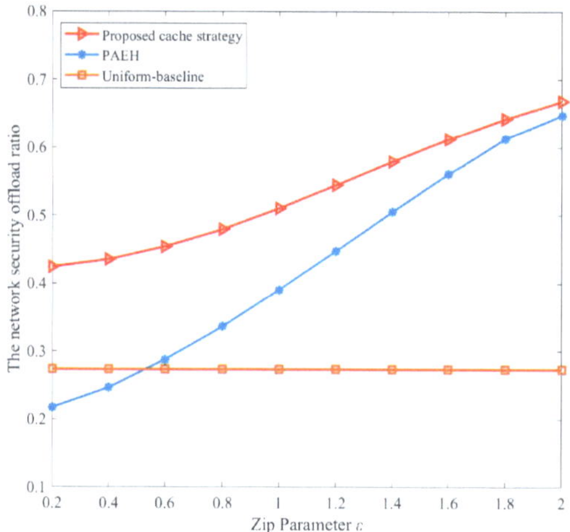

Figure 4. The network security offloading ratio of the caching strategies varies with Zipf parameters.

In Figure 5, we investigate the effect of different CV densities and EV densities ($\lambda_e = 1 \times 10^{-3}/\text{m}^2$, $\lambda_e = 4 \times 10^{-3}/\text{m}^2$, $\lambda_e = 10 \times 10^{-3}/\text{m}^2$) on the security offloading ratio of the MEC-enabled air-to-ground integrated IoV. Monte Carlo method is used to obtain the simulation results. From Figure 5, we can see that the simulation results match well with the theoretical values. This indicates that the theoretical derivation of this paper is reasonable. By analyzing the abscissa in Figure 5, we can conclude that the network security offloading ratio increases with the increase of CV density. This is because the increase in CV density also increases the probability of the RVs finding the surrounding required files. In addition, with the increase of CV density, the network security offloading ratio increases slowly and gradually tends to balance. This indicates that when the CV density reaches a certain value, the CV's cache capacity will become the main influencing factor of the network security offloading ratio. Furthermore, it can be seen from the three EV density curves that the network security offloading ratio decreases as the EV density increases. The reason for this phenomenon may be that the proposed caching strategy considers the factors of successful V2V communication connection and secure transmission. When the EV density increases, the risk of the file secure transmission also increases, which may lead to the decrease of the network security offloading ratio. In addition, with the increase in EV density, the network security offloading ratio will decrease slowly. The main reason is that the proposed caching strategy takes into account the impact of self-caching, which can ensure that the files can be obtained confidentially through self-caching even when the communication conditions are very risky.

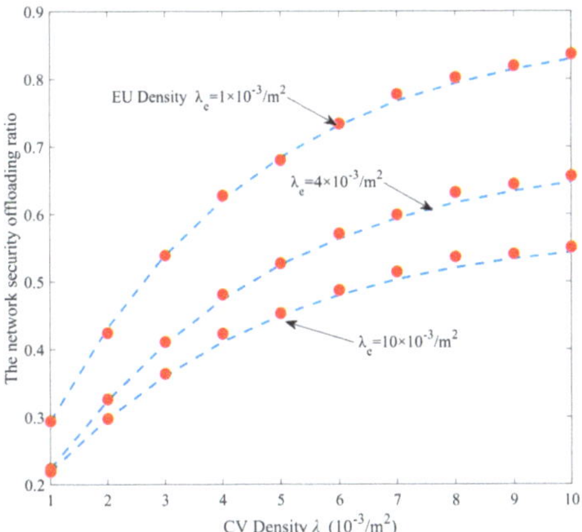

Figure 5. The network security offloading ratio varies with EV density and CV density.

In Figure 6, we compare the data security offloading performance of the proposed caching strategy, the PAEH strategy and the uniform-baseline caching strategy with different CV densities and Zipf parameters. It is obvious that all the caching strategies involved in the comparison will increase rapidly with the increase of CV density. The proposed caching strategy significantly outperforms the PAEH strategy and the uniform-baseline caching strategy in terms of the network security offloading ratio. This is the same conclusion as in Figure 5, in which the increase in the density of CV gives the requester a greater chance of obtaining the desired file. With the increase of Zipf parameters, the proposed caching strategy and the PAEH caching strategy will be significantly improved. Although both the proposed caching strategy and the PAEH caching strategy consider the influence of self-caching, it can be seen from the distribution of caching strategies in Figure 2 that the caching probability of the proposed caching strategy is strongly correlated with the request probability. However, the network security offloading ratio curves of the uniform-baseline caching strategy under the two Zipf parameters are coincident, because the uniform-baseline caching strategy does not consider the influence of content popularity.

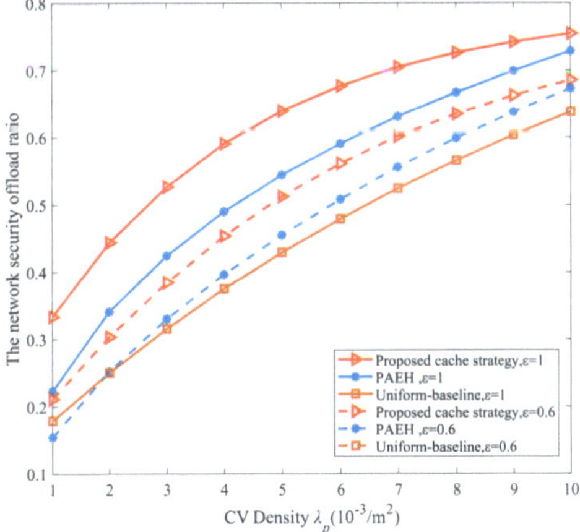

Figure 6. The network security offload ratio of different caching strategies varies with CV densities.

In Figure 7, we compare the network security offloading ratio of the proposed caching strategy, the PAEH strategy, and the uniform-baseline caching strategy under different

CV cache capacities. It can be seen from Figure 7 that the proposed caching strategy, the PAEH strategy, and the uniform-baseline caching strategy all increase with the increase of CV cache capacity. Due to the consideration of self-caching by all caching strategies and our assumption of a limited file library $F = 10$, when the CV cache capacity is $S = 10$, the network security offloading ratio of all caching strategies can reach the maximum value. In addition, we can also observe that with the increase of CV capacity, all cache strategies gradually narrow the gap in network security offloading ratio. Because when the cache capacity of the CV is large enough (compared with the file library), the probability that the requester obtains the required file through the self-cache is increased. At this time, the proportion of caching strategy and file popularity distribution to network data offloading will decrease. Therefore, the network security offloading ratio curve is gradually approaching. Furthermore, the proposed caching strategy is better than the PAEH strategy and the uniform-baseline caching strategy due to the consideration of the successful transmission of V2V communication. Since the uniform-baseline caching strategy does not fully utilize the cache space (all files are cached with the same probability), its network offloading ratio is the worst.

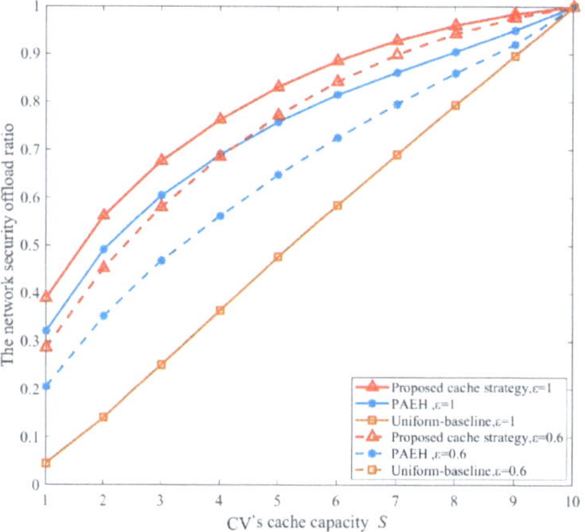

Figure 7. The network security offloading ratio of different caching strategies varies with CV cache capacity.

6. Conclusions

In this paper, we propose a novel mobile edge caching strategy to improve the security offloading ratio of the MEC-enabled air-to-ground integrated IoV, which comprehensively considers the effects of self-caching, the successful connection of V2V communication, and the secure transmission. On the basis of stochastic geometry theory and Laplace transform, we calculate the accurate expression for the network security offloading ratio. Based on the network security offloading ratio, we construct a joint optimization problem of the caching strategy and the secure transmission rate. Due to the complexity of the optimization problem, it is difficult to directly obtain the joint optimal solution of the caching strategy and the secure transmission rate. We propose an alternating optimization algorithm to jointly optimize the caching strategy and the secure transmission rate. Through the limited number of alternate optimizations, we can obtain a set of the optimal caching strategy and secure transmission rate that maximize the network security offloading ratio. Finally, we verify the superiority and feasibility of the proposed caching strategy through simulation experiments.

In addition, this paper focuses on considering a single line and a single ABS. If multiple lines and ABS are considered, road layout should be further considered. In this scenario, the network model should meet the Cox process or doubly stochastic Poisson point process.

However, the network will become more complicated, which is out of the scope of this paper and will be our future work. This paper focuses on highrise urban scenarios, consisting of many ground obstructions. Furthermore, studies regarding the comprehensive impact on LoS and NLoS groups will be conducted in the future.

Author Contributions: Methodology, Formal analysis, and Writing—Reviewing, W.W.; Supervision and Project administration, H.L.; Formal analysis, Y.L.; Project administration and Investigation, W.C.; Editing, R.L. All authors have read and agreed to the published version of the manuscript.

Funding: This work was supported by the National Natural Science Foundation of China (No. 62271395) and (No. 61401360), Natural Science Foundation of Shaanxi Province (No. 2021JM-076).

Institutional Review Board Statement: Not applicable.

Informed Consent Statement: Not applicable.

Data Availability Statement: Not applicable.

Conflicts of Interest: The authors declare no conflict of interest.

Appendix A

We set $\psi\left(q_f\right) = \exp\left(-k_b P_{V2V} 2^{2(R_s+R_v)/\alpha}\right) \times \left(1 - \exp\left(-k_e 2^{2R_v/\alpha}\right)\right)$, so the objective function can be rewritten as $H_{total} = \sum_{f=1}^{F} p_f\left[q_f + \left(1 - q_f\right)(1 - e^{-\pi q_f \lambda_p r^2})\psi\left(q_f\right)\right]$. The first derivative of H_{total} with respect to q_f can be calculated as

$$
\begin{aligned}
\frac{\partial H_{total}}{\partial q_f} = 1 - (1 - e^{-\pi q_f \lambda_p z^2})\psi\left(q_f\right) + \left(1 - q_f\right)\psi\left(q_f\right) \\
\times \left[\pi \lambda_p r^2 e^{-\pi q_f \lambda_p r^2} - P_{V2V}' k_b 2^{2(R_s+R_v)/\alpha}\left(1 - e^{-\pi q_f \lambda_p z^2}\right)\right]
\end{aligned}
\tag{A1}
$$

Furthermore, we can also calculate the second derivative of H_{total} with respect to q_f as

$$
\begin{aligned}
\frac{\partial^2 H_{total}}{\partial q_f^2} &= \psi\left(q_f\right)\left\{
\begin{array}{l}
-2\pi \lambda_p r^2 e^{-\pi q_f \lambda_p r^2} + 2P_{V2V}'\zeta\left(q_f\right) \\
+ \left(1 - q_f\right)\left[
\begin{array}{l}
-k_b 2^{2(R_s+R_v)/\alpha}\left(P_{V2V}'\right)^2 \zeta\left(q_f\right) \\
-\left(\pi \lambda_p r^2\right)^2 e^{-\pi q_f \lambda_p r^2} - P_{V2V}''\zeta\left(q_f\right)
\end{array}
\right]
\end{array}
\right\} \\
&= \psi\left(q_f\right)\left\{
\begin{array}{l}
-\left(1 - q_f\right)\zeta\left(q_f\right)\left(P_{V2V}'\right)^2 k_b 2^{2(R_s+R_v)/\alpha} - \left(2 + \left(1 - q_f\right)\pi \lambda_p r^2\right)\pi \lambda_p r^2 e^{-\pi q_f \lambda_p r^2} \\
+ k_b 2^{2(R_s+R_v)/\alpha}\left[2(1 - e^{-\pi q_f \lambda_p z^2})P_{V2V}' - \left(1 - e^{-\pi q_f \lambda_p r^2}\right)\left(1 - q_f\right)P_{V2V}''\right]
\end{array}
\right\}
\end{aligned}
\tag{A2}
$$

where $\zeta\left(q_f\right) = k_b 2^{2(R_s+R_v)/\alpha}\left(1 - e^{-\pi q_f \lambda_p z^2}\right)$. Obviously, the other terms of $\partial^2 H_{total}/\partial q_f^2$ are negative, so we just need to judge the positive and negative of $G\left(q_f\right) = 2(1 - e^{-\pi q_f \lambda_p z^2})P_{V2V}' - \left(1 - e^{-\pi q_f \lambda_p r^2}\right)\left(1 - q_f\right)P_{V2V}''$. The first derivative of $G\left(q_f\right)$ with respect to q_f can be calculated as

$$
\begin{aligned}
G\left(q_f\right)' &= 2\pi \lambda_p r^2 e^{-\pi q_f \lambda_p z^2} P_{V2V}' - \pi \lambda_p r^2 e^{-\pi q_f \lambda_p r^2}\left(1 - q_f\right)P_{V2V}'' \\
&\quad - \left(1 - e^{-\pi q_f \lambda_p r^2}\right)\left(1 - q_f\right)P_{V2V}''' + 3(1 - e^{-\pi q_f \lambda_p z^2})P_{V2V}'' \\
&= -(1 - e^{-\pi q_f \lambda_p z^2})\pi \lambda_p r^2 e^{-\pi q_f \lambda_p r^2}\left[8 + 3\pi \lambda_p r^2\left(1 - q_f\right)\right] \\
&\quad + 3(1 - e^{-\pi q_f \lambda_p z^2})P_{V2V}''
\end{aligned}
\tag{A3}
$$

Through the calculation of Equation (A4), we can easily judge the positive and negative of the second derivative $p_{V2V}{}''$.

$$p_{V2V}{}'' = -2\pi\lambda_p r^2 e^{-\pi q_f \lambda_p z^2} - \left(\pi\lambda_p r^2\right)^2 e^{-\pi q_f \lambda_p z^2}\left(1 - q_f\right) < 0. \tag{A4}$$

Therefore, we can get the conclusion $G\left(q_f\right)' < 0$, which represents $G\left(q_f\right)$ as a decreasing function within $0 \leq q_f \leq 1$. So, we can judge that the function $G\left(q_f\right) \leq G(0) = 0$. Furthermore, we can get $\partial^2 H_{total}/\partial q_f^2 < 0$. Therefore, the objective function H_{total} is a concave function on the convex set $0 \leq q_f \leq 1$, then the optimization problem **P1 − b** is proved to be a standard convex optimization problem.

References

1. Cisco. *Cisco Annual Internet Report (2018–2023) White Paper*; Cisco: San Jose, CA, USA, 2020.
2. Israr, A.; Ali, Z.A.; Alkhammash, E.H.; Jussila, J.J. Optimization Methods Applied to Motion Planning of Unmanned Aerial Vehicles: A Review. *Drones* **2022**, *6*, 126. [CrossRef]
3. You, X.; Wang, C.X.; Huang, J.; Gao, X.; Zhang, Z.; Wang, M.; Huang, Y.; Zhang, C.; Jiang, Y.; Wang, J.; et al. Towards 6G wireless communication networks: Vision, enabling technologies, and new paradigm shifts. *Sci. China Inf. Sci.* **2021**, *64*, 1–74. [CrossRef]
4. Sun, W.; Li, S.; Zhang, Y. Edge caching in blockchain empowered 6G. *China Commun.* **2021**, *18*, 1–17. [CrossRef]
5. Wang, D.; Zhou, F.; Lin, W.; Ding, Z.; Al-Dhahir, N. Cooperative hybrid nonorthogonal multiple access-based mobile-edge computing in cognitive radio networks. *IEEE Trans. Cogn. Commun. Netw.* **2022**, *8*, 1104–1117. [CrossRef]
6. Wang, D.; Wu, M.; He, Y.; Pang, L.; Xu, Q.; Zhang, R. An HAP and UAVs Collaboration Framework for Uplink Secure Rate Maximization in NOMA-Enabled IoT Networks. *Remote Sens.* **2022**, *14*, 4501. [CrossRef]
7. He, Y.; Wang, D.; Huang, F.; Zhang, R.; Pan, J. Trajectory optimization and channel allocation for delay sensitive secure transmission in UAV-relayed VANETs. *IEEE Trans. Veh. Technol.* **2022**, *71*, 4512–4517. [CrossRef]
8. Thandavarayan, G.; Sepulcre, M.; Gozalvez, J. Generation of cooperative perception messages for connected and automated vehicles. *IEEE Trans. Veh. Technol.* **2020**, *69*, 16336–16341. [CrossRef]
9. Chen, Y.; Liu, Y.; Zhao, J.; Zhu, Q. Mobile edge cache strategy based on neural collaborative filtering. *IEEE Access* **2020**, *8*, 18475–18482. [CrossRef]
10. Wang, D.; He, T.; Zhou, F.; Cheng, J.; Zhang, R.; Wu, Q. Outage-driven link selection for secure buffer-aided networks. *Sci. China Inf. Sci.* **2022**, *65*, 1–6. [CrossRef]
11. He, Y.; Nie, L.; Guo, T.; Kaur, K.; Hassan, M.M.; Yu, K. A NOMA-enabled framework for relay deployment and network optimization in double-layer airborne access VANETs. *IEEE Trans. Intell. Transp. Syst.* **2022**, *23*, 22452–22466. [CrossRef]
12. He, Y.; Zhai, D.; Huang, F.; Wang, D.; Tang, X.; Zhang, R. Joint task offloading, resource allocation, and security assurance for mobile edge computing-enabled UAV-assisted VANETs. *Remote Sens.* **2021**, *13*, 1547. [CrossRef]
13. Grlj, C.G.; Krznar, N.; Pranjić, M. A Decade of UAV Docking Stations: A Brief Overview of Mobile and Fixed Landing Platforms. *Drones* **2022**, *6*, 17. [CrossRef]
14. Narang, M.; Xiang, S.; Liu, W.; Gutierrez, J.; Chiaraviglio, L.; Sathiaseelan, A.; Merwaday, A. UAV-assisted edge infrastructure for challenged networks. In Proceedings of the 2017 IEEE Conference on Computer Communications Workshops (INFOCOM WKSHPS), Atlanta, GA, USA, 1–4 May 2017; pp. 60–65.
15. Dai, Y.; Xu, D.; Maharjan, S.; Zhang, Y. Joint load balancing and offloading in vehicular edge computing and networks. *IEEE Internet Things J.* **2018**, *6*, 4377–4387. [CrossRef]
16. Ning, Z.; Zhang, K.; Wang, X.; Obaidat, M.S.; Guo, L.; Hu, X.; Hu, B.; Guo, Y.; Sadoun, B.; Kwok, R.Y. Joint computing and caching in 5G-envisioned Internet of vehicles: A deep reinforcement learning-based traffic control system. *IEEE Trans. Intell. Transp. Syst.* **2020**, *22*, 5201–5212. [CrossRef]
17. Anjum, N.; Yang, Z.; Khan, I.; Kiran, M.; Wu, F.; Rabie, K.; Bahaei, S.M. Efficient algorithms for cache-throughput analysis in cellular-d2d 5g networks. *Comput. Mater. Contin.* **2021**, *67*, 1759–1780. [CrossRef]
18. Ma, Z.; Nuermaimaiti, N.; Zhang, H.; Zhou, H.; Nallanathan, A. Deployment model and performance analysis of clustered D2D caching networks under cluster-centric caching strategy. *IEEE Trans. Commun.* **2020**, *68*, 4933–4945. [CrossRef]
19. Lee, M.C.; Molisch, A.F. Caching policy and cooperation distance design for base station-assisted wireless D2D caching networks: Throughput and energy efficiency optimization and tradeoff. *IEEE Trans. Wirel. Commun.* **2018**, *17*, 7500–7514. [CrossRef]
20. Cai, J.; Wu, X.; Liu, Y.; Luo, J.; Liao, L. Network coding-based socially-aware caching strategy in D2D. *IEEE Access* **2020**, *8*, 12784–12795. [CrossRef]
21. Kafıloğlu, S.S.; Gür, G.; Alagöz, F. Cooperative Caching and Video Characteristics in D2D Edge Networks. *IEEE Commun. Lett.* **2020**, *24*, 2647–2651. [CrossRef]
22. Li, M.; Cheng, N.; Gao, J.; Wang, Y.; Zhao, L.; Shen, X. Energy-efficient UAV-assisted mobile edge computing: Resource allocation and trajectory optimization. *IEEE Trans. Veh. Technol.* **2020**, *69*, 3424–3438. [CrossRef]

23. Wyner, A.D. The wire-tap channel. *Bell Syst. Tech. J.* **1975**, *54*, 1355–1387. [CrossRef]
24. Irram, F.; Ali, M.; Naeem, M.; Mumtaz, S. Physical layer security for beyond 5G/6G networks: Emerging technologies and future directions. *J. Netw. Comput. Appl.* **2022**, *206*, 103431. [CrossRef]
25. Wang, H.M.; Zheng, T.X. *Physical Layer Security in Random Cellular Networks*; Springer: Berlin/Heidelberg, Germany, 2016.
26. Liu, Y.; Qin, Z.; Elkashlan, M.; Gao, Y.; Hanzo, L. Enhancing the Physical Layer Security of Non-Orthogonal Multiple Access in Large-Scale Networks. *IEEE Trans. Wirel. Commun.* **2017**, *16*, 1656–1672. [CrossRef]
27. Zheng, T.X.; Wang, H.M.; Yuan, J. Physical-layer security in cache-enabled cooperative small cell networks against randomly distributed eavesdroppers. *IEEE Trans. Wirel. Commun.* **2018**, *17*, 5945–5958. [CrossRef]
28. Ren, D.; Gui, X.; Zhang, K.; Wu, J. Mobility-aware traffic offloading via cooperative coded edge caching. *IEEE Access* **2020**, *8*, 43427–43442. [CrossRef]
29. Meng, Y.; Zhang, Z.; Huang, Y. Cache-and energy harvesting-enabled d2d cellular network: Modeling, analysis and optimization. *IEEE Trans. Green Commun. Netw.* **2021**, *5*, 703–713. [CrossRef]
30. Farooq, M.J.; ElSawy, H.; Alouini, M.S. A stochastic geometry model for multi-hop highway vehicular communication. *IEEE Trans. Wirel. Commun.* **2015**, *15*, 2276–2291. [CrossRef]
31. Steinmetz, E.; Wildemeersch, M.; Quek, T.Q.; Wymeersch, H. A stochastic geometry model for vehicular communication near intersections. In Proceedings of the 2015 IEEE Globecom Workshops (GC Wkshps), San Diego, CA, USA, 6–10 December 2015; pp. 1–6.
32. Wu, Y.; Zheng, J. Modeling and Analysis of the Local Delay in an MEC-Based VANET for an Urban Area. *IEEE Trans. Veh. Technol.* **2022**, *71*, 13266–13280. [CrossRef]
33. Sial, M.N.; Deng, Y.; Ahmed, J.; Nallanathan, A.; Dohler, M. Stochastic geometry modeling of cellular V2X communication over shared channels. *IEEE Trans. Veh. Technol.* **2019**, *68*, 11873–11887. [CrossRef]
34. Andrews, J.G.; Baccelli, F.; Ganti, R.K. A tractable approach to coverage and rate in cellular networks. *IEEE Trans. Commun.* **2011**, *59*, 3122–3134. [CrossRef]
35. Chen, Z.; Pappas, N.; Kountouris, M. Probabilistic caching in wireless D2D networks: Cache hit optimal versus throughput optimal. *IEEE Commun. Lett.* **2017**, *21*, 584–587. [CrossRef]
36. Liu, H.W.; Zheng, T.X.; Wen, Y.; Feng, C.; Wang, H.M. Performance Analysis of Uplink mmWave Communications in C-V2X Networks. In Proceedings of the GLOBECOM 2020—2020 IEEE Global Communications Conference, Taipei, Taiwan, 7–11 December 2020; pp. 1–6.
37. Ullah, A.; Choi, W. Massive MIMO Assisted Aerial-Terrestrial Network: How Many UAVs Need to Be Deployed? *TechRxiv* **2022**, *10*, 36227.
38. Zhang, C.; Wei, Z.; Feng, Z.; Zhang, W. Spectrum sharing of drone networks. In *Handbook of Cognitive Radio*; Springer: Berlin/Heidelberg, Germany, 2019; pp. 1279–1304.
39. Zhang, S.; Zhu, Y.; Liu, J. Multi-UAV Enabled Aerial-Ground Integrated Networks: A Stochastic Geometry Analysis. *IEEE Trans. Commun.* **2022**, *70*, 7040–7054. [CrossRef]
40. Malak, D.; Al-Shalash, M. Device-to-device content distribution: Optimal caching strategies and performance bounds. In Proceedings of the 2015 IEEE International Conference on Communication Workshop (ICCW), London, UK, 8–12 June 2015; pp. 664–669.
41. Li, S.; Sun, W.; Zhang, H.; Zhang, Y. Physical Layer Security for Edge Caching in 6G Networks. In Proceedings of the GLOBECOM 2020—2020 IEEE Global Communications Conference, Taipei, Taiwan, 7–11 December 2020; pp. 1–6.
42. Stoyan, D.; Kendall, W.S.; Chiu, S.N.; Mecke, J. *Stochastic Geometry and Its Applications*; John Wiley & Sons: Hoboken, NJ, USA, 2013.
43. Chai, R.; Li, Y.; Chen, Q. Joint cache partitioning, content placement, and user association for D2D-enabled heterogeneous cellular networks. *IEEE Access* **2019**, *7*, 56642–56655. [CrossRef]
44. Vu, T.X.; Chatzinotas, S.; Ottersten, B.; Trinh, A.V. Full-duplex enabled mobile edge caching: From distributed to cooperative caching. *IEEE Trans. Wirel. Commun.* **2019**, *19*, 1141–1153. [CrossRef]
45. Mozaffari, M.; Saad, W.; Bennis, M.; Debbah, M. Unmanned aerial vehicle with underlaid device-to-device communications: Performance and tradeoffs. *IEEE Trans. Wirel. Commun.* **2016**, *15*, 3949–3963. [CrossRef]
46. Okabe, A.; Boots, B.; Sugihara, K.; Chiu, S.N. *Concepts and Applications of Voronoi Diagrams*; John Wiley: Chichester, UK, 2000.
47. Wang, C.; Li, Z.; Xia, X.G.; Shi, J.; Si, J.; Zou, Y. Physical layer security enhancement using artificial noise in cellular vehicle-to-everything (C-V2X) networks. *IEEE Trans. Veh. Technol.* **2020**, *69*, 15253–15268. [CrossRef]
48. Zheng, T.X.; Wen, Y.; Liu, H.W.; Ju, Y.; Wang, H.M.; Wong, K.K.; Yuan, J. Physical-Layer Security of Uplink mmWave Transmissions in Cellular V2X Networks. *IEEE Trans. Wirel. Commun.* **2022**, *21*, 9818–9833. [CrossRef]
49. Gradshteyn, I.S.; Ryzhik, I.M. *Table of Integrals, Series, and Products*; Academic Press: Cambridge, MA, USA, 2014.
50. Bai, T.; Wang, J.; Ren, Y.; Hanzo, L. Energy-efficient computation offloading for secure UAV-edge-computing systems. *IEEE Trans. Veh. Technol.* **2019**, *68*, 6074–6087. [CrossRef]
51. Su, Z.; Feng, W.; Tang, J.; Chen, Z.; Fu, Y.; Zhao, N.; Wong, K.K. Energy efficiency optimization for D2D communications underlaying UAV-assisted industrial IoT networks with SWIPT. *IEEE Internet Things J.* **2022**, *10*, 1990–2002. [CrossRef]

52.   Boyd, S.; Boyd, S.P.; Vandenberghe, L. *Convex Optimization*; Cambridge University Press: Cambridge, UK, 2004.
53.   Amer, R.; Baza, M.; Salman, T.; Butt, M.M.; Alhindi, A.; Marchetti, N. Optimizing joint probabilistic caching and channel access for clustered D2D networks. *J. Commun. Netw.* **2021**, *23*, 433–441. [CrossRef]

 *drones*

Article

Dynamic Robust Spectrum Sensing Based on Goodness-of-Fit Test Using Bilateral Hypotheses

Shaoyang Men [1], Pascal Chargé [2] and Zhe Fu [3,*]

1   School of Medical Information Engineering, Guangzhou University of Chinese Medicine, Guangzhou 510006, China
2   Institut d'Electronique et Télécommunications de Rennes (IETR), Université de Nantes, UMR CNRS 6164, Rue Christian Pauc BP 50609, 44306 Nantes, France
3   School of Integrated Circuits, Guangdong University of Technology, Guangzhou 510006, China
*   Correspondence: zhe_fu@foxmail.com; Tel.: +86-155-2115-4913

Abstract: Dynamic spectrum detection has attracted increasing interest in drone or drone controller detection problems. Spectrum sensing as a promising solution allows us to provide a dynamic spectrum map within the target frequency band by estimating the occupied sub-bands in a specific period. In this paper, a robust Student's t-distribution model is built to tackle the scenario with a small number of observed samples. Then, relying on the characteristics of the statistical model, we propose an appropriate goodness-of-fit (GoF) test statistic regarding a small number of samples. Moreover, to obtain a reliable sensing, bilateral hypotheses of the test statistic are both used to make a decision. Numerical simulations show the superiority of the proposed method compared with other schemes, including the unilateral hypothesis-based GoF testing and the conventional energy detection, in a small number of sample cases.

Keywords: spectrum sensing; Student's t-distribution; powerful goodness-of-fit test; cognitive drone network

Citation: Men, S.; Chargé, P.; Fu, Z. Dynamic Robust Spectrum Sensing Based on Goodness-of-Fit Test Using Bilateral Hypotheses. *Drones* **2023**, *7*, 18. https://doi.org/10.3390/drones7010018

Academic Editor: Diego González-Aguilera

Received: 28 November 2022
Revised: 18 December 2022
Accepted: 23 December 2022
Published: 27 December 2022

1. Introduction

Nowadays, due to the rapid development of the wireless communication, the number of civil unmanned aerial vehicles (UAVs) has increased significantly in recent years, which could cause many problems for city administration [1]. Reliable detection of UAVs or their controllers is a prerequisite for further administration [2–4]. Cognitive radio (CR) [5], which enables dynamic detection of surrounding signal spectrum, becomes a promising solution for frequency detection. More specifically, the whole spectrum can be divided into sub-bands and different signal occupancies can be estimated [6–8]. It has been widely applied to drone networks in order to create promising infrastructures of cognitive drone networks, in which multiple resource-constrained sensor nodes are equipped with cognitive ability [9–12]. As the fundamental prerequisite for CR, namely spectrum sensing (SS) [13,14], reliable and quick detection of signal existence is the key for further strategy and decision.

To dynamically estimate the existing spectrum, many algorithms have been developed, including cyclostationary feature, matched filter, waveform-based detection [15–17], etc. However, these algorithms need to acquire prior knowledge of primary user (PU), which is difficult in practice, i.e., illegal quad-rotor drone intrusion. Therefore, blind detection techniques that do not need prior knowledge about PU are developed, for instance, the energy detection (ED) scheme [18–20] and the eigenvalue-based estimation [21–23]. ED is one commonly adopted method due to its simplicity for implementation, but the noise uncertainty in practice significantly degrades its detection performance. Thus, the eigenvalue-based blind detection is proposed to settle the disadvantage of ED by analyzing the covariance matrix. The corresponding eigenvalues are utilized to increase the robustness against noise uncertainty. Unfortunately, it needs a large number of samples to obtain a good performance, and has relatively high complexity.

Some GoF test-based strategies have been proposed to achieve better estimation performance given a small number of available samples [24–30]. In [31], the Anderson–Darling (AD) test is exploited to achieve reliable detection under a small number of samples. On the basis of this, the authors in [32] make use of a Student's *t*-distribution test for fully blind detection with noise uncertainty. In [33], the Kolmogorov–Smirnov (KS) test, a non-parametric GoF method, is used to perform a fast and reliable spectrum sensing, which is also robust to non-Gaussian noise and channel uncertainty. In [34], the authors utilize a powerful GoF test to achieve non-parametric sensing for Middleton noise scenario. In [35], the characteristic of non-symmetrical differences is exploited to construct the unilateral right-tail AD test. In [36], the ratio between maximum eigenvalue relative to the trace is exploited to achieve blind detection. However, only one side of the binary hypothesis information is used in all of the above works. That is to say, only the null hypothesis is considered. Therefore, the reliability of the decision may be improved by using the bilateral hypotheses.

To address the above problems, an enhanced detection method based on the GoF test using bilateral hypotheses for a small number of samples in cognitive drone network is proposed in this work. One of the main objectives is to obtain a short signal processing and a real-time decision. Considering the case that there is only a single-radio module at SU, the sampling period of the observation sensors is expected to be as short as possible. Moreover, we consider a special information environment where there is only few steady state receptions available. First of all, we propose to utilize the Student's *t*-distribution in order to address a small number of sample problems, which are collected by the low power sensor nodes (SU) in the cognitive drone network. In fact, the performance of ED using Gaussian approximation becomes good only when sufficiently large sample size is available [37]. It has been shown that the Student's *t*-test is the optimal test in spectrum sensing given a small number of samples [38,39]. Then, taking into account the limitations of the traditional GoF test (e.g., AD test and KS test) under a small number of samples, the powerful GoF test [40–44] is introduced to precisely evaluate the distance between common cumulative distribution and the empirical distribution of observation. As in the proposed method in [45], the statistic based on the likelihood ratio is used, which is substantially more powerful than the traditional statistic. The main contribution stands in the proposition of the powerful GoF test to accommodate the small samples situation. Finally, two new statistics based on bilateral hypotheses are calculated based on the statistical characteristic of Student's *t*-distribution, and a high reliability sensing decision is obtained based on bilateral hypotheses.

The rest of the paper is organized as follows. The traditional unilateral hypothesis-based GoF test is introduced in Section 2. The proposed scheme is illustrated in Section 3, where the Student's *t*-distribution-based statistical model is provided. A powerful goodness-of-fit test statistic Z_c is introduced for computing the distance between the common cumulative distribution of the observations and the empirical distribution, and the bilateral hypotheses information is utilized for high reliability decision. Numerical simulations are discussed in Section 4 and the conclusions are provided in Section 5.

2. Traditional GoF Test Based on Unilateral Hypothesis

The traditional sensing scheme on the basis of GoF test using unilateral hypothesis is presented in this section. Spectrum sensing aims to detect the existence of PU signal in a specific frequency band for a given set of observed samples. This can be expressed as a traditional GoF test problem, which can be written as:

$$
\begin{aligned}
H_0 &: \quad F_n(x) = F_0(x) \\
H_1 &: \quad F_n(x) \neq F_0(x)
\end{aligned}
\tag{1}
$$

where H_0 is the null hypothesis and H_1 is the alternative hypothesis. $F_0(x)$ denotes the cumulative distribution function (CDF) of noise distribution of H_0 hypothesis, while $F_n(x)$ represents the CDF of collected samples, which can be calculated by using the empirical CDF

$$F_n(x) = |i : x_i \leq x, 1 \leq i \leq n|/n \tag{2}$$

where $|S|$ represents the cardinality of a given set S, and n denotes the number of samples utilized for acquiring the statistical distribution.

In other words, the detection problem is turned to a problem of testing the null hypothesis against the alternative hypothesis. Assuming that Z is a statistic for testing H_0 against H_1, which is defined as [46]

$$Z = \int_{-\infty}^{\infty} Z_x \mathrm{d}w(x). \tag{3}$$

Here, $w(x)$ represents some weight function and large values of Z and reject the null hypothesis H_0. The power of Z depends on Z_x and $w(x)$, and the natural candidate for Z_x is generally considered as the Pearson χ^2 test statistic defined as follows [46]:

$$P_x^2 = \frac{n(F_n(x) - F_0(x))^2}{F_0(x)(1 - F_0(x))}. \tag{4}$$

For a traditional GoF test, Z_x in Equation (3) is firstly replaced by P_x^2. Then, various traditional GoF tests have been proposed to evaluate the distance between $F_0(x)$ and $F_n(x)$, and how to choose different weight functions. For instance, the AD test, Kolmogorov–Smirnov (KS) test, and Cramér-von Mises (CM) test [47,48]. They belong to the one-side hypothesis test for H_0. A one-side hypothesis is utilized for determining whether the collected samples meet the distribution with CDF $F_0(x)$ or not. These GoF tests are illustrated as follows.

(A) KS test: To evaluate the relative distance, the empirical CDF of collected samples and the reference CDF are considered in KS test and $w(x) = n^{-1}F_0(x)(1 - F_0(x))$ is chosen. Then, the GoF test statistic can be obtained using the largest absolute distance between the two CDFs, which can be written as

$$D^2 = \{\sup|F_n(x) - F_0(x)|\}^2. \tag{5}$$

Here, $\sup\{\cdot\}$ represents the supremum function denoting the maximum value in a given set. In a practical scenario, it can be rewritten as [49]

$$D^2 = \left(\max(D^+, D^-)\right)^2 \tag{6}$$

$$D^+ = \max_{1 \leq i \leq n} \{\frac{i}{n} - F_0(x_i)\} \tag{7}$$

$$D^- = \max_{1 \leq i \leq n} \{F_0(x_i) - \frac{i-1}{n}\}. \tag{8}$$

(B) CM test: In the CM test, the term $\mathrm{d}w(x)$ is set to $\mathrm{d}w(x) = F_0(x)(1 - F_0(x))\mathrm{d}F_0(x)$. In other words, CM test is an alternative to the KS test. The statistic of the CM test is defined by

$$W^2 = n \int_{-\infty}^{\infty} (F_n(x) - F_0(x))^2 \mathrm{d}F_0(x). \tag{9}$$

The integral can be divided into n parts as provided in [47]. Then, W^2 can be approximately rewritten by

$$W^2 = \frac{1}{12n} + \sum_{i=1}^{n} (F_0(x_i) - \frac{2i-1}{2n})^2.$$

(10)

(C) AD test: It can be seen from Equation (10) for distribution $F_0(x)$, there is not enough weight to the tails included in W^2. Thus, Anderson and Darling generalized the CM test statistic in order to enhance the difference between the lower and upper tails of the distribution. By choosing $w(x) = F_0(x)$, the AD test statistic is given as follows [48]:

$$A^2 = n \int_{-\infty}^{\infty} \frac{(F_n(x) - F_0(x))^2}{F_0(x)(1 - F_0(x))} \mathrm{d}F_0(x).$$

(11)

For an efficient implementation, the simplified formula of the AD statistic can be denoted as [47]:

$$A^2 = -n - \frac{1}{n} \sum_{i=1}^{n} (2i-1)(\ln Z_i + \ln(1 - Z_{n+1-i}))$$

(12)

with $Z_i = F_0(x_i)$. n is the number of the collected samples.

The above traditional GoF test statistics are derived considering the unilateral hypothesis. The spectrum sensing can be reformulated as

$$\begin{aligned} H_0: &\quad \mathcal{T} \leq \eta \\ H_1: &\quad \mathcal{T} > \eta \end{aligned}$$

(13)

where $\mathcal{T}$ is one of the GoF test statistics D^2, W^2 and A^2, and η is a threshold which can be found in [47] or be calculated using the Monte Carlo approach. Hence, when $\mathcal{T} \leq \eta$, the null hypothesis H_0 can be considered to be accepted and the licensed frequency band is assumed to be available (not used by the PU).

3. Proposed Enhanced GoF Test-Based Spectrum Sensing Using Bilateral Hypotheses

In this section, an enhanced GoF test-based spectrum sensing technique using bilateral hypotheses is proposed. Firstly, a statistical model of the collected data is provided considering the Student's t-distribution, then a powerful GoF test statistic Z_c is introduced. Moreover, in order to obtain an improved decision, bilateral hypothesis information is utilized and a final decision is made by comparing them.

In order to use bilateral hypotheses for GoF tests, the traditional sensing scheme based on hypothesis test in Equation (1) can be rewritten as:

$$\begin{aligned} H_0: &\quad X_i = W_i \\ H_1: &\quad X_i = hs_t + W_i \end{aligned}$$

(14)

where H_0 indicates the absence hypothesis of a PU signal while H_1 denotes the presence hypothesis of the PU signal, respectively. X_i denotes the received samples at time slot i ($i = 1, 2, \ldots, l$), W_i represents the sample noise contribution. Here, the noise is assumed to be additive white Gaussian noise (AWGN) with zero mean and variance σ^2, and h denotes the channel gain between PU and SU, and s_t is the PU signal component. In addition, considering that the distribution of the PU signal power spectral density is unavailable in practice, we can make a reasonable, fair, and neutral assumption that the PU signal distributes uniformly within the entire bandwidth. For example, in many multi-carrier signals scenarios, the signal is assumed to be a constant in both frequency and time domains. The received signal is assumed to pass a down converter to a baseband frequency bandwidth for presentation convenience in this paper. Then, the samples are acquired with a sampling rate several times faster than the baseband frequency. Thus, s_t can be considered as a constant, $s_t = 1$, as shown in [39]. The sensing problem in this case can be considered as a standard scenario with same variance and different mean value Gaussian distributions with corresponding hypotheses.

3.1. Statistical Model of the Collected Samples

We consider the case with a small number of samples and need to establish the test statistic model based on Student's t-distribution. Hence, we further assume that the detected signal is a wide-band signal. As shown in [39], the bandwidth to be detected can be divided into n subbands, where each subband has equal bandwidth. For each subband, the m collected samples are limited to small numbers. Then, after receiving the samples at an SU, the samples $X = \{X_i\}_{i=1}^{l}$ are divided into n groups, with m ($m > 1$) samples in each group, which indicates that $n = l/m$. The mean of the samples is defined as $\bar{X}_j$ and the variance is S_j^2 for the j-th group, respectively. Consequently, we can have

$$\bar{X}_j \triangleq \sum_{k=0}^{m-1} \frac{X_{mj-k}}{m}, \quad S_j^2 \triangleq \sum_{k=0}^{m-1} \frac{(X_{mj-k} - \bar{X}_j)^2}{m-1} \tag{15}$$

where $j = 1, 2, \cdots, n$ and $k = 0, 1, \cdots, m-1$. Let

$$Y_j \triangleq \frac{\bar{X}_j}{S_j/\sqrt{m}}, \quad j = 1, 2, \cdots, n. \tag{16}$$

Note that in order to calculate the following test statistic, the sequence $\{Y_j\}_{j=1}^{n}$ is sorted in increasing order and we assume that $Y_1 \leq Y_2 \leq \cdots \leq Y_n$.

Under H_0 hypothesis, PU transmits signal and the received samples follow $X_i \sim \mathcal{N}(0, \sigma^2)$. Then, Y_j follows a $v = m - 1$ degree Student's t-distribution. In the case of the H_1 hypothesis, the transmitted signal and noise are both included in the received signal, it results that $X_i \sim \mathcal{N}(\mu, \sigma^2)$, where $\mu = h$. In this case, Y_j is proved to follow a non-central Student's t-distribution with degree of freedom given by $v = m - 1$ and $\delta = \sqrt{m\mu^2/\sigma^2}$, where μ^2/σ^2 represents the signal-to-noise ratio (SNR) [32,50]. The histograms for different scenarios and GoFs of Y_j for H_0 as well as H_1 hypotheses are shown in Figure 1, where SNR = -2 dB and the number of samples $l = 64$. It can be observed from Figure 1 that the case with noise only fits the Student's t-distribution well. Meanwhile, the case including both signal and noise matches the noncentral t-distribution curve. Moreover, for the same degree of freedom, we can notice that the noncentral t-distribution curve shifts slightly to the right side of the red curve of the Student's t-distribution.

In addition, the curve shape of the Student's t-distribution tends to approach a zero mean normal distribution with variance equal to 1. For parameter m, it can also be derived that the student's t-distribution is closer to a standard normal distribution if m becomes larger. In contrast, if m gets smaller, the tails of the Student's t-distribution tend to locate at a higher level as shown in Figure 2. Tails with different m values distribute at a higher level than that of the normal distribution,. Therefore, for small m, it indicates that variables Y_j in Equation (16) tend to take values that deviate from their statistical mean. This could potentially result in inaccurate computation of the distance between the common cumulative distribution function and empirical distribution of the observation. Therefore, we need to accurately estimate the above distance in the next section.

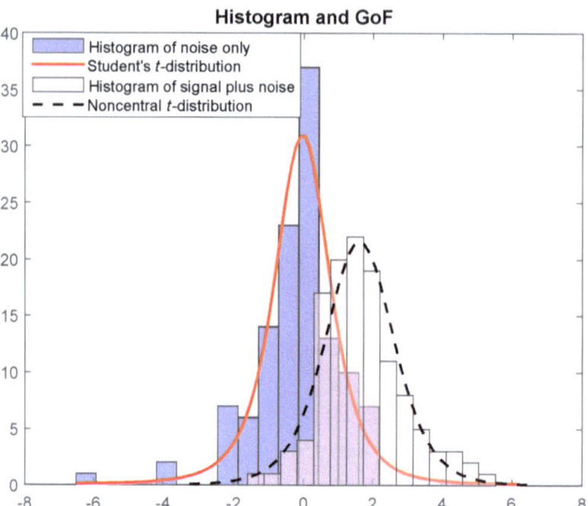

Figure 1. Histogram for scenarios with signal and without signal, GoF of Y_j under different hypotheses.

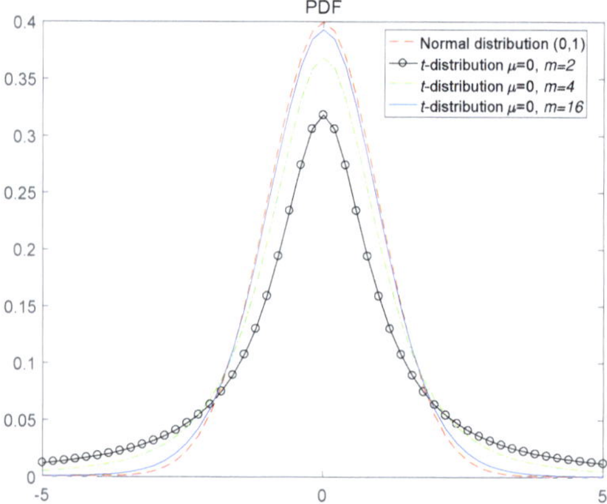

Figure 2. Probability density function (PDF) with different degrees of freedom $v = m - 1$ in Student's t-distribution.

It should be emphasized that the collected sample X_i as Y_j is reformulated and the Student's t-distribution is satisfied with H_0 hypothesis while noncentral t-distribution is met for the H_1 hypothesis. Notice that for the H_0 hypothesis, the CDF $F_0(y)$ relates only with degrees of freedom v, while $F_1(y)$ depends on the term $\delta = \sqrt{m}\mathrm{SNR}$. The noise variance σ^2 is also assumed to be known, which is the same assumption as ED-based methods. In addition, taking into account the limitations of the traditional GoF test (e.g., AD test and KS test) under a small number of samples, we propose a likelihood ratio-based powerful statistic instead of the traditional statistic in the following section.

3.2. Powerful GoF Test

To precisely evaluate the distance between common CDF and the empirical distribution of the observation, a novel GoF test statistic Z_c on the basis of the likelihood ratio is proposed. It is asymptotically equivalent to the Pearson χ^2-statistic in Equation (4) under large sample situations. For obtaining the test statistic Z_c, two kinds of statistic for testing H_0 with H_1 are defined by

$$Z = \int_{-\infty}^{\infty} Z_t \mathrm{d}w(t) \tag{17}$$

$$Z_{\max} = \sup_{t \in (-\infty,\infty)} \{Z_t w(t)\} \tag{18}$$

where Z_t is the statistic for comparing $H_0(t)$ with $H_1(t)$ such that its large values reject $H_0(t)$ and $w(t)$ is one of the weight functions. Note that

$$H_0 = \bigcap_{t \in (-\infty,\infty)} H_0(t) \tag{19}$$

$$H_1 = \bigcap_{t \in (-\infty,\infty)} H_1(t) \tag{20}$$

with $H_0(t) : F_n(t) = F_0(t)$ and $H_1(t) : F_n(t) = F_1(t)$.

In [45], authors present a natural candidate for Z_t, which is the likelihood ratio test statistic defined as follows:

$$G_t^2 = 2n[F_n(t)\log\{\frac{F_n(t)}{F_0(t)}\} + (1 - F_n(t))\log\{\frac{1 - F_n(t)}{1 - F_0(t)}\}] \tag{21}$$

Setting Z_t in Equation (17) equal to G_t^2 and setting the weight function to a proper value $\mathrm{d}w(t) = F_0(t)^{-1}\{1 - F_0(t)\}^{-1}\mathrm{d}F_0(t)$, we have

$$Z = \sum_{j=1}^{n} [log\{F_0(y)^{-1} - 1\} - b_{i-1} + b_i]^2 + C_n, \tag{22}$$

where $b_i = ilog(i/n) + (n - i)log(1 - i/n)$ and C_n is a constant value.

Since $b_{i-1} - b_i \approx log\{(n - \frac{1}{2})/(i - \frac{3}{4}) - 1\}$, we can derive the powerful GoF test statistic Z_{c0} compared with the traditional GoF test. It is approximately obtained in the following

$$Z_{c0} = \sum_{j=1}^{n} [log\{\frac{F_0(y)^{-1} - 1}{(n - 1/2)/(j - 3/4) - 1}\}]^2 \tag{23}$$

where $F_0(y)$ denotes the CDF of Y_j under H_0 hypothesis and it can be calculated by [50]:

$$F_0(y) = \begin{cases} \frac{1}{2} + \frac{1}{\pi}\tan^{-1}(y), & v = 1, \\ \frac{1}{2} + \frac{y}{2\sqrt{v+y^2}}\sum_{j=0}^{(v-2)/2}\frac{h_j}{(1+\frac{y^2}{v})^j}, v & even, \\ \frac{1}{2} + \frac{1}{\pi}\tan^{-1}(\frac{y}{\sqrt{v}}) \\ + \frac{y\sqrt{v}}{\pi(v+y^2)}\sum_{j=0}^{(v-3)/2}\frac{a_j}{(1+\frac{y^2}{v})^j}, & v \quad odd, \end{cases} \tag{24}$$

where $a_j = \frac{2j}{2j+1}a_{j-1}, a_0 = 1, b_j = \frac{2j-1}{2j}b_{j-1}, b_0 = 1$. In this situation, the statistic Z_{c0} denotes the distance between the CDF of Y_j under H_0 hypothesis and the empirical CDF of the collected samples. A large Z_{c0} means that H_0 hypothesis is rejected with a large probability. Otherwise, a small Z_{c0} means that the H_0 hypothesis is accepted. This is just the traditional GoF test, which is only based on the null hypothesis. However, in the proposed method, H_1 hypothesis is also considered to improve the reliability of the decision.

Therefore, corresponding to Equation (23), the other GoF test statistic Z_{c1} based on H_1 hypothesis, is given as follows:

$$Z_{c1} = \sum_{j=1}^{n} [\log\{\frac{F_1(y)^{-1} - 1}{(n - 1/2)/(j - 3/4) - 1}\}]^2 \tag{25}$$

where $F_1(y)$ denotes the CDF of Y_j under H_1 hypothesis and it can be calculated by [50].

$$F_1(y) = \begin{cases} \frac{1}{2} \sum_{j=0}^{\infty} \frac{1}{j!}(-\delta\sqrt{2})^j e^{-\frac{\delta^2}{2}} \frac{\Gamma(\frac{j+1}{2})}{\sqrt{\pi}} \\ \qquad \times I(\frac{v}{v+y^2}; \frac{v}{2}, \frac{j+1}{2}), \quad y \geq 0, \\ 1 - \frac{1}{2} \sum_{j=0}^{\infty} \frac{1}{j!}(-\delta\sqrt{2})^j e^{-\frac{\delta^2}{2}} \frac{\Gamma(\frac{j+1}{2})}{\sqrt{\pi}} \\ \qquad \times I(\frac{v}{v+y^2}; \frac{v}{2}, \frac{j+1}{2}), \quad y < 0. \end{cases} \tag{26}$$

where Γ represents the gamma function and I denotes the regularized incomplete beta function. In this situation, the statistic Z_{c1} denotes the distance between the CDF of Y_j under H_1 hypothesis and the empirical CDF of the collected samples. A large Z_{c1} means that H_1 hypothesis is rejected with a large probability. Otherwise, a small Z_{c1} means that H_1 hypothesis is accepted.

In this section, due to adapting a more appropriate weight function, we derive a new GoF test statistic Z_{c0} that is substantially more powerful than the traditional GoF test statistic under a small sample situation. Moreover, we further derive the other powerful GoF test statistic Z_{c1}, which utilizes the CDF of noncentral t-distribution under H_1 hypothesis. At the end, in order to enhance the reliability of final decision, we propose to make a final decision based on bilateral hypotheses (Z_{c0} and Z_{c1}) in the next section.

3.3. Final Decision Based on Bilateral Hypotheses

In this section, we propose to make use of the information from bilateral hypotheses in order to more accurately detect the PU signal given a small number of received data. According to the statistical characteristic of the Student's t-distribution and the noncentral t-distribution with different hypothesis in Section 3.1, and the two new powerful GoF test statistics Z_{c0} in Equation (23) and Z_{c1} in Equation (25) in Section 3.2, we make a final decision by comparing these two new GoF test statistics. Moreover, the normalization of the two GoF test statistics Z_{c0} and Z_{c1} can be written as $\mathcal{T}_0 = Z_{c0}/(Z_{c0} + Z_{c1})$ and $\mathcal{T}_1 = Z_{c1}/(Z_{c0} + Z_{c1})$. Finally, the final decision can be determined based on the following rule:

$$\begin{aligned} H_0 &: \quad \mathcal{T}_0 \leq \mathcal{T}_1 \\ H_1 &: \quad \mathcal{T}_1 < \mathcal{T}_0, \end{aligned} \tag{27}$$

where the information of bilateral hypotheses are both utilized to make a final decision, which significantly enhances the reliability of detection with small samples compared with the decision rule in Equation (13).

4. Simulation Results

In this section, the traditional GoF test (AD test, KS test, and CM test) based spectrum sensing methods and ED are considered for comparison. We assume that the PU signal is unknown while the noise power σ^2 is available in these methods.

First of all, to present the advantage of the proposed method, Figure 3 shows the detection probability providing different number of samples of the proposed method, AD test-based, KS test-based, and CM test-based method with the increasing of number of samples from 12 to 100 when $P_{fa} = 0.1$ and the SNR is -5 dB. The parameter m is set to 4 for the proposed method. As shown in Figure 3, the proposed method surpasses the traditional GoF test-based spectrum sensing methods. Particularly, if the number of samples is smaller than 48, the proposed method has a big improvement of the detection probability compared to other techniques. In addition, Table 1 shows the detection probability of compared methods when the number of samples are 32, 40, and 48. We can see that the proposed method can achieve a detection probability of 0.9514 for SNR = −5 dB. It can also validate the robustness of proposed method.

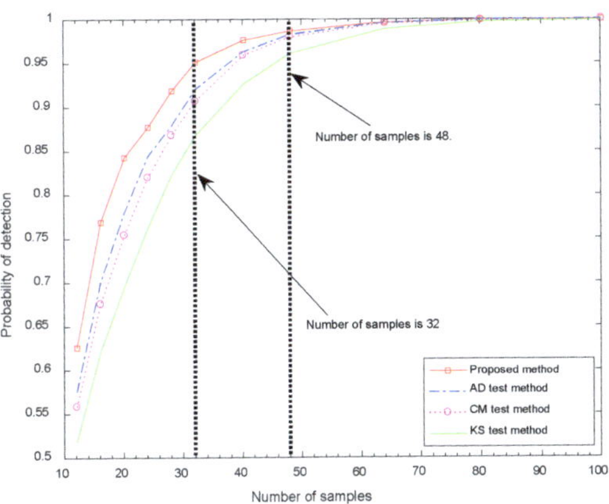

Figure 3. Probability of detection with different number of samples over AWGN channels with $P_{fa} = 0.1$.

Table 1. The probability of detection for different methods when the number of samples are 32, 40, and 48.

Number of Samples	KS Test Method	CM Test Method	AD Test Method	Proposed Method
32	0.8688	0.9070	0.9206	**0.9514**
40	0.9268	0.9592	0.9632	**0.9764**
48	0.9614	0.9796	0.9826	**0.9866**

Moreover, to compare the proposed method and other methods under different SNR conditions, the detection probability is provided in Figure 4 corresponding to the proposed method, the traditional GoF test based spectrum sensing methods and ED method with the increasing of SNR when $P_{fa} = 0.1$ and the number of samples $l = 32$. The parameter m is set to 4 for the proposed method. It can be seen from Figure 4 that the proposed scheme greatly surpasses the ED method. Importantly, the proposed method also has a better performance at low SNR region than the traditional GoF test (AD test, KS test, and CM test) based spectrum sensing methods. A specific example provided in Table 1 shows the detection probabilities corresponding to the $l = 32$ dotted lines in Figure 4 (SNR = −5 dB), which are 0.9514, 0.9206, 0.9070, and 0.8688 for the proposed method, AD test, CM test, and KS test, respectively. This also verifies that the proposed powerful GoF test method can achieve more reliable test statistic, leading to a better detection performance.

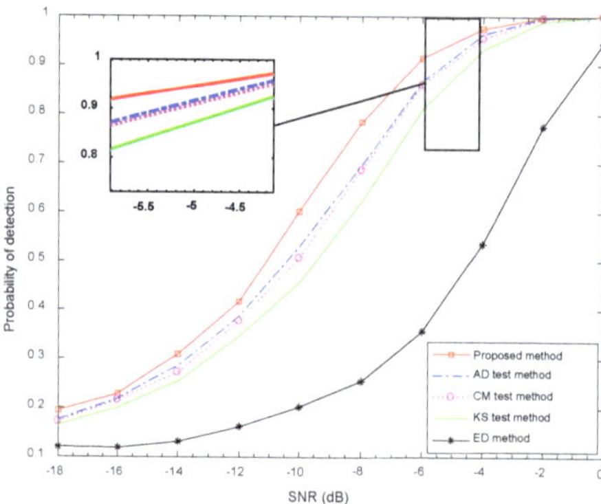

Figure 4. Detection probability with different SNR over AWGN channels with $P_{fa} = 0.1$.

The receiver operating characteristic (ROC) curves are compared in Figure 5. It is obvious that the proposed method has better performance than the ED method. Given the probability of false alarm set to 0.1, SNR of -5 dB and $l = 32$, the proposed method with $m = 4$ has about a 5% and 10% improvement relative to the AD test-based method and KS test-based method. That is because the proposed method utilizes the powerful GoF test statistic Z_c that outperforms the traditional GoF test. For D^2, W^2, and A^2 in KS test, CM test, and AD test, it is difficult to find their exact null distributions for finite sample cases. In the powerful GoF test statistic Z_c, we can use the sample mean and the sample variance to estimate μ and σ^2, respectively, and it outperforms the best tests in the literature, including the KS test, CM test, and AD test [45].

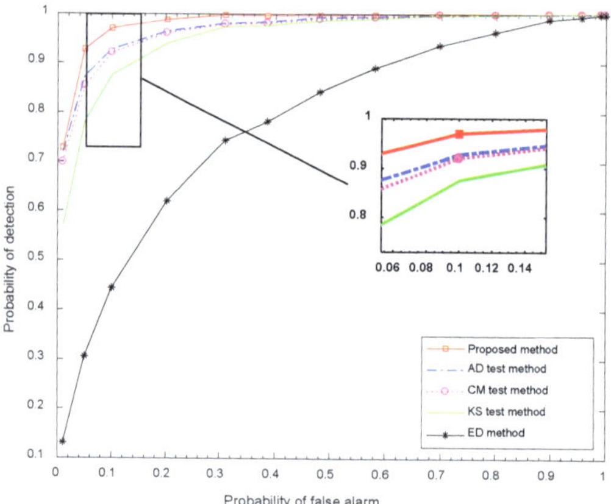

Figure 5. ROC curves comparison with SNR = -5 dB.

In addition, in order to show the advantages of using the bilateral hypotheses to make the decision, we compare the proposed method with and without bilateral hypotheses, and a similar study using the GoF [27] in Figure 6. As shown in Figure 6, the performances of the proposed method with and without bilateral hypotheses are superior to the method in [27] and the AD test method. Moreover, the proposed method with bilateral hypotheses makes full use of the bilateral hypotheses information, which increases the utilization of the distribution properties. Thus, it has a higher probability of detection compared to the proposed method without bilateral hypotheses.

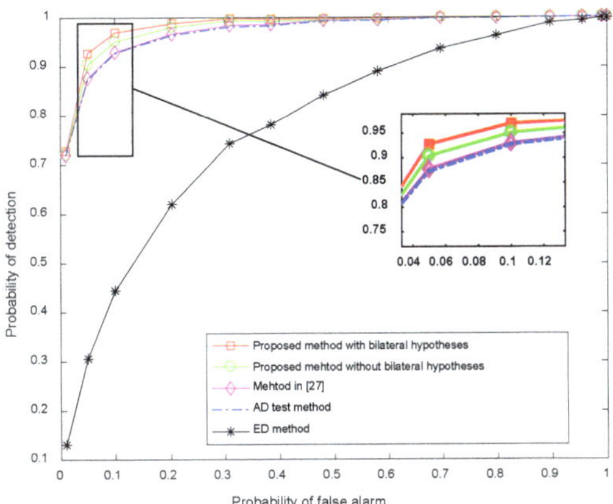

Figure 6. ROC curves comparison with SNR = −5 dB.

5. Conclusions

In this paper, an enhanced spectrum sensing method is proposed on the basis of a GoF test using bilateral hypotheses in a cognitive drone network. Only a small number of samples is required by the proposed scheme compared to the traditional ED, which is attractive for a dynamic weak signal scenario, including illegal drone detection. More specifically, samples at the SU with statistical model are thoroughly exploited, which strengthen its capacity for dealing with the small sample size case. Then, a powerful GoF test statistic Z_c is proposed to obtain a better measurement, and bilateral hypotheses GoF test Z_{c0} and Z_{c1} are both used for making a reliable decision. Finally, simulations validate the superiority of the proposed method compared with the traditional GoF test-based methods (AD test, KS test, and CM test) and ED method provided a small number of samples. The capability of proposed method for settling small sample size problem without sacrificing the detection performance could bring several potential benefits, including sensing time, energy consumption, and computational burden to the whole drone network.

Author Contributions: Conceptualization and methodology, S.M. and Z.F., writing and validation, S.M., Z.F. and P.C. All authors have read and agreed to the published version of the manuscript.

Funding: This research was funded in part by the National Natural Science Foundation of China under Grant 82004259, in part by the GuangDong Basic and Applied Basic Research Foundation under Grant 2020A1515110503, in part by the Guangzhou Basic and Applied Basic Research Project under Grant 202102020674.

Institutional Review Board Statement: Not applicable

Informed Consent Statement: Not applicable

Data Availability Statement: Not applicable

Conflicts of Interest: No conflict of interest exits in the submission of this manuscript, and the manuscript is approved by all authors for publication.

References

1. Adamopoulos, E.; Rinaudo, F. UAS-Based Archaeological Remote Sensing: Review, Meta-Analysis and State-of-the-Art. *Drones* **2020**, *4*, 46. [CrossRef]
2. Kaplan, B.; Kahraman, I.; Gorcin, A.; Çırpan, H.A.; Ekti, A.R. Measurement based FHSS–type Drone Controller Detection at 2.4GHz: An STFT Approach. In Proceedings of the 2020 IEEE 91st Vehicular Technology Conference (VTC2020-Spring), Antwerp, Belgium, 25–28 May 2020.
3. Wang, D.; He, T.; Zhou, F.; Cheng, J.; Zhang, R.; Wu, Q. Outage-driven link selection for secure buffer-aided networks. *Sci. China Inf. Sci.* **2022**, *65*, 182303. [CrossRef]

4. Wang, D.; Wu, M.; He, Y.; Pang, L.; Xu, Q.; Zhang, R. An HAP and UAVs Collaboration Framework for Uplink Secure Rate Maximization in NOMA-Enabled IoT Networks. *Remote Sens.* **2022**, *14*, 4501. [CrossRef]
5. Ahmad, A.; Ahmad, S.; Rehmani, M.H.; Hassan, N.U. A Survey on Radio Resource Allocation in Cognitive Radio Sensor Networks. *IEEE Commun. Surv. Tuts.* **2015**, *17*, 888–917. [CrossRef]
6. Chen, X.; Chen, H.; Meng, W. Cooperative communications for cognitive radio networks from theory to applications. *IEEE Commun. Surv. Tuts.* **2014**, *16*, 1180–1192. [CrossRef]
7. Kakalou, I.; Psannis, K.E.; Krawiec, P.; Badea, R. Cognitive radio network and network service chaining toward 5g: Challenges and requirements. *IEEE Commun. Mag.* **2017**, *55*, 145–151. [CrossRef]
8. Liang, W.; Ng, S.X.; Hanzo, L. Cooperative overlay spectrum access in cognitive radio networks. *IEEE Commun. Surv. Tuts.* **2017**, *19*, 1924–1944. [CrossRef]
9. Hefnawi, M. Large-Scale Multi-Cluster MIMO Approach for Cognitive Radio Sensor Networks. *IEEE Sens. J.* **2016**, *16*, 4418–4424. [CrossRef]
10. Akan, O.B.; Karli, O.B.; Ergul, O. Cognitive radio sensor networks. *IEEE Netw.* **2009**, *23*, 34–40. [CrossRef]
11. Joshi, G.P.; Nam, S.Y.; Kim, S.W. Cognitive radio wireless sensor networks: Applications, challenges and research trends. *Sensors* **2013**, *13*, 11196–11228. [CrossRef]
12. Wang, D.; Zhou, F.; Lin, W.; Ding, Z.; Dhahir, N.A. Cooperative Hybrid Non-Orthogonal Multiple Access Based Mobile-Edge Computing in Cognitive Radio Networks. *IEEE Trans. Cogn. Commun. Netw.* **2022**, *8*, 1104–1117. [CrossRef]
13. Ali, A.; Hamouda, W. Advances on Spectrum Sensing for Cognitive Radio Networks: Theory and Applications. *IEEE Commun. Surv. Tuts.* **2017**, *19*, 1277–1304. [CrossRef]
14. Axell, E.; Leus, G.; Larsson, E.G.; Poor, H.V. Spectrum sensing for cognitive radio: State-of-the-art and recent advances. *IEEE Signal Process. Mag.* **2012**, *29*, 101–116. [CrossRef]
15. Cicho, K.; Kliks, A.; Bogucka, H. Energy-Efficient Cooperative Spectrum Sensing: A Survey. *IEEE Commun. Surv. Tuts.* **2016**, *18*, 1861–1886. [CrossRef]
16. Yücek, T.; Arslan, H. A survey of spectrum sensing algorithms for cognitive radio applications. *IEEE Commun. Surv. Tuts.* **2009**, *11*, 116–130. [CrossRef]
17. Wang, B.; Liu, K.J. Advances in cognitive radio networks: A survey. *IEEE J. Sel. Top. Signal Process.* **2011**, *5*, 5–23. [CrossRef]
18. Urkowitz, H. Energy detection of unknown deterministic signals. *Proc. IEEE* **1967**, *55*, 523–531. [CrossRef]
19. Sofotasios, P.C.; Rebeiz, E.; Tsiftsis, L.Z.T.A.; Cabric, D.; Freear, S. Energy Detection Based Spectrum Sensing Over $\kappa - \mu$ and $\kappa - \mu$ Extreme Fading Channels. *IEEE Trans. Veh. Technol.* **2013**, *62*, 1031–1040. [CrossRef]
20. Chatziantonious, E.; Allen, B.; Velisavljevic, V.; Karadimas, P.; Coon, J. Energy Detection Based Spectrum Sensing Over Two-Wave With Diffuse Power Fading Channels. *IEEE Trans. Veh. Technol.* **2017**, *66*, 868–874.
21. Zeng, Y.; Liang, Y.C. Eigenvalue-based spectrum sensing algorithms for cognitive radio. *IEEE Trans. Commun.* **2009**, *57*, 1784–1793. [CrossRef]
22. Tsinos, C.G.; Berberidis, K. Decentralized Adaptive Eigenvalue-Based Spectrum Sensing for Multiantenna Cognitive Radio Systems. *IEEE Trans. Wireless Commun.* **2015**, *14*, 1703–1715. [CrossRef]
23. Bouallegue, K.; Dayoub, I.; Gharbi, M.; Hassan, K. Blind Spectrum Sensing Using Extreme Eigenvalues for Cognitive Radio Networks. *IEEE Commun. Lett.* **2018**, *2*, 1386–1389. [CrossRef]
24. Nguyen-Thanh, N.; Kieu-Xuan, T.; Koo, I. Comments and Corrections Comments on "Spectrum Sensing in Cognitive Radio Using Goodness-of-Fit Testing". *IEEE Trans. Wirel. Commun.* **2012**, *11*, 3409–3411. [CrossRef]
25. Teguig, D.; Nir, V.L.; Scheers, B. Spectrum sensing method based on goodness of fit test using chi-square distribution. *Electron. Lett.* **2014**, *50*, 713–715. [CrossRef]
26. Teguig, D.; Nir, V.L.; Scheers, B. Spectrum sensing Method Based on the Likelihood Ratio Goodness of Fit test under noise uncertainty. *Electron. Lett.* **2015**, *51*, 253–255. [CrossRef]
27. Scheers, B.; Teguig, D.; Nir, V.L. Wideband spectrum sensing technique based on Goodness-of-Fit testing. In Proceedings of the 2015 International Conference on Military Communications and Information Systems (ICMCIS), Cracow, Poland, 16 July 2015; pp. 1–6. [CrossRef]
28. Jin, M.; Guo, Q.; Xi, J.; Yu, Y. Spectrum sensing based on goodness of fit test with unilateral alternative hypothesis. *Electron. Lett.* **2014**, *50*, 1645–1646. [CrossRef]
29. Kockaya, K.; Develi, I. Spectrum sensing in cognitive radio networks: Threshold optimization and analysis. *J. Wirel. Com Netw.* **2020**, *2020*, 255. [CrossRef]
30. Gai, J.; Zhang, L.; Wei, Z. Spectrum Sensing Based on STFT-ImpResNet for Cognitive Radio. *Electronics* **2022**, *11*, 2437. [CrossRef]
31. Wang, H.; Yang, E.H.; Zhao, Z.; Zhang, W. Spectrum sensing in cognitive radio using goodness of fit testing. *IEEE Trans. Wirel. Commun.* **2009**, *8*, 5427–5430. [CrossRef]
32. Shen, L.; Wang, H.; Zhang, W.; Zhao, Z. Blind spectrum sensing for cognitive radio channels with noise uncertainty. *IEEE Trans. Wirel. Commun.* **2011**, *10*, 1721–1724. [CrossRef]
33. Zhang, G.; Wang, X.; Liang, Y.C.; Liu, J. Fast and robust spectrum sensing via Kolmogorov-Smirnov test. *IEEE Trans. Commun.* **2010**, *58*, 3410–3416. [CrossRef]
34. Patel, D.K.; Trivedi, Y.N. Goodness-of-fit-based non-parametric spectrum sensing under Middleton noise for cognitive radio. *Electron. Lett.* **2015**, *51*, 419–421. [CrossRef]

35. Ye, Y.; Lu, G.; Li, Y.; Jin, M. Unilateral right-tail Anderson-Darling test based spectrum sensing for cognitive radio. *Electron. Lett.* **2017**, *53*, 1256–1258. [CrossRef]
36. Liu, C.; Wang, J.; Liu, X.; Liang, Y. Maximum Eigenvalue-Based Goodness-of-Fit Detection for Spectrum Sensing in Cognitive Radio. *IEEE Trans. Veh. Technol.* **2019**, *68*, 7747–7760. [CrossRef]
37. Rugini, L.; Banelli, P.; Leus, G. Small sample size performance of the energy detector. *IEEE Commun. Lett.* **2013**, *17*, 1814–1817. [CrossRef]
38. Arshad, K.; Moessner, K. Robust spectrum sensing based on statistical tests. *IET Commun.* **2013**, *7*, 808–817. [CrossRef]
39. Men, S.; Chargé, P.; Wang, Y.; Li, J. Wideband signal detection for cognitive radio applications with limited resources. *Eurasip J. Adv. Signal Process.* **2019**, *2019*, 1–10. [CrossRef]
40. Rostami, S.; Arshad, K.; Moessner, K. Order-statistic based spectrum sensing for cognitive radio. *IEEE Commun. Lett.* **2012**, *16*, 592–595. [CrossRef]
41. Denkovski, D.; Atanasovski, V.; Gavrilovska, L. HOS Based Goodness-of-Fit Testing Signal Detection. *IEEE Commun. Lett.* **2012**, *16*, 310–313. [CrossRef]
42. Pakyari, R.; Balakrishnan, N. A General Purpose Approximate Goodness-of-Fit Test for Progressively Type-II Censored Data. *IEEE Trans. Reliab.* **2012**, *61*, 238–244. [CrossRef]
43. Noughabi, H.A.; Balakrishnan, N. Goodness of Fit Using a New Estimate of Kullback-Leibler Information Based on Type II Censored Data. *IEEE Trans. Reliab.* **2015**, *64*, 627–635. [CrossRef]
44. Qiu, Y.; Liu, L.; Lai, X.; Qiu, Y. An Online Test for Goodness-of-Fit in Logistic Regression Model. *IEEE Access* **2019**, *7*, 107179–107187. [CrossRef]
45. Zhang, J. Powerful goodness-of-fit tests based on the likelihood ratio. *J. Roy. Statist. Soc. Ser. B* **2002**, *64*, 281–294. [CrossRef]
46. Terry, K.; Agostino, R.B.; Stephens, M.A. Goodness-of-Fit Techniques. *Technometrics* **1987**, *29*, 493.
47. Stephens, M.A. EDF statistics for goodness of fit and some comparisons. *J. Amer. Statist. Assoc.* **1974**, *69*, 730–737. [CrossRef]
48. Anderson, T.W.; Darling, D.A. Asymptotic theory of certai "goodness of fit" criteria based on stochastic processes. *Ann. Math. Stat.* **1952**, *23*, 193–212. [CrossRef]
49. Stephens, M.A. Use of the Kolmogorov-Smirnov, Cramer-Von Mises and Related Statistics Without Extensive Tables. *J. R. Stat. Soc. Ser. B* **1970**, *32*, 115–122. [CrossRef]
50. Forbes, C.; Evans, M.; Hastings, N.; Peacock, B. *Statistical Distributions*; John Wiley: Hoboken, NJ, USA, 2011.

Article

Vehicle to Everything (V2X) and Edge Computing: A Secure Lifecycle for UAV-Assisted Vehicle Network and Offloading with Blockchain

Abdullah Ayub Khan [1,2], **Asif Ali Laghari** [3], **Muhammad Shafiq** [4,*], **Shafique Ahmed Awan** [2] **and Zhaoquan Gu** [5]

[1] Department of Computer Science, Sindh Madressatul Islam University, Karachi 74000, Pakistan
[2] Department of Computing Science and Information Technology, Benazir Bhutto Shaheed University Lyari, Karachi 75660, Pakistan
[3] School of Software Engineering, Shenyang Normal University, Shenyang 110034, China
[4] Cyberspace Institute of Advanced Technology, Guangzhou University, Guangzhou 510006, China
[5] School of Computer Science and Technology, Harbin Institute of Technology (Shenzhen), Shenzhen 518055, China
* Correspondence: srsshafiq@gmail.com

Citation: Khan, A.A.; Laghari, A.A.; Shafiq, M.; Awan, S.A.; Gu, Z. Vehicle to Everything (V2X) and Edge Computing: A Secure Lifecycle for UAV-Assisted Vehicle Network and Offloading with Blockchain. *Drones* **2022**, *6*, 377. https://doi.org/10.3390/drones6120377

Academic Editors: Dawei Wang and Ruonan Zhang

Received: 1 November 2022
Accepted: 20 November 2022
Published: 25 November 2022

Publisher's Note: MDPI stays neutral with regard to jurisdictional claims in published maps and institutional affiliations.

Abstract: Due to globalization and advances in network technology, the Internet of Vehicles (IoV) with edge computing has gained increasingly more attention over the last few years. The technology provides a new paradigm to design interconnected distributed nodes in Unmanned Aerial Vehicle (UAV)-assisted vehicle networks for communications between vehicles in smart cities. The process hierarchy of the current UAV-assisted networks is also becoming more multifaceted as more vehicles are connected, requiring accessing and exchanging information, performing tasks, and updating information securely. This poses serious issues and limitations to centralized UAV-assisted vehicle networks, directly affecting computing-intensive tasks and data offloading. This paper bridges these gaps by providing a novel, transparent, and secure lifecycle for UAV-assisted distributed vehicle communication using blockchain hyperledger technology. A modular infrastructure for Vehicle-to-Everything (V2X) is designed and 'B-UV2X', a blockchain hyperledger fabric-enabled distributed permissioned network-based consortium structure, is proposed. The participating nodes of the vehicle are interconnected with others in the chain of smart cities and exchange different information such as movement, etc., preserving operational logs on the blockchain-enabled immutable ledger. This automates IoV transactions over the proposed UAV-assisted vehicle-enabled consortium network with doppler spread. Thus, for this purpose, there are four different chain codes that are designed and deployed for IoV registration, adding new transactions, updating the ledger, monitoring resource management, and customized multi-consensus of proof-of-work. For lightweight IoV authentication, B-UV2X uses a two-way verification method with the defined hyperledger fabric consensus mechanism. Transaction protection from acquisition to deliverance and storage uses the NuCypher threshold proxy re-encryption mechanism. Simulation results for the proposed B-UV2X show a reduction in network consumption by 12.17% compared to a centralized network system, an increase in security features of up to 9.76%, and a reduction of 7.93% in the computational load for computed log storage.

Keywords: Vehicle-to-Everything (V2X); edge computing; UAV-assisted vehicle network; blockchain; smart contract; cost-effective scheduling

1. Introduction

The future development of the Internet of Vehicles (IoV) depends on Vehicle to Everything (V2X) applicational maturity [1,2]. However, applications include intent sharing, distributed environment design, bi-directional intercommunication, interactive gaming, and wireless network-based coordinated driving. These developments are anticipated to make the system more efficient, reliable, secure, and diverse, allowing execution of vehicular transactions autonomously [3]. However, in recent advances, the success of these vehicular applications relies on a large amount of generated data, a fusion of data handling, and processing from wireless sensor-based distributed networks that are associated with the vehicles. Therefore, deployed on-road infrastructure guarantees data management in real time while precisely perceiving the environment. Perception and related computation for each vehicle increases the continuity of learning in the smart environment. In fact, it leads to low resource consumption in terms of computing energy, network bandwidth, and storage, due to the high price of precision sensors and the powerful computational processing units (CPUs) used [4]. The usage of low-priced equipment in the IoV domain impacts the performance ratio of systems. Furthermore, it limits local perception capabilities, which directly affects centralized network-based V2X applications. Resource usage, especially energy consumption for computation and high-precision sensor management, substantially reduces the efficiency of vehicular systems in terms of battery life, reducing mileage. In order to improve the use of computational resources, different applications of the IoV have adopted different artificial intelligence (AI)-based methods for dynamic monitoring. This has led to various significant challenges that pose resource-management-related problems in IoV terminals [5].

Beyond these limitations, each IoV node is able to directly interconnect with other nodes in the chain of smart cities, as shown in Figure 1. This is possible because the centralized network-enabled infrastructure provides intercommunication facilities with no repudiation [6]. However, the technology is more robust with vehicular edge computing, fifth-generation networks (5G), and fog/cloud-based system integration. These developments aim to provide a vigorous computational environment, high storage, sensor connectivity, smart sharing and exchanging, and service orientation by leveraging distributed IoV energy usage with low-cost distributed communication [7]. This differs from cloud/fog-enabled technologies, which use an old paradigm where physical proximity between information systems and computing services promised distinct advantages. These advantages include low throughput and latency, high power efficiency, security and privacy protection reliability, reduced network bandwidth usage, and storage-related context awareness. On the other hand, vehicular edge computing integration depends on distributed wireless networks to divide a large number of computing tasks and sensor-based corresponding records/details over the IoV and the edge network. This leads to a quick response by enabling smart vehicles to perform all applications for a distributed IoV network.

The current standard of V2X (UAV-assisted network) is categorized into two different parts: short-range dedicated communication and a cellular V2X. Dedicated short-range communication is used as a standard protocol of IEEE 802.11p, while cellular communication follows 5G protocols [8,9], as shown in Figure 1. The combination of short-range and cellular V2X on the existing system is used to gain broader coverage for pilot-distributed applicational facilities with distance transmission, steadier channels, and UAV network-related deployment. The recent of UAV-assisted infrastructure system uses 5G cellular V2X to offload a large amount of data to IoV edge nodes with new frequencies, such as millimeter-wave frequency bands.

However, the offloading problem in a UAV-assisted network creates different challenges in the V2X environment. These include problem segmentation, IoV edge selection and offloading, problem mitigation, and privacy protection, which are becoming widely researched concerns [10]. For instance, edge computing offloading strongly depends on a large-scale ubiquitous base station to handle data and coverage for transmission. It assumes that the current communication resources of the UAV-assisted network are not sufficient for vehicular edge offloading [11]. Therefore, deploying a dense-base station reduces the load of vehicular edge computing-enabled traffic by providing a simple structure to manage all the transactions. However, it is considered a cost-inefficient method for IoV services and locations. To manage these complexities, a middle base station/middle infrastructure with the standard process hierarchy is required to evaluate the requests of IoV devices concurrently. Only a valid service request can pass; otherwise, it is discarded at an initial level.

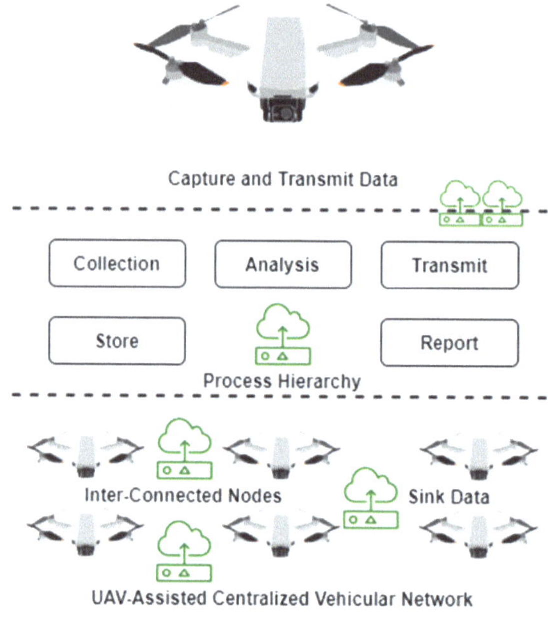

Figure 1. The current Vehicle to Everything (V2X) environment.

Major actors are analyzing distributed UAV-assisted vehicular network solutions related to connectivity and offloading for futuristic transportation. One of the main reasons for adopting blockchain hyperledger-enabled distributed architecture in a UAV-assisted vehicular network is to eliminate the dependency on certificate authority [12–16]. The decentralized nature of blockchain ledger technology integrates with different domains of computing to allow new designs in IoV transaction processing, privacy, and security. The modular infrastructure of blockchain technology provides automated transaction execution facilities via distributed applications (DApp). Thus, it ensures information security against malicious attacks during inter/outer-communication between nodes. However, the current system of UAV-assisted vehicle network consumes significant computing resources because no standard protocol for request management has been proposed [17]. For this reason, hyperledger technology is used to provide a customized design for consensus, chaincode execution, privacy and security procedures, and network communication-related facilities, directly reducing resource consumption. In addition, another advantage of this technology is that it provides ledger (log records) preservation and protection in a serverless environment through the proxy threshold re-encryption mechanism.

1.1. Objectives and Contributions

This paper addresses the current issues, challenges, and limitations involved in centralized UAV-assisted vehicle networks. It highlights the changes in the evolution of Vehicle to Everything (V2X) and presents the role of V2X in futuristic transportation development. A number of enhancements/improvements when deploying a UAV-assisted vehicular network with blockchain, including in record scheduling, managing, organizing, optimizing, and offloading in a secure and protected manner in a decentralized environment. By enabling the current design of UAV-assisted V2X to be integrated with a consortium channel's blockchain structure, the load on vehicular network resources is reduced drastically compared to previous infrastructures. The major contributions of this research paper are as follows:

- This paper proposes B-UV2X, a novel and secure distributed UAV-assisted vehicular network infrastructure for IoV interconnectivity. The designed system realizes interoperable communication between devices in the V2X environment with blockchain;
- A blockchain-enabled standardized lifecycle is designed. The main objective is to maintain the process hierarchy throughout transactions acquisition towards deliverance in a secure manner;
- A consortium network with doppler spread is deployed for edge-enabled IoV systems to handle requests related to permissioned or permissionless environments;
- To protect individual transactions of the IoV, B-UV2X uses a proxy re-encryption threshold mechanism. Furthermore, a multi-consensus protocol is created with the predefined method of the digital signature of the hyperledger to schedule the list of node transaction executions, which helps in the management of resources;
- In this paper, three different types (IoV connectivity and data management, record updates, and exchanging) of smart contracts are created and deployed;
- Finally, this paper highlights the implementation challenges faced in the process of B-UV2X deployment, with future open research questions. Possible solutions are discussed.

1.2. Section Distribution

The remainder of this paper is structured and organized as follows. In Section 2, various related works are studied and investigated to find the current gaps in vehicle-to-everything, UAV-assisted vehicle networks, the IoV, and seamless edge computing services for centralized networks. The problem description, formulations, and related working objectives of the proposed B-UV2X are discussed in Section 3. The experimental results of B-UV2X and related comparisons with other state-of-the-art methods are presented in Section 4. In Section 5, the paper describes different implementation challenges, issues, and limitations, and highlights futuristic objectives as well. Finally, the conclusion of this research is discussed in Section 6.

2. Related Work

2.1. Vehicle-to-Everything (V2X) and UAV-Assisted Vehicle Network

Recently, drone-enabled technology has been widely adopted in different industrial, manufacturing, and production units to smartly enhance working objectives in terms of scheduling, managing, and monitoring. The open nature of centralized vehicular networks threatens privacy of information [18]. This may also lead to privacy leakage of personal information, posing various tampering- and forgery-related issues. In this regard, several artificial intelligence, machine learning, deep learning, federated learning, and blockchain-enabled distributed modular architectures of UAV-assisted vehicular networks with doppler spread have been presented [19]. These address different kinds of dependent centralized aggregative servers, which are designed to maintain system objectives and a crash-less environment. In addition, unauthorized participation also drives positioning attack, reducing the usability of the system and creating communication barriers that hinder integration in the large number of cross-domain IoVs. The research gaps in previously published state-of-the-art methods are discussed as follows (as shown in Table 1).

Table 1. Related literature on blockchain, UAV-assisted vehicular networks, and vehicle to everything.

Title of the Article	Proposed Method/Procedure	Research Gaps in the Study	Similarities and Differences with the Proposed B-UV2X
Internet of Drones (IoD) applications with blockchain [20]	The authors of this paper discussed the role of blockchain and its integration in the improvement of IoD connectivity and security, as well as the importance of distributed applications for drone-based data management and monitoring in a protected manner, especially in smart-city environments.	• Scope of data privacy security issues • Explored commercial applicational problems • Derived blockchain mechanism proposed • Data optimization and offloading issues	• Internet-of-Vehicles (IoV) connectivity • Blockchain permissionless network • Platform interoperability limitation • Security and privacy concerns
A decentralized machine learning framework for intrusion detection in UAV using blockchain distributed ledger modular infrastructure [21]	This paper presents a distributed framework for intrusion detection using integrated machine learning and blockchain technologies. In this design, the system is potentially able to significantly enhance the integrity, transparency, and storage of information for smart decision-making among multiple UAVs.	• Conventional UAVs • Complex machine learning algorithm used • Predictive analysis • Multi-UAV intercommunication	• Cross chain platform-based challenges • Intercommunication node integrity • Permissionless network structure
Drone-based delivery scheme for industrial healthcare using blockchain technology [22]	This paper highlights the list of current blockchain-based drone-enabled industrial healthcare applicational challenges and limitations. These include harsh environmental conditions, rough terrain, war-prone areas, congested traffic, remote location, etc.	• Integrated IoD delivery scheme • Data driven analytics • Two-way verification and validation process • Cross-chaining platform	• Blockchain distributed ledger technology • Permissioned architecture • Hash-encryption • Cloud-enabled storage
Internet of Drones (IoD): communication leveraging with blockchain [23]	The authors of this paper presented a security approach for drone-to-everything communication, in which the locations of drones are traced by segment divisions of the areas in which they are deployed.	• Fifth generation network (5G) connectivity • Deployed across remote sides • Remote cloud for storage	• Cryptographic hash-encryption mechanism used • Advanced sensors and GPS used • Segment division by areas
Internet of Vehicles (IoV) security [24]	In this paper, the authors defined the taxonomy of IoD security and privacy along with access to the controlled airspace to provide an inter-location navigation service using AI, machine learning, blockchain, and federated learning.	• Federated learning architecture used • Proposed IoD paradigm (standardized) • Level-of-security category	• Blockchain integrates with AI • Permissionless architecture • On-chain and off-chain intercommunication channels designed • Distributed interconnected node hierarchy
A lightweight assisted secure routing scheme for the IoV using blockchain Ethereum [25]	A secure routing algorithm for IoT-enabled drone management swarm UAS networking is proposed in this research. The benefits are as follows: • Swarm UAS orientation; • Customized consensus using blockchain; • Estimate traffic status/dynamic monitoring; • Lookup table and scheduling.	• Customized protocols and policies • Improved predefined consensus • Blockchain permissionless network • Distributed ledger preservation and digital signature	• Monitoring resource usage • Reduce network bandwidth consumption • Data security and preservation • Interoperability issues between inner and outer chain connectivity

2.2. Internet of Vehicles and Mobile Edge Computing with Blockchain

By virtue of intelligence in UAV-assisted vehicular networks, the IoV performs a primary role for transport systems in dynamic-time information exchange, which improves data processing and traffic management, especially in smart cities. In addition, to ease the computational energy and preservation load, which is increased by a large number of IoV nodes requesting to connect, edge-enabled computing resources are introduced to reduce the load of computing tasks, offload data and management, and optimize the local vehicular network with low latency [26–30]. Data integrity and privacy are still challenging prospects in these proposed systems. To address these problems, various researchers apply different methods in their proposed architectures, along with conditional privacy-preservation authentication protocols to enable the IoV with edge computing using blockchain distributed ledger technology.

The use of mobile edge computing provides enormous storage resources with a powerful computing network infrastructure. The IoV with mobile edge computing ensures that the paradigm can handle a large amount of data storage, sharing and exchanging, and processing capabilities close to the devices. However, the system is unable to share data when the architectural approach is based on a centralized server. With these potential risks of data leakage, an IoV node faces difficulty when evaluating the credibility of a message; this is also because it receives requests for transactions from an untrusted centralized environment [2–30]. To enhance security, blockchain hyperledger technology with a consortium network structure with doppler spread is proposed.

3. Preliminary Knowledge of the Proposed B-UV2X

This section discusses the fundamentals and critical assumptions of blockchain-enabled distributed technologies in UAV-assisted vehicular networks to create a new paradigm in the IoV. Related problem formulations are discussed as follows:

3.1. Notation, Problem Formulation, and Description

First, to design a UAV-assisted distributed vehicular network lifecycle, the state of the data item is requested by the IoV nodes in a transactional manner; then, the IoV builds encoded data point broadcasts based on the high clique (priority) of the transactional request. Second, these transactional requests must be secure and protected while being transmitted. To ensure this, we present a standardized process hierarchy: (i) data generation (transaction request sent from the IoV devices/nodes), (ii) capturing, (iii) examining, (iv) analyzing, (v) preserving, (vi) sharing/exchanging, and (vii) reporting (details of log recordings), as shown in Figure 2. Finally, the IoV receives encoded data points via the process hierarchy of the proposed B-UV2X lifecycle. In the decoding procedure, the broadcasted logs of UAV-assisted vehicular data points are accepted by the IoV itself. In the design and development, the main objective is to reduce resource usage in terms of IoV-enabled computational energy, network bandwidth consumption, and memory, which directly affect battery life. For the sake of standardization and simplification, we use IoV_1, IoV_2, IoV_3, ..., IoV_n to denote Internet of Vehicles (IoV) devices; the position coordinates of IoVs in smart cities are as follows: (a_x, b_x) of IoV_n. d_1, d_2, d_3, ... , d_n represent the transactional requests and related data broadcasted over the UAV-assisted vehicular network. 'r' represents the radius of interconnected IoV nodes for communication (or sharing information) in the designed distributed vehicular network.

In order to check that the IoV nodes in the proposed B-UV2X-enabled distributed vehicular network are receiving or sending data points/transactional requests at the same time, the interoperable platform provides a distanced measurement structure that calculates the distance between devices before transmitting requests. The distance between IoV devices is represents as '2r', 'r', and 'r_0', showing the coverage radius of two nodes in the UAV-assisted distributed network. The maximum radius of the distributed IoV network is equal to the distance between locations, which must be less than or equal to the coverage area ('r').

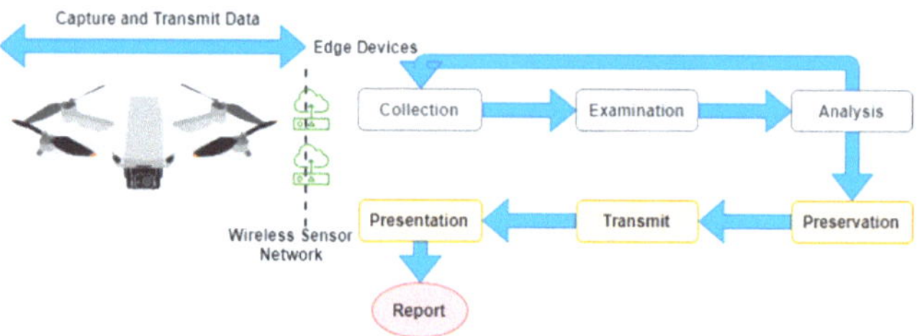

Figure 2. Block diagram of the proposed lifecycle of B-UV2X.

With the goal of reducing network bandwidth usage in the distributed environment, we tune the maximum radius of multiple IoVs (maximum in the four interconnected pairs of shared information/transactional requests). Given the maximum area ('r') of the distributed network circle, we can obtain the coverage area of a circle of the inscribed graphical domain (such as an equilateral triangle). The length of the sides of the equilateral triangle (which means the area of IoVs/drones) is $\sqrt{}$four-sides ($\sqrt{}$4s).

If the size of the IoVs is = $\sqrt{}$4s or >= $\sqrt{}$4s, the maximum distance between IoVs is 'r', which is directly proportional to $\sqrt{}$4s. The exterior radius of the designed UAV-assisted distributed vehicular network is 2r. In another case, the exterior is >= 2r.

The vehicle IoV_n sends or receives transactional requests for data points d_n; then, the UAV-assisted distributed vehicular network schedules transactions t_n. In this way, the system can identify from which IoV devices the requested and scheduled transactions originate, as well as where they are shared. In addition, the integration of edge-enabled computation with the proposed B-UV2X lifecycle reduces the load of data offloading. With request/transaction scheduling, the computational processing of the IoVs is reduced by increasing the rate of execution and transmission. In this manner, the cost of information preservation is also reduced.

However, duplicate and redundant scheduled transactions can be discarded before execution. This is because the system verifies and validates request automatically by the use of deployed chain codes (and functions such as IoVReg(), UAVAVLC(), AddNTD(), update() and InfoPre()). The role of a hyperledger expert, the person responsible for initiating the proposed B-UV2X chain and handling the request for participating IoV registrations, is also highlighted in this scenario.

3.2. Proposed Architecture

The operation of the proposed B-UV2X is divided into four phases. First, B-UV2X registers the IoV node in the designed consortium chain after proper verification and validation while obtaining the enrollment request from the nodes and related device stakeholders. For a complete analysis of the registered request, the role of a blockchain hyperledger expert is crucial; this person is responsible for initiating chain transactions, handling a number of requests and related executions, and managing information preservation (records logs), as shown in Figure 3. The data movement hierarchy of this phase is bi-directional in nature. Second, in the distributed vehicular environment, the IoV-enabled captured data is received via a wireless sensor network (which is placed between the first and second phases); the captured data points are processed through the proposed B-UV2X standardized lifecycle, as shown in Figure 3. Initially, data points are collected and data are preprocessed and filtered for different types of noises, such as duplication, shallow data, etc. After that, the system examines and analyzes the data (schedules to execute); if the data have necessary details that require further investigation, then the system preserves the data, transmits them towards execution, and presents a report (category of information of further investigation).

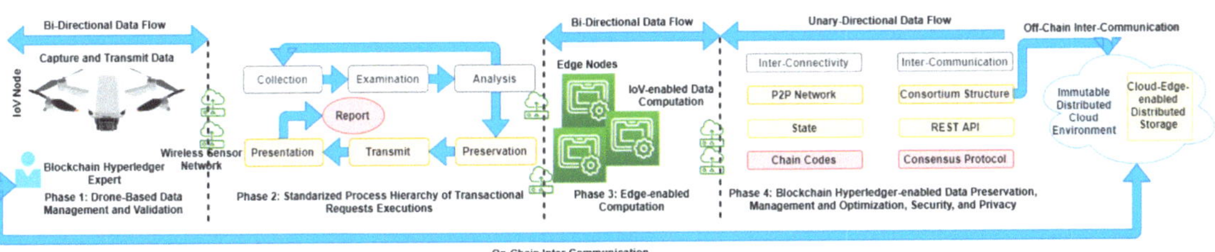

Figure 3. Proposed B-UV2X.

In the third phase, the computational node is placed between the lifecycle and blockchain hyperledger-enabled distributed ledger technology. There, it receives scheduled transactions/data points for execution. Therefore, the performance of B-UV2X is robust while reducing usage of computational energy. This is because edge nodes consume fewer computing resources compared to the fog, cloud, and other customized computational units. The data movement hierarchy of this phase is bidirectional. The fourth phase includes security and privacy operations of the proposed B-UV2X, as shown in Figures 3 and 4. A blockchain hyperledger-enabled consortium modular infrastructure is proposed; the main purpose of this design is to protect data from malicious attackers, provide data integrity, transparency, provenance. organization, and management, and to prevent forgery of and tampering with data, thereby maintaining privacy and security.

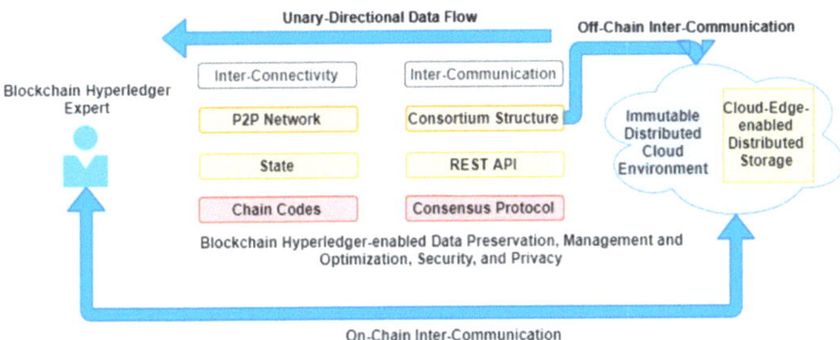

Figure 4. Working operation of the proposed B-UV2X security and privacy hierarchy.

In order to maintain privacy and security, we design two channels of inter-communication between inter-connected nodes of the IoV over the distributed UAV-assisted vehicular network, as shown in Figure 4. In this process, a transaction processor is placed, the main objective of which is to handle requests for transactions in the deployed B-UV2X consortium Peer-to-Peer network (P2P). For instance, to reset the stack of transactional requests and schedule transactions and exchanges, we use the REST API and state facilities of hyperledger technology. Multi-proof-of-stack (MPoS), along with chain code (with different functions ()), is designed, created, and deployed to automate verification, validation, transaction execution, and preservation. For storage of IoV-based logs (transactional requests), cloud-edge-enabled distributed immutable storage is utilized, which is considered one of the most customizable and cost-efficient distributed information preservation methods in the domain of blockchain-enabled ledger technology.

3.3. Smart Contracts Implementation

In this section, we discuss the procedure to automate transactional requests of IoVs and process each request through the designed lifecycle of a UAV-assisted vehicular network, responding to these transactions via DApp and records (in the cloud–edge-enabled immutable storage, as mentioned in Table 2 (InfoPre())). For execution (request verification and validation) automation, we designed, created, and deployed chain code with five different functions, multi-consensus protocols, and digital signature, as shown in Table 2.

The main objectives of these codes/functions are to provide operations automation in terms of IoV registration (IoVReg()), monitoring stakeholder registration in accordance with the designed lifecycle (UAVAVLC()), schedule the number of transactions, perform related requests executions (AddNTD()), conduct record management (UpdTr()), and preserve (InfoPre()) and exchange information.

Table 2. Chain codes, consensus, and digital signature implementation.

Input Variables: The engineer of the blockchain hyperledger is the person to initiate chain/transactional requests.
Manages events of node (IoV) transactions executions and preservation.
Stakeholder registration (verification and validation).
Exchange information between the participating nodes.
Updates logs/records and sharing.
Assumptions and Declaration:
int main().File[x].X:
IoV node/device registration,
IoVReg();
Stakeholder registration (smart cities),
StkReg();
UAV-assisted vehicular lifecycle,
UAVAVLC();
Add new transaction/request details,
AddNTD();
Resource management and monitoring,
ResMM();
Consortium channels,
Ccha();
Update transactions,
UpdTr();
Exchange information,
ExInfo();
Data/information preservation,
InfoPre();
Blockchain fabric timestamp [run];
Blockchain hyperledger expert schedule list of requests and executions,
Counter + 1;
Count(request/executed);
Executions:
if   IoV is not in IoVReg(),
then, AddNTD() and exchange;
if transactions initiated/requested passes through UAVAALC(),
then, AddNTD(), ResMM(), Ccha(), and ExInfo();
Multi-Proof-of-Stack (MPoS()),
Digital signature (after receiving 51% consensus votes),
Consensus();
Counter + 1, updTr(), and InfoPre();
else
check error, change state, share, exchange, and preserve,
terminate;
else
check error, change state, share, exchange, and preserve,
terminate;
Outputs: IoVReg(); UAVAVLC(); AddNTD(); UpdTr(); and InfoPre();

4. Simulations, Results, and Discussion

This section discusses the simulation of the proposed B-UV2X. The results are based on a blockchain hyperledger-enabled consortium modular infrastructure designed for connected nodes of IoVs for a UAV-assisted distributed vehicular network. The proposed B-UV2X was tested on a Core i7 VPro CPU (2.8 base clock speed—3.4 Turbo Boost) with

16 GB RAM, 8 GB shared Iris Xe Graphics, and 1 TB storage/internal SSD. The network connectivity between IoVs and the proposed architecture was 24 MB/s with dedicated channels of distribution. With some assumptions (discussed as follows), the docker of the blockchain hyperledger included:

- Heterogeneous node connectivity;
- 4MB size of transactional nodes;
- Single network bandwidth used
- Cloud-edge enabled customized distributed storage deployed;
- Blockchain hyperledger expert initiates a chain of the transactional requests of IoVs, as shown in Figure 5 (the test code of smart contract/chain codes with MPoS consensus is presented, along with the parameters of simulations executions).

```go
package abac

import (
    "encoding/base64"
    "encoding/json"
    "fmt"

    "github.com/hyperledger/fabric-contract-api-go/contractapi"
)

// SmartContract provides functions for managing an Asset
type SmartContract struct {
    contractapi.Contract
}

// Asset describes basic details of what makes up a simple asset
type Asset struct {
    ID             string `json:"ID"`
    Color          string `json:"color"`
    Size           int    `json:"size"`
    Owner          string `json:"owner"`
    AppraisedValue int    `json:"appraisedValue"`
}
```

Figure 5. Chain code with MPoS. Test code for simulation of the proposed B-UV2X.

In Figure 6(1–3), the simulation results of the proposed B-UV2X show that it decreases the cost of computing resources by 7.93%, which allows edge-enabled computation (as shown in Figure 3). After evaluation of the computation of B-UAV2X, it is considered to be a good candidates for real-time industrial implementation and scheduling. The operation of this simulation shows that the process initiates when the request management of the proposed lifecycle executes (as shown in Figure 6(3)), the matrices of which are the number of data requests received and the data examined for further executions.

However, the IoV-enabled self-data-capturing capability was enhanced after the tuning of lifecycle hierarchy, as shown in Figure 4(1–3) and Figure 6(1–3), which is most importantly used for the sake of heuristic reconciliation in smart cities. This capability is also needed because the process hierarchy sends a request to preserve captured records for heuristic investigation. To do this, we manipulate the design of the lifecycle to examine and analyze the collected details in terms of binary color transformation, entropy, extracted critical features, and preserve optimized records, as shown in Figures 7 and 8 (also discussed in the proposed architecture section and highlighted in Figure 3).

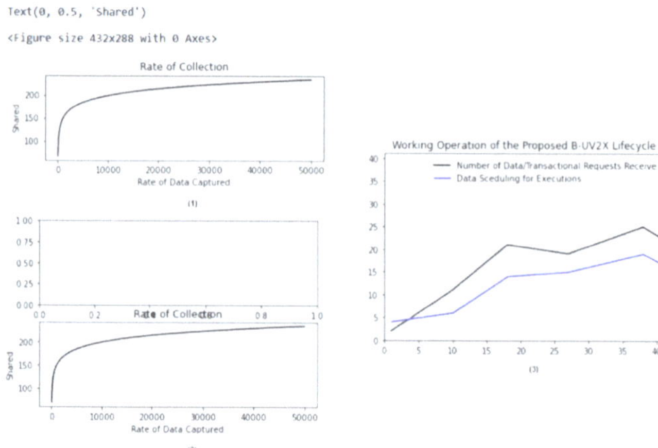

Figure 6. The working operation of the proposed B-UV2X lifecycle: (**1**) shows the rate of data captured by the IoVs, (**2**) shows the rate of data collected, and (**3**) shows that the fluctuation between the data received and the data scheduled for execution.

Figure 7. IoV-enabled data capture: (**1**) shows the original image, (**2**) shows the transformation in the binary color format, (**3**) shows the extracted features for heuristic purposes, and (**4**) shows the optimized record preservation.

Figure 8. IoV-enabled data capture: (**1**) shows the original image, (**2**) shows the transformation in the binary color format, (**3**) shows the extracted features for heuristic purposes, and (**4**) shows the optimized record preservation.

In the entire process of B-UV2X simulations, the blockchain hyperledger expert and registered participating stakeholders can observe the resources utilized in the complete process execution. The consumption of network bandwidth is reduced by 12.17% throughout the execution of each transaction, and security capability is increased by up to 9.76% by protecting individual ledger transactions of UAV-assisted vehicles using the NuCypher threshold proxy re-encryption mechanism (as shown in Figure 9).

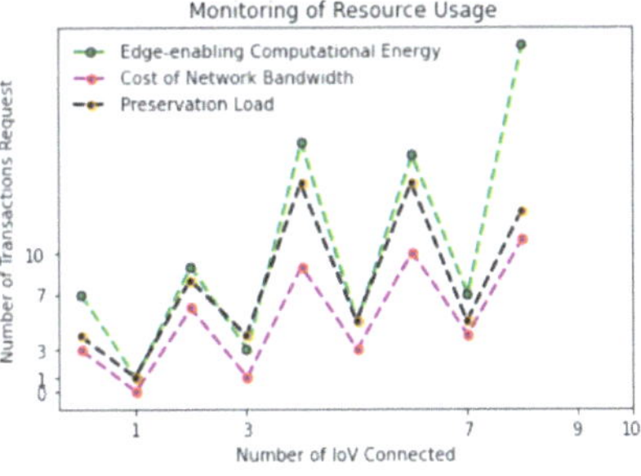

Figure 9. Monitoring of resource usage: shows the consumption fluctuation in number of IoVs connected and number of requested transactions.

Observe that the deployment of the chaincodes and related functions (such as IoVReg(), UAVAVLC(), AddNTD(), UpdTr(), and InfoPre()) with MPoS customized consensus policy and edge computing technology decreases the cost of IoV-enabled data scheduling, organization, management, optimization, and preservation. Figure 10 illustrates the fluctuations between the predefined hyperledger consensus and the proposed B-UV2X over the UAV-assisted consortium distributed vehicular network in terms of the number of transactional requests and the number of connected IoVs' for data verification and validation.

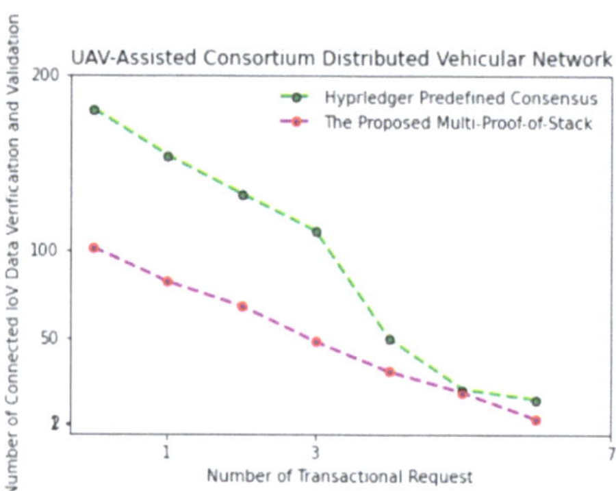

Figure 10. UAV-assisted consortium distributed vehicular network shows the fluctuation between the predefined hyperledger consensus and the proposed B-UV2X.

The evaluation matrices of the proposed B-UV2X are compared with newly published methods (previous state-of-the-art methods) such as "edge intelligence for IoV" [31] and "blockchain-based conditional privacy preservation" [32]. The metrics for this analytical procedure are based on the usage of resources, preservation, protection efficiency (proxy threshold re-encryption level), reliability, privacy, and security. In Table 3, a few more analytical comparisons are discussed, indicating the superiority of the proposed work compared to other state-of-the-art methods. A comparative parameter of the evaluation is presented (as mentioned in Table 3, attribute 3), which helps to measure the fluctuation/improvement of the proposed B-UAV2X compared to other methods.

Table 3. Comparison table: state-of-the-art methods.

Methodology of Other State-of-the-Art Methods	Main Contributions	Analytical Matrices of Other State-of-the-Art Methods	Proposed B-UV2X
A resource trading, computational offloading, and management approach for enhanced drone-to-drone assisted environment using blockchain distributed ledger [33]	• Decentralized resource sharing system • One ledger multi follower strategy • KKT-based algorithm	• Blockchain: yes • Hyperledger: no • Network type: permissionless • Encryption mechanism: hash-encryption • Block size: variable • Intercommunication channels: two • Consensus: predefined • Digital signature: predefined • Efficiency: not applicable • Accuracy: not applicable	The analytical matrices of the proposed B-UV2X are as follows: • Blockchain: yes; • Hyperledger: yes; • Network type: consortium; • Encryption mechanism: NuCypher proxy threshold re-encryption; • Block size: 4 MB–6 MB; • Intercommunication channels: on and off-chain; • Consensus: multi-proof-of-stack (MPoS); • Digital signature: customized (51% vote based); • Efficiency: 12.17%, 7.93%; • Accuracy: not applicable.
Edge-enabled mobile server deployment scheme for IoVs with blockchain [34]	• Edge server deployment • Roadside node management • Distributed application uses for resource monitoring	• Blockchain: yes • Hyperledger: no • Network type: permissionless • Encryption mechanism: defined hash encryption • Block size: variable • Intercommunication channels: not defined • Consensus: predefined • Digital signature: [redefined • Efficiency: not applicable • Accuracy: not applicable	

Table 3. *Cont.*

Methodology of Other State-of-the-Art Methods	Main Contributions	Analytical Matrices of Other State-of-the-Art Methods	Proposed B-UV2X
A multi-access edge computing for vehicular network using a deep neural approach [35]	• Multiple multi-access edge is designed • VANET ecosystem deployment • Distributed permissionless structure is proposed for IoV interconnectivity	• Blockchain: yes • Hyperledger: no • Network type: permissioned • Encryption mechanism: hash encryption SHA-256 • Block size: variable • Intercommunication channels: not applicable • Consensus: predefined • Digital signature: predefined • Efficiency: not defined • Accuracy: not defined	
A resource efficient framework for IoVs using blockchain, AI, and edge computing [36]	• Proof-of-lottery consensus mechanism is proposed • ETCZ: the edge terminal consensus zone • Resource efficient distributed modular framework is presented	• Blockchain: yes • Hyperledger: no • Network type: permissioned • Encryption mechanism: hash SHA-256 • Block size: variable • Intercommunication channels: two • Consensus: predefined • Digital signature: predefined • Efficiency: not applicable • Accuracy: not applicable	

5. Current Status of Edge Computing and Related Implementation Issues

The use of edge-enabled computing integrated with UAV-assisted vehicular network technology to design a secure intercommunication channel for IoV interconnectivity and related advantages are discussed. With blockchain distributed technologies, edge networks provide a cost-efficient manner to share and exchange information in a distributed environment. However, there are various implementation challenges that impact resource usage, such as computing power, storage, and network bandwidth.

5.1. Edge Computing Integrated with Outsourced Computation

The edge computing-enabled Internet of Things (IoT) plays a significant role in outsourcing computation and related management, such as providing participating node proximity [37]. In the UAV-assisted vehicular network environment, nodes receive a reward for computational task executions. For instance, the technology uses with a blockchain hyperledger to verify the integrity of arbitrary deterministic functions and restricts illegal authentications of false negative contractors that try to maximize their activities in the distributed environment [38,39]. The verification mechanisms of blockchain-enabled technology with IoV creates a challenging problem when the pre-trained models are designed for validation purpose. However, all the nodes consume less than 1 milli-second (ms) computational overhead with minimum network bandwidth (almost 80 bytes/frame). This may lead to another limitation regarding resource management and parallel usage.

5.2. Vehicle to Everything-Enabled Distributed Node Interconnectivity

In the V2X environment, one of the biggest issues to design and develop an efficient and secure distributed node architecture. For instance, when applying a blockchain distributed consortium mechanism over a Peer-to-Peer (P2P) network, there are different node scaling challenges that arise, while the lack of cost-efficiency requirements is also considerable [37,38]. However, with the introduction of a hyperledger-enabled modular framework, we can meet various integrity, transparency, provenance, and trustworthiness requirements [39]. By constantly stimulating the ledger, every request for IoV transactions is incorporated, with the details of information acquisition towards deliverance and exchange. However, the participating stakeholders can see the movement of individual IoVs through the dynamic monitoring capability/traceability using the blockchain hyperledger, regardless of the particular stakeholder that initiates activities.

5.3. Role of Blockchain Hyperledger Technology in Edge Computing Environment

Edge-enabled technology brings computational resources close to end devices (IoT-enabled devices), allowing edge computing, preservation, operation and control, and analysis of related data [38,39]. Blockchain distributed ledger technology has the potential to provide a platform to solve privacy-, protection-, and security-related problems associated with edge computing, including access control, authentication, verification, and validation. In a blockchain-enabled edge network, the system provides UAV-assisted vehicular intercommunication channel facilities, from which on-chain and off-chain channels are derived. These interconnected node channels are designed to handle the list of implicitly and explicitly transactional requests more efficiently and reliably.

5.4. Drone-Based Data Management and Monitoring

In the domain of data management and monitoring, there are major limitations to providing data integrity and transparency, most importantly in the distributed ledger environment [37,38]. At present, most hyperledger technology cannot provide a customized data integrity policy and consensus management, only allowing moderate predefined validator processors for distributed verification of consensus, such as PoET, PoW, PoS, etc. [39]. However, a robust structure of privacy protection has been proposed by the Linux community to allow construction of an infrastructure to preserve information and chain-of-records with data traceability. This modular improvement of the hyperledger effectively tracks information management at every step of the transactional request schedule. In addition, it enhances the dynamic monitoring facilities by providing a better transaction/drone registration (IoV registry) hierarchy, with more efficient control compared to previously state-of-the-art methods [38,39].

6. Conclusions

This paper addresses current problems involving centralized UAV-assisted vehicular networks such as scheduling, offloading, management, optimization, privacy, and security. The key objective of this paper is to address gaps in the design, development, and deployment of distributed vehicular networks for the IoV in smart cities using the blockchain hyperledger. The existing protocol/process hierarchy of IoV-enabled request execution via distributed applications (DApp) is also highlighted. This paper proposes B-UV2X, a secure and novel lifecycle of UAV-assisted vehicular data processing for the IoV, using blockchain consortium architecture. It includes a customized consensus mechanism for multi-proof-of-stack (MPoS), where data offloading can be managed, directly impacting the management of resources as well. Transactional executions of the proposed B-UV2X are fully protected by the NuCypher threshold proxy re-encryption algorithm. The individual ledger/records of the node's transactions are preserved in immutable storage, such as edge network-enabled cloud storage. The participating stakeholders of the proposed B-UV2X receive details of ledger traceability for the sake of dynamic monitoring of resource management and related IoV node activities in smart city environments. The simulation results of B-UV2X show that it reduces network consumption by 17%, reduces the computing load with preservation by 7.93%, and increases security by 9.76% compared to other state-of-the-art methods.

Author Contributions: A.A.K. wrote the original draft and was responsible for organization, preparation, and analysis. A.A.K., A.A.L., S.A.A., M.S. and Z.G. reviewed and rewrote the draft, performed part of the literature survey, investigated and designed the architecture/framework, were responsible for lifecycle design, code, and presentation, and explored software tools. All authors have read and agreed to the published version of the manuscript.

Funding: This work was supported by the National Natural Science Foundation of China (grant nos. 62250410365, 61902082), and the Guangzhou Science and Technology Planning Project (no. 202102010507).

Institutional Review Board Statement: Not applicable.

Informed Consent Statement: Not applicable.

Data Availability Statement: Not applicable.

Conflicts of Interest: The authors of this paper declare that there are no conflict of interest.

References

1. Lv, Z.; Qiao, L.; Hossain, M.S.; Choi, B.J. Analysis of Using Blockchain to Protect the Privacy of Drone Big Data. *IEEE Netw.* **2021**, *35*, 44–49. [CrossRef]
2. Gumaei, A.; Al-Rakhami, M.; Hassan, M.M.; Pace, P.; Alai, G.; Lin, K.; Fortino, G. Deep Learning and Blockchain with Edge Computing for 5G-Enabled Drone Identification and Flight Mode Detection. *IEEE Netw.* **2021**, *35*, 94–100. [CrossRef]
3. Cheema, M.A.; Ansari, R.I.; Ashraf, N.; Hassan, S.A.; Qureshi, H.K.; Bashir, A.K.; Politis, C. Blockchain-based secure delivery of medical supplies using drones. *Comput. Netw.* **2022**, *204*, 108706. [CrossRef]
4. Hasan, M.K.; Islam, S.; Shafiq, M.; Ahmed, F.R.A.; Ataelmanan, S.K.M.; Babiker, N.B.M.; Abu Bakar, K.A. Communication Delay Modeling for Wide Area Measurement System in Smart Grid Internet of Things Networks. *Wirel. Commun. Mob. Comput.* **2021**, *2021*, 9958003. [CrossRef]
5. Khan, A.A.; Laghari, A.A.; Awan, S.; Jumani, A.K. Fourth Industrial Revolution Application: Network Forensics Cloud Security Issues. In *Security Issues and Privacy Concerns in Industry 4.0 Applications*; John Wiley & Sons, Inc.: Hoboken, NJ, USA, 2021; pp. 15–33. [CrossRef]
6. Feng, C.; Yu, K.; Bashir, A.K.; Al-Otaibi, Y.D.; Lu, Y.; Chen, S.; Zhang, D. Efficient and Secure Data Sharing for 5G Flying Drones: A Blockchain-Enabled Approach. *IEEE Netw.* **2021**, *35*, 130–137. [CrossRef]
7. Li, T.; Liu, W.; Liu, A.; Dong, M.; Ota, K.; Xiong, N.N.; Li, Q. BTS: A Blockchain-Based Trust System to Deter Malicious Data Reporting in Intelligent Internet of Things. *IEEE Internet Things J.* **2021**, *9*, 22327–22342. [CrossRef]
8. Khan, A.A.; Laghari, A.A.; Shaikh, A.A.; Dootio, M.A.; Estrela, V.V.; Lopes, R.T. A blockchain security module for brain-computer interface (BCI) with Multimedia Life Cycle Framework (MLCF). *Neurosci. Inform.* **2021**, *2*, 100030. [CrossRef]
9. Alsamhi, S.H.; Almalki, F.A.; Afghah, F.; Hawbani, A.; Shvetsov, A.V.; Lee, B.; Song, H. Drones' Edge Intelligence Over Smart Environments in B5G: Blockchain and Federated Learning Synergy. *IEEE Trans. Green Commun. Netw.* **2021**, *6*, 295–312. [CrossRef]
10. Aloqaily, M.; Bouachir, O.; Boukerche, A.; Al Ridhawi, I. Design Guidelines for Blockchain-Assisted 5G-UAV Networks. *IEEE Netw.* **2021**, *35*, 64–71. [CrossRef]
11. Shafiq, M.; Tian, Z.; Bashir, A.K.; Cengiz, K.; Tahir, A. SoftSystem: Smart Edge Computing Device Selection Method for IoT Based on Soft Set Technique. *Wirel. Commun. Mob. Comput.* **2020**, *2020*, 8864301. [CrossRef]
12. Khan, A.A.; Wagan, A.A.; Laghari, A.A.; Gilal, A.R.; Aziz, I.A.; Talpur, B.A. BIoMT: A State-of-the-Art Consortium Serverless Network Architecture for Healthcare System Using Blockchain Smart Contracts. *IEEE Access* **2022**, *10*, 78887–78898. [CrossRef]
13. Luo, S.; Li, H.; Wen, Z.; Qian, B.; Morgan, G.; Longo, A.; Rana, O.; Ranjan, R. Blockchain-Based Task Offloading in Drone-Aided Mobile Edge Computing. *IEEE Netw.* **2021**, *35*, 124–129. [CrossRef]
14. Khan, A.A.; Shaikh, A.A.; Shaikh, Z.A.; Laghari, A.A.; Karim, S. IPM-Model: AI and metaheuristic-enabled face recognition using image partial matching for multimedia forensics investigation with genetic algorithm. *Multimedia Tools Appl.* **2022**, *81*, 23533–23549. [CrossRef]
15. Abualigah, L.; Diabat, A.; Sumari, P.; Gandomi, A.H. Applications, Deployments, and Integration of Internet of Drones (IoD): A Review. *IEEE Sens. J.* **2021**, *21*, 25532–25546. [CrossRef]
16. Khan, A.A.; Shaikh, Z.A.; Belinskaja, L.; Baitenova, L.; Vlasova, Y.; Gerzelieva, Z.; Laghari, A.A.; Abro, A.A.; Barykin, S. A Blockchain and Metaheuristic-Enabled Distributed Architecture for Smart Agricultural Analysis and Ledger Preservation Solution: A Collaborative Approach. *Appl. Sci.* **2022**, *12*, 1487. [CrossRef]
17. Shaikh, Z.A.; Khan, A.A.; Teng, L.; Wagan, A.A.; Laghari, A.A. BIoMT Modular Infrastructure: The Recent Challenges, Issues, and Limitations in Blockchain Hyperledger-Enabled E-Healthcare Application. *Wirel. Commun. Mob. Comput.* **2022**, *2022*, 813841. [CrossRef]
18. Wang, D.; Wu, M.; He, Y.; Pang, L.; Xu, Q.; Zhang, R. An HAP and UAVs Collaboration Framework for Uplink Secure Rate Maximization in NOMA-Enabled IoT Networks. *Remote Sens.* **2022**, *14*, 4501. [CrossRef]
19. Wang, D.; Zhou, F.; Lin, W.; Ding, Z.; Al-Dhahir, N. Cooperative Hybrid Non-Orthogonal Multiple Access Based Mobile-Edge Computing in Cognitive Radio Networks. *IEEE Trans. Cogn. Commun. Netw.* **2022**, *8*, 1104–1117. [CrossRef]
20. Singh, M.P.; Aujla, G.S.; Bali, R.S. Blockchain for the Internet of Drones: Applications, Challenges, and Future Directions. *IEEE Internet Things Mag.* **2021**, *4*, 47–53. [CrossRef]
21. Khan, A.A.; Khan, M.M.; Khan, K.M.; Arshad, J.; Ahmad, F. A blockchain-based decentralized machine learning framework for collaborative intrusion detection within UAVs. *Comput. Netw.* **2021**, *196*, 108217. [CrossRef]

22. Gupta, R.; Bhattacharya, P.; Tanwar, S.; Kumar, N.; Zeadally, S. GaRuDa: A Blockchain-Based Delivery Scheme Using Drones for Healthcare 5.0 Applications. *IEEE Internet Things Mag.* **2021**, *4*, 60–66. [CrossRef]
23. Aujla, G.S.; Vashisht, S.; Garg, S.; Kumar, N.; Kaddoum, G. Leveraging Blockchain for Secure Drone-to-Everything Communications. *IEEE Commun. Stand. Mag.* **2021**, *5*, 80–87. [CrossRef]
24. Yahuza, M.; Idris, M.Y.I.; Bin Ahmedy, I.; Wahab, A.W.A.; Nandy, T.; Noor, N.M.; Bala, A. Internet of Drones Security and Privacy Issues: Taxonomy and Open Challenges. *IEEE Access* **2021**, *9*, 57243–57270. [CrossRef]
25. Wang, J.; Liu, Y.; Niu, S.; Song, H. Lightweight blockchain assisted secure routing of swarm UAS networking. *Comput. Commun.* **2021**, *165*, 131–140. [CrossRef]
26. Mei, Q.; Xiong, H.; Zhao, Y.; Yeh, K.-H. Toward Blockchain-Enabled IoV with Edge Computing: Efficient and Privacy-Preserving Vehicular Communication and Dynamic Updating. In Proceedings of the 2021 IEEE Conference on Dependable and Secure Computing (DSC), Aizuwakamatsu, Japan, 30 January–2 February 2021; pp. 1–8. [CrossRef]
27. Wang, D.; He, T.; Zhou, F.; Cheng, J.; Zhang, R.; Wu, Q. Outage-driven link selection for secure buffer-aided networks. *Sci. China Inf. Sci.* **2022**, *65*, 182303. [CrossRef]
28. Al-Hourani, A.; Kandeepan, S.; Lardner, S. Optimal LAP Altitude for Maximum Coverage. *IEEE Wirel. Commun. Lett.* **2014**, *3*, 569–572. [CrossRef]
29. He, Y.; Wang, D.; Huang, F.; Zhang, R.; Pan, J. Trajectory Optimization and Channel Allocation for Delay Sensitive Secure Transmission in UAV-Relayed VANETs. *IEEE Trans. Veh. Technol.* **2022**, *71*, 4512–4517. [CrossRef]
30. Islam, S.; Badsha, S.; Sengupta, S.; La, H.; Khalil, I.; Atiquzzaman, M. Blockchain-Enabled Intelligent Vehicular Edge Computing. *IEEE Netw.* **2021**, *35*, 125–131. [CrossRef]
31. Jiang, X.; Yu, F.R.; Song, T.; Leung, V.C. Edge Intelligence for Object Detection in Blockchain-Based Internet of Vehicles: Convergence of Symbolic and Connectionist AI. *IEEE Wirel. Commun.* **2021**, *28*, 49–55. [CrossRef]
32. Yang, J.; Liu, J.; Song, H.; Liu, J.; Lei, X. Blockchain-based Conditional Privacy-Preserving Authentication Protocol with Implicit Certificates for Vehicular Edge Computing. In Proceedings of the 2022 7th International Conference on Cloud Computing and Big Data Analytics (ICCCBDA) 2022, Chengdu, China, 22–24 April 2022; pp. 210–216. [CrossRef]
33. Jing, W.; Fu, X.; Liu, P.; Song, H. Joint resource trading and computation offloading in blockchain enhanced D2D-assisted mobile edge computing. *Clust. Comput.* **2022**, 1–15. [CrossRef]
34. Xu, L.; Ge, M.; Wu, W. Edge Server Deployment Scheme of Blockchain in IoVs. *IEEE Trans. Reliab.* **2022**, *71*, 500–509. [CrossRef]
35. Zhang, D.; Yu, F.R.; Yang, R. Blockchain-Based Multi-Access Edge Computing for Future Vehicular Networks: A Deep Compressed Neural Network Approach. *IEEE Trans. Intell. Transp. Syst.* **2021**, *23*, 12161–12175. [CrossRef]
36. Wang, K.; Tu, Z.; Ji, Z.; He, S. Faster Service with Less Resource: A Resource Efficient Blockchain Framework for Edge Computing. 2022. Available online: https://assets.researchsquare.com/files/rs-1719287/v1_covered.pdf?c=1654894162 (accessed on 31 October 2022).
37. Shaikh, Z.A.; Khan, A.A.; Baitenova, L.; Zambinova, G.; Yegina, N.; Ivolgina, N.; Laghari, A.A.; Barykin, S.E. Blockchain Hyperledger with Non-Linear Machine Learning: A Novel and Secure Educational Accreditation Registration and Distributed Ledger Preservation Architecture. *Appl. Sci.* **2022**, *12*, 2534. [CrossRef]
38. Ahmed, M.M.; Hasan, M.K.; Shafiq, M.; Qays, O.; Gadekallu, T.R.; Nebhen, J.; Islam, S. A peer-to-peer blockchain based interconnected power system. *Energy Rep.* **2021**, *7*, 7890–7905. [CrossRef]
39. Khan, A.A.; Laghari, A.A.; Gadekallu, T.R.; Shaikh, Z.A.; Javed, A.R.; Rashid, M.; Estrela, V.V.; Mikhaylov, A. A drone-based data management and optimization using metaheuristic algorithms and blockchain smart contracts in a secure fog environment. *Comput. Electr. Eng.* **2022**, *102*, 108234. [CrossRef]

Article

Enhanced Artificial Gorilla Troops Optimizer Based Clustering Protocol for UAV-Assisted Intelligent Vehicular Network

Hadeel Alsolai [1], Jaber S. Alzahrani [2], Mohammed Maray [3], Mohammed Alghamdi [3,4], Ayman Qahmash [3], Mrim M. Alnfiai [5], Amira Sayed A. Aziz [6] and Anwer Mustafa Hilal [7,*]

[1] Department of Information Systems, College of Computer and Information Sciences, Princess Nourah bint Abdulrahman University, Riyadh 11671, Saudi Arabia
[2] Department of Industrial Engineering, College of Engineering at Alqunfudah, Umm Al-Qura University, Makkah 24382, Saudi Arabia
[3] Department of Information Systems, College of Computer Science, King Khalid University, Abha 62529, Saudi Arabia
[4] Department of Information and Technology Systems, College of Computer Science and Engineering, University of Jeddah, Jeddah 23218, Saudi Arabia
[5] Department of Information Technology, College of Computers and Information Technology, Taif University, Taif 21944, Saudi Arabia
[6] Department of Digital Media, Faculty of Computers and Information Technology, Future University in Egypt, New Cairo 11835, Egypt
[7] Department of Computer and Self Development, Preparatory Year Deanship, Prince Sattam bin Abdulaziz University, Al-Kharj 16278, Saudi Arabia
*   Correspondence: a.hilal@psau.edu.sa

Citation: Alsolai, H.; Alzahrani, J.S.; Maray, M.; Alghamdi, M.; Qahmash, A.; Alnfiai, M.M.; Aziz, A.S.A.; Mustafa Hilal, A. Enhanced Artificial Gorilla Troops Optimizer Based Clustering Protocol for UAV-Assisted Intelligent Vehicular Network. *Drones* **2022**, *6*, 358. https://doi.org/10.3390/drones6110358

Academic Editors: Dawei Wang and Ruonan Zhang

Received: 12 October 2022
Accepted: 14 November 2022
Published: 16 November 2022

Publisher's Note: MDPI stays neutral with regard to jurisdictional claims in published maps and institutional affiliations.

Abstract: The increasing demands of several emergent services brought new communication problems to vehicular networks (VNs). It is predicted that the transmission system assimilated with unmanned aerial vehicles (UAVs) fulfills the requirement of next-generation vehicular network. Because of its higher flexible mobility, the UAV-aided vehicular network brings transformative and far-reaching benefits with extremely high data rates; considerably improved security and reliability; massive and hyper-fast wireless access; much greener, smarter, and longer 3D communications coverage. The clustering technique in UAV-aided VN is a difficult process because of the limited energy of UAVs, higher mobility, unstable links, and dynamic topology. Therefore, this study introduced an Enhanced Artificial Gorilla Troops Optimizer–based Clustering Protocol for a UAV-Assisted Intelligent Vehicular Network (EAGTOC-UIVN). The goal of the EAGTOC-UIVN technique lies in the clustering of the nodes in UAV-based VN to achieve maximum lifetime and energy efficiency. In the presented EAGTOC-UIVN technique, the EAGTO algorithm was primarily designed by the use of the circle chaotic mapping technique. Moreover, the EAGTOC-UIVN technique computes a fitness function with the inclusion of multiple parameters. To depict the improved performance of the EAGTOC-UIVN technique, a widespread simulation analysis was performed. The comparison study demonstrated the enhancements of the EAGTOC-UIVN technique over other recent approaches.

Keywords: vehicular networks; unmanned aerial vehicles; clustering; gorilla troops optimizer; fitness function

1. Introduction

Transport commuting becomes a ubiquitous part of day to day lives; the vehicular network (VN) plays a positive role and increases the quality of life [1]. VN technology is advantageous in information applications, automatic toll collection, public safety, automatic driving, traffic coordination, event-driven safety message broadcasting, and so on. VN accesses the location where a traffic accident occurs, timely notifies the pertinent vehicle to adopt security measurement, and then offers quality of service of multimedia information for the tourist during the journey. Regarding the data services, VN is applied for the remainder of services with respect to infectious diseases, carbon emissions, pollution levels of the haze, and other related services that could enhance the living environment of humans [2]. Moreover, traffic-flow coordination could be benefitted by VN. People timely obtain present traffic data on the road of the vehicle and select the optimum path. These data are particularly effective for the best travel experience for the passenger and for preventing congested sections. Intelligent transport systems benefited considerably from VN [3]. It provides data with respect to restaurants, petrol stations, weather information, navigation, service areas, and every desirable datum regarding the neighboring environments. VN is employed to automated driving, namely the distance detection among velocity estimation, vehicles, road-condition perception, self-parking, and location service [4]. VN is employed to realize automated charging of the vehicle. In V2I transmission, RSU could automatically sense the journey mileage of a vehicle, the entrance location, and exit location, later realizing the automated charging that could decrease the congestion at the charging place and enhance the charging efficiency [5]. Figure 1 depicts the framework of UAV-assisted vehicular network.

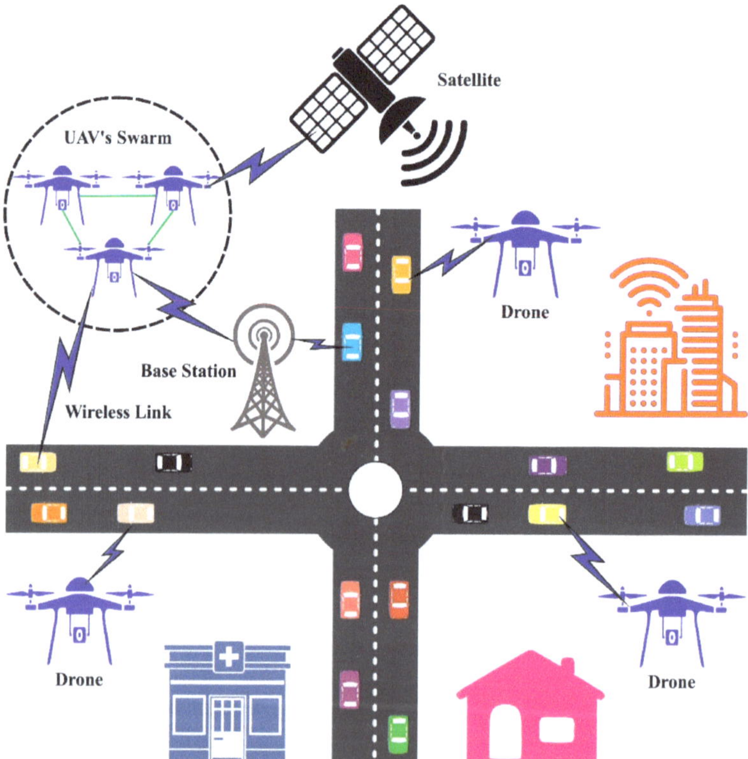

Figure 1. Structure of UAV-assisted vehicular network.

A data-distribution technique is broadly employed in different circumstances, namely emergency collision avoidance [6], the data acquirement of public entertainment, and traffic-flow management that could decrease the number of traffic accidents, which promote the urban building of a smart city and discharge urban traffic congestion [7]. In certain scenarios, the timeliness of data dissemination is crucial. To accomplish the objective, unmanned aerial vehicle (UAV) is applied to help with data distribution. Due to better maneuverability of the UAV, it is widely used for completing the data dissemination task in certain scenarios [8]. For instance, in certain locations, once the transmission framework is damaged, the UAV could be deployed rapidly as the mobile base station to help the transmission network. Network lifetime is a significant parameter in UAV network that is based on whether a specific number of nodes die due to energy consumption [9]. In mobile UAV networks, topology control is the major aspect for extending network lifetime and reducing communication interference. A hierarchical network based on clustering model is widely employed in mobile networks. The cluster head (CH) selection and clustering process are major factors in the hierarchical network [10–12].

This study introduces an Enhanced Artificial Gorilla Troops Optimizer based Clustering Protocol for UAV-Assisted Intelligent Vehicular Network (EAGTOC-UIVN). The goal of the EAGTOC-UIVN technique lies in the clustering of the nodes in UAV-based VN to achieve maximum lifetime and energy efficiency. In the presented EAGTOC-UIVN technique, the EAGTO algorithm is primarily designed by the use of circle chaotic mapping technique. Moreover, the EAGTOC-UIVN technique computes a fitness function with the inclusion of multiple parameters. To depict the improved performance of the EAGTOC-UIVN technique, a widespread simulation analysis was performed.

2. Literature Review

In Reference [13], an efficient routing technique depending upon a flooding method was developed for robust route identification. It assures an alternate path during path-failure scenarios. In addition, a forecasting approach is employed for anticipating the expiration time of every discovered route. The authors in Reference [14] considered the issue of content distribution to the vehicles on roadways with overloaded or no available communication structure. Incoming vehicles demand service from a library of content which is partially cached at the UAV; the content of the library is also considered for modifying new vehicles carrying more popular content. A non-orthogonal multiple access (NOMA)-enabled double-layer airborne access vehicular ad hoc networks (DLAA-VANETs) architecture was designed in Reference [15], which consists of a high-altitude platform (HAP), multiple unmanned aerial vehicles (UAVs), and vehicles. For the designed DLAA-VANETs, the UAV deployment and network optimization problem is addressed. Particularly, the UAV deployment method, depending upon particle swarm optimization, is presented. Next, the NOMA model is introduced into the designed model for improving the transmission rate.

Khabbaz et al. [16] aimed at enhancing the ground vehicle connectivity in the framework of an alternating vehicle-to-UAV (V2U) transmission condition, whereas vehicles create time-limited connectivity with transient by UAV served as flying BSs responsible to route arriving vehicle information on backbone network or Internet. Zheng et al. [17] used cyclic-flight UAVs for assisting RSU by offering video download services to vehicles. With the utilization of UAV, seamless communication coverage and stable broadcast connections ensure the optimum quality of services to vehicle. Moreover, the authors present a model-free technique dependent upon DQN for determining an optimum UAV decision procedure for achieving the minimization of stalling time. Raza et al. [18] examined a UAV-assisted VANET communication structure, whereas UAVs fly over the used region and offer communication service to basic coverage region. UAV-assisted VANET aims for the benefits of line-of-sight (LOS) communication, flexibility, load balancing (LB), and cost-effectual deployment.

Wu et al. [19] examined a deep supervised learning system to enable intelligence edge for making decisions on the extremely dynamic vehicular network. Specifically, the authors initially presented a clustering-based two-layered (CBTL) technique for solving the JCTO problem offline. Afterward, they planned a deep supervised learning structure of CNN for making fast decisions online. Ghazzai et al. [20] established a mobility- and energy-aware data routing protocol for UAV-supported VANETs. Most UAVs perform as a flying RSU, gathering information in ground vehicles, but another UAV role is the play of relays for providing the information to mobility service center (MSC). The UAV is modifying its 3D places in an existing range if required for ensuring reliable communication links.

A UAV-helped data dissemination system dependent upon network coding was presented in Reference [21]. Initially, the graph concept for modeling the occurrence of data loss of the vehicles was utilized; the data dissemination issue was changed to the maximal clique issue of graphs. With the coverage of directional antenna being restricted, a parallel system for determining the maximal clique dependent upon the area separation was presented. Alioua et al. [22] examine a new distributed SDN-related structure for UAV-support-structure-less vehicular networks. An important purpose is to fill the gap in which no SDN-based infrastructure was presented for these networks. The author's concentrated mostly on a road safety use-case that integrated UAVs for assisting emergency vehicles in the exploration of affected regions from crucial emergency conditions. In addition, the authors examined an effectual data processing strategy with shared decision-making or computation-offloading problems. Though several models are available in the literature, the network efficiency in UAV-assisted VNs still needs to be improved. In addition, the inclusion of multiple parameters for optimal UAV selection is important as CHs become essential.

3. The Proposed Model

In this study, a new EAGTOC-UIVN technique was developed for clustering the UAV-assisted VN. The major aim of the EAGTOC-UIVN technique exists in the grouping of the nodes in the UAV-based VN to achieve maximum lifetime and energy efficiency. In the presented EAGTOC-UIVN technique, the EAGTO algorithm is primarily designed by the use of the circle chaotic mapping technique.

3.1. System Model

In the presented model, the types of UAVs considered could be middle-size drones or mini drones. A simple collision process is utilized for collision avoidance [23]. In this work, the UAV changed altitude for possible collision. The UAV's maximum speed could reach up to 30 m/s. Every UAV device relies on a location-aware component. This location-conscious mechanism allows the routing method to function efficiently and precisely. In general, location data can be attained from an alternative scheme. In the presented method, inertial measurement units and GPS are given for the motion sensing and positioning of the UAV. Each UAV is aware of its ground station and neighbors' positions. Each UAV is equipped with long- and short-range wireless transmission. Long-range wireless transmission can be utilized for inter-cluster transmission with the ground station and other CHs. Short-range wireless transmission is utilized for intra-transmission with its peers in the cluster.

3.2. Design of EAGTO Technique

With other metaheuristics, AGTO's stability and convergence accuracy suffer as the optimization problem to be resolved grows in variety and complexity. This flaw requires the further development of novel mechanisms to perform exploitation and exploration and help accomplish improved performance [24]. A troop comprises a dominant adult male gorilla (silverback), numerous dominant adult females, and their offspring. A silverback gorilla is over 12 years old and obtains the name from the distinct hairs that grow on his back while he attains puberty. Furthermore, the silverback is the leader of the entire troop and is accountable for ensuring everyone's safety, planning and executing group travel, allocating food and other resources, and making each decision, mediating any conflicts that arise. Male gorillas between the ages of 8 and 12 are considered "black" since the silver fur is not fully grown. It is common for gorillas to leave the birth group for joining a third. However, some male gorillas decide to stick around and keep following the silverback. Such males might fight viciously for controlling the group and accessing adult females when the silverback is killed. The idea of group behaviors in wild gorillas acts as the motivation for the AGTO algorithm. Initialization, local exploitation, and global exploration are the three phases that make up AGTO, the same as they are in other intelligent techniques.

3.2.1. Initialization Phase

Consider the D-dimension space has N gorillas. To specify where i-th gorillas are in the universe, we could formulate $X_i = (x_{i,1}, x_{i,2}, \ldots x_{i,D})$, whereas $I = 1, 2, \ldots N$, and it can be defined as follows:

$$X_{N \times D} = rand(N, D) \times (ub - lb) + lb \quad (1)$$

where rand () lies between 0 and 1. The search range can be determined by using the upper and lower limits, ub and lb, respectively; and the matrix, X, has a random value, A, within [0, 1] that is allocated to all the elements of the N rows and D columns in the matrix represented as $rand\ (N, D)$.

3.2.2. Exploration Phase

$$GX(t + 1) = (ub - lb) \times r2 + lb,\ r1 < p$$

$$(r3 - C) \times XA(t) + L \times Z \times X(t),\ r1 \geq 0.5$$

$$X(t) - L \times (L \times (X(t) - XB(t)) + r4 \times (X(t) - XB(t)))r1 < 0.5 \quad (2)$$

In the above equations, t signifies iteration times, $X(t)$ indicates the gorilla's existing location vector, and $GX(t + 1)$ denotes the potential search agent position for the following iteration. Furthermore, the random numbers $r1$, $r2$, $r3$, and $r4$ denote a number value between zero and one. Two locations among the existing population of gorillas, $XA(t)$ and $B(t)$, are selected randomly; p is a predetermined value. By utilizing the problem dimension as an index, Z denotes the row vector where the component value was derived randomly from $[-C, C]$. Additionally, C can be defined as follows:

$$C = (\cos(2 \times r5) + 1) \times \left(1 - \frac{t}{Maxiter}\right) \quad (3)$$

where $\cos(\bullet)$ denotes the cosine function, r_5 indicates positive real numbers amongst [0, 1], and $Maxiter$ denotes the maximal iteration number. It is possible to evaluate L, the value of variable, as follows:

$$L = C \times l \quad (4)$$

where l indicates the arbitrary value within $[-1, 1]$. Afterward, every probable $GX(t + 1)$ solution is produced, owing to the exploration, and the fitness value is compared. If GX outperformes X, it is kept and utilized in the location of X. This is represented as the condition $(GX) < F(X)$, whereas F indicates the fitness function for the problem in question (t). Additionally, the better option available at the time is now considered to be the silverback.

3.2.3. Exploitation Phase

Once the new troop of gorillas is formed, the silverback is the dominant male and is at the peak of his health and strength. They follow the silverback gorilla since they forage for food. Unavoidably, the silverback will age and die, and in his location, a younger blackback in the troop might engage in fighting over mating and leadership with other males. AGTO's exploitation stage follows the silverback and competes for adult female gorillas. W is presented for controlling these transitions. When C in Equation (4) is higher than W, this follows the silverback's initial model:

$$GX(t + 1) = L \times M \times (X(t) - Xsilverback) + X(t) \quad (5)$$

In such cases, the optimum solution found so far is indicated as X silverback, the existing location vector is represented as (t), and L is estimated by means of Equation (5). The values of M are defined as follows:

$$M = \left(\frac{\sum_{i=1}^{n} xi(t)}{N\left|\frac{xi(t)}{N}\right|^{2l}} \right) \frac{1}{2l}$$ (6)

where N denotes the overall individual number, and $Xi(t)$ indicates a vector demonstrating the gorilla's position:

$$GX(t+1) = Xsilverback - (Xsilverback \times Q - X(t) \times Q) \times A$$ (7)

$$Q = 2 \times r6 - 1$$ (8)

$$A = \phi \times E,$$ (9)

$$E = \begin{cases} N1, & r7 \geq 0.5 \\ N2, & r7 < 0.5 \end{cases}$$ (10)

It is the existing location, represented as (t), and the impact force, Q, that are evaluated by Equations (7) and (8). A random value within zero and one is utilized for r6 in Equation (4). Additionally, Equation (9) is utilized for assessing the efficiency of the coefficient. A is utilized for stimulating the level of violence in the game. With the equation denoting a constant, we could define what number represents Equation (10). Equation (6) involves r7, which is a value selected randomly within zero and one. Standard distribution, $E(1, D)$, is when r70.5 is a coincidental event, and D indicates the number of spatial dimensions. However, if r7 is less than half, E is equivalent to the random quantity that fits neatly into the standard distribution. Afterward, the exploitation stage is complete, and the value of candidate fitness for the recently generated $GX(t+1)$ problems is calculated. GX is preserved if $F(GX)F$. Figure 2 depicts the flowchart of GTOA.

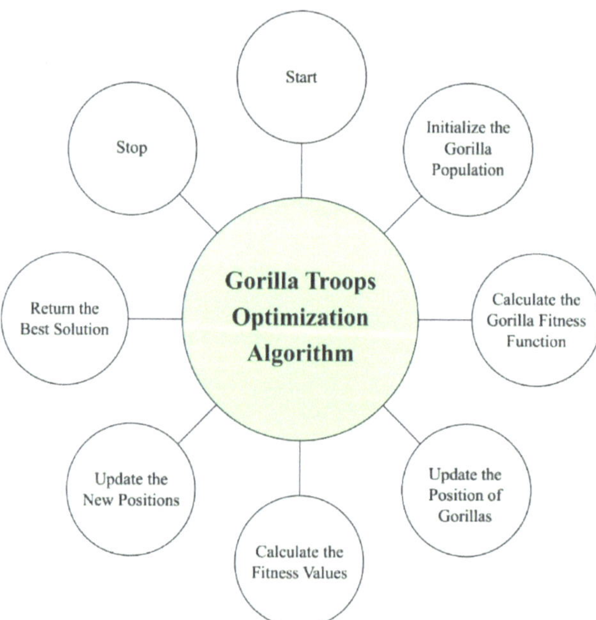

Figure 2. Flowchart of GTOA.

In this work, the EAGTO algorithm was primarily designed by the use of the circle chaotic mapping technique. To increase the population diversity and exploit the data in the solution space, the circle chaotic function is proposed to increase the initialization mode of the GTOA. Additionally, it can be mathematically expressed as follows:

$$z_{k+1} = z_k + b - \frac{a}{2\pi} \cdot \sin(a\pi z_k) mod(1), z_k \in (0,1) \tag{11}$$

whereas $a = 0.5$ and $b = 0.2$, the circle mapping and random search mechanism are chosen to be independently implemented 300 times. The traversal of circle chaotic mapping is more homogeneously distributed and wider in the range of [0, 1]. Thus, after integrating circle chaotic mapping, the presented technique has a robust global exploration capability.

3.3. Clustering Process Involved in EAGTOC-UIVN Technique

The EAGTOC-UIVN technique computes a fitness function with the inclusion of multiple parameters. The EAGTOC-UIVN method is proposed with the existence of 4 fitness variables, namely energy efficacy of cluster node density, UAV nodes, distance in CH to sink, and average distance of UAV for CH enclosed by their sensing series [25]. The data on fitness parameters was provided by the following:

Energy efficiency: The *CH* executes various events, such as gathered, sense, data broadcast, aggregation, and so on; hence, *CH* intakes the greatest amount of energy when compared to other nodes. Then it is vital for defining an *FF* that shared the load among each UAV from the network. The fitness variable for effective deployment of network energy is given below:

$$R_e = e(n_i)$$

$$Avg_e = \frac{1}{n} \sum_{i=0}^{n} e(n_i)$$

$$f_1 = CH_{opt} * \frac{R_e}{Avg_e} = \frac{CH_{opt} * e(n_i)}{\frac{1}{n}\sum_{i=0}^{n} e(n_i)} \forall CH_{opt} = 5\% \ of \ n, \ e(n_i)$$

$$= 0.5J \ or \ 1.25J \ or \ 1.75J \tag{12}$$

In Equation (12), R_e, Avg_e, and n_i denote the node RE, network average energy, and whole quantity of UAV nodes, respectively. CH_{opt} shows the optimum percentage of CHs.

Cluster node density: In intra-cluster communication, the cost is a crucial parameter for the high energy effectiveness of the network. Next, the network energy deployment was larger when the cost function of cluster was determined as follows:

$$f_2 - \max \left(n(CH_1), n(CH_2), n(CH_3) n(CH_j) \right) \forall n = 2 \ To \ 95, \ j = 1 \ to \ 15 \tag{13}$$

From the expression, $n(CH_j)$ denotes the number of UAVs from the range of jth CH (CH_j). The values of objective function f_2 are greater than able choice of *CH* and exploits from reducing the energy reduction.

The average distance of UAVs to the *CH* within its sensing range: In intra-cluster communication, UAVs transmit information to the *CH*. When the *CH* is farther from the CM, the energy of the UAV diminishes; when the *CH* is closer to the member UAV nodes afterward, there is a deployment of minimal energy.

$$f_3 = \frac{1}{n_{s\tau}} \sum_{i=0}^{n_{sr}} disT(CH, \ i) \ \forall dist(CH, \ i) = 1 \ to \ 35 \ m, n_{sr} = 1 \ to \ 100 \tag{14}$$

where n_{sr} and $dist(CH, \ i)$ indicate the number of UAVs from the sensing sequence and Euclidean distance in node and *CH* from the sensing series of the cluster. Thus, the value of f_3 is minimal; however, the intra-cluster communication power is lessened.

Distance from *CH* to *BS*: The distance between the *BS*s and *CH*s takes a basic function, as if the *CHS* is farther from the sink and exploits energy quickly that is evaluated by the following:

$$f_4 = \frac{1}{CH} \sum_{i=0}^{CH} dist(BS,\ CH_i)\ \forall dist(BS,\ CH_i) = 1\ to\ 70m,\ CH = 1\ to\ 15 \tag{15}$$

In Equation (15), $dist(BS,\ CH_i)$ indicates the Euclidean distance among the *BS* and CH_i. Minimizing the f_4 objective function specified that the *CHS* is not farther from the *BS*.

Once the f_1, f_2, f_3, and f_4 function parameters were evaluated, the objective function was also named FF and calculated as follows:

$$F = Maximize\ Fitness = \alpha * f_1 + \beta * f_2 + \gamma * \frac{1}{f_3} + \delta * \frac{1}{f_4} \tag{16}$$

In Equation (16), α, β, γ, and δ denote the weight coefficient for the f_1, f_2, f_3, and f_4 FF parameters, correspondingly. The range of weight coefficient ranges from 0 to 1.

4. Results and Discussion

The proposed model was simulated by using MATLAB R2019a. The simulation parameters are listed in Table 1. In this section, a detailed experimental validation of the EAGTOC-UIVN approach is investigated under distinct UAVs. Table 2 and Figure 3 report an overall PDR examination of the EAGTOC-UIVN model under several UAVs, with existing models such as swarm-intelligence-based clustering (SIC), EALC, ant-colony optimization (ACO), GBLADSR, and genetic algorithm (GA) [23].

Table 1. Parameter settings.

Parameter	Value
Network area	1000 m*1000 m
UAV transmission range	250–300 m
Number of UAVs	150
Number of ground station	1
Traffic type	CBR
CBR rate	2 Mbps
Speed	10–30 m/s
UAV transmission power	5 W

Table 2. PDR analysis of EAGTOC-UIVN approach with other systems under varying UAVs.

Number of UAVs	Packet Delivery Ratio (%)					
	EAGTOC-UIVN	SIC	EALC	ACO	GBLADSR	GA
30	90.15	85.00	71.21	67.81	64.83	64.00
60	94.37	88.19	77.28	74.09	68.74	66.27
90	96.12	90.46	84.49	79.96	73.47	69.87
120	98.18	92.52	87.89	82.84	77.39	73.58
150	98.90	95.09	87.78	84.49	80.89	78.52

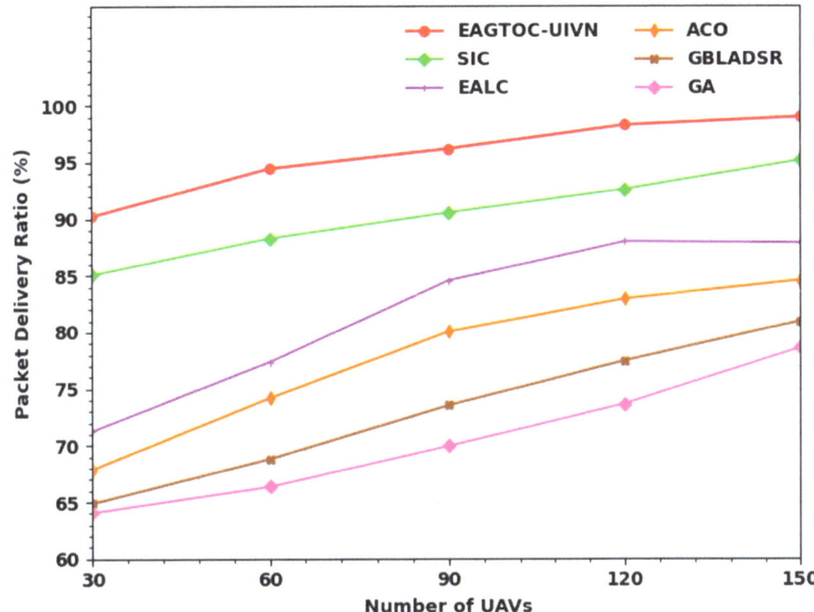

Figure 3. PDR analysis of EAGTOC-UIVN system under varying UAVs.

The experimental value implies that the EAGTOC-UIVN technique obtained a better performance under all UAVs. For example, on 30 UAVs, the EAGTOC-UIVN method reached an increased PDR value of 90.15%. On the other hand, the SIC, EALC, ACO, GBLADSR, and GA approaches accomplished decreased PDR values of 85%, 71.21%, 67.81%, 64.83%, and 64%, correspondingly. Meanwhile, on 150 UAVs, the EAGTOC-UIVN technique attained an improved PDR value of 98.90%. In contrast, the SIC, EALC, ACO, GBLADSR, and GA techniques attained reduced PDR values of 95.09%, 87.78%, 84.49%, 80.89%, and 78.52%, correspondingly.

In Table 3 and Figure 4, a brief average end-to-end delay (AETED) assessment of the EAGTOC-UIVN with recent techniques is given. The results implied that the GA model failed to portray effectual outcomes with maximum values of AETED. At the same time, the EALC, ACO, and GBLADSR models reached closer AETED values. Although the SIC model tried to show a reasonable AETED value, the EAGTOC-UIVN model gained effectual outcomes with minimal AETED values. Notice that the EAGTOC-UIVN model reached an AETED value of at least 0.078 s under 30 UAVs.

Table 3. AETED analysis of EAGTOC-UIVN approach with other systems under varying UAVs.

	Average End-to-End Delay (s)					
Number of UAVs	**EAGTOC-UIVN**	**SIC**	**EALC**	**ACO**	**GBLADSR**	**GA**
30	0.078	0.091	0.110	0.111	0.111	0.122
60	0.093	0.105	0.135	0.152	0.171	0.179
90	0.102	0.129	0.166	0.182	0.194	0.216
120	0.121	0.143	0.227	0.229	0.242	0.269
150	0.136	0.174	0.262	0.268	0.286	0.302

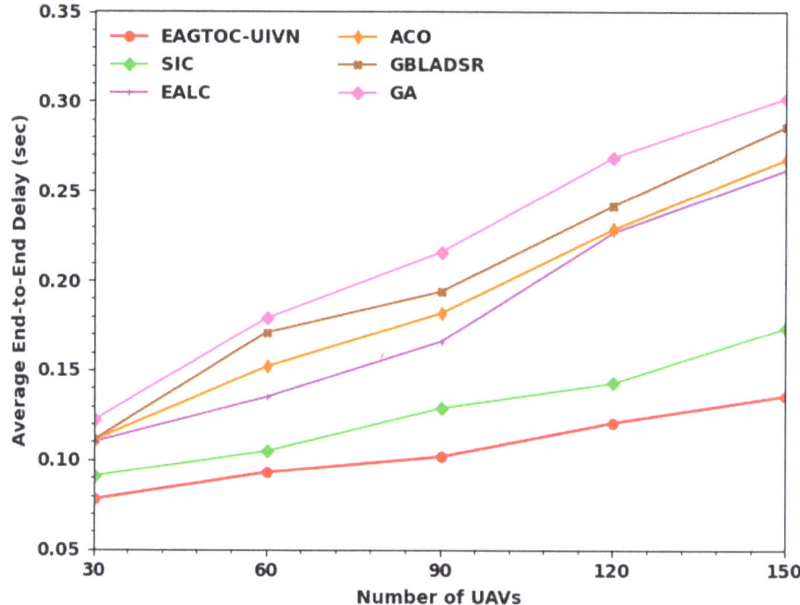

Figure 4. AETED analysis of EAGTOC-UIVN algorithm under varying UAVs.

In Table 4 and Figure 5, a brief cluster overhead (COH) assessment of the EAGTOC-UIVN with recent techniques is given. The result implies that the GA technique failed to represent effectual outcomes with a maximal value of COH. Simultaneously, the EALC, ACO, and GBLADSR techniques attained closer COH values. Even though the SIC method tried to demonstrate a reasonable COH value, the EAGTOC-UIVN technique obtained effectual outcomes with minimal COH values. Note that the EAGTOC-UIVN methodology attained a minimum COH value of 0.153 under 30 UAVs.

Table 4. COH analysis of EAGTOC-UIVN technique with other systems under varying UAVs.

	Cluster Overhead					
Number of UAVs	EAGTOC-UIVN	SIC	EALC	ACO	GBLADSR	GA
30	0.153	0.160	0.168	0.174	0.177	0.195
60	0.179	0.197	0.209	0.218	0.228	0.247
90	0.195	0.219	0.240	0.250	0.258	0.274
120	0.198	0.227	0.268	0.278	0.291	0.309
150	0.216	0.253	0.290	0.300	0.307	0.329

In Table 5 and Figure 6, a brief cluster building time (CBT) assessment of the EAGTOC-UIVN with recent approaches is given. The result implies that the GA approach failed to represent effectual outcomes with maximal value of CBT. Simultaneously, the EALC, ACO, and GBLADSR techniques attained closer CBT values. Even though the SIC approach tried to demonstrate a reasonable CBT value, the EAGTOC-UIVN technique obtained effectual outcomes with the lowest CBT values. Note that the EAGTOC-UIVN method has attained a minimum CBT value of 0.51 s under 30 nodes.

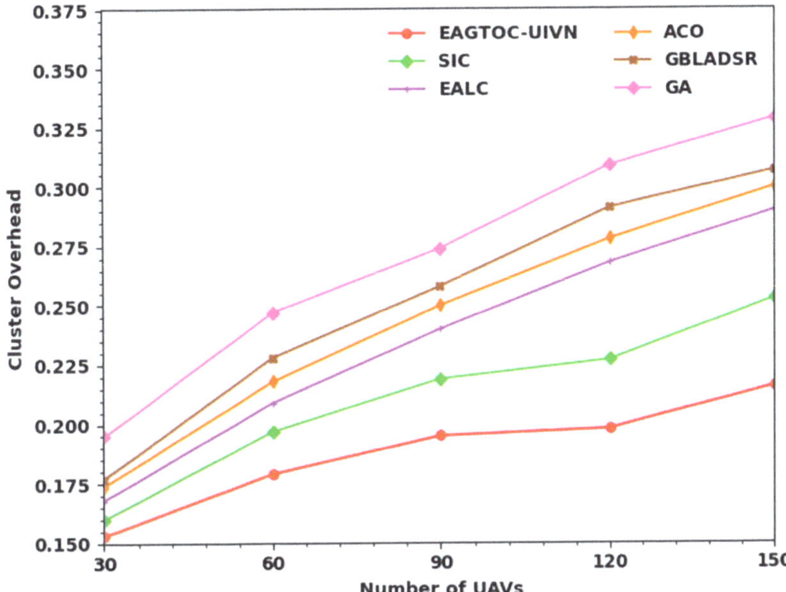

Figure 5. COH analysis of EAGTOC-UIVN algorithm under varying UAVs.

Table 5. CBT analysis of EAGTOC-UIVN technique with other systems under varying nodes.

	Cluster Building Time (s)					
Number of Nodes	**EAGTOC-UIVN**	**SIC**	**EALC**	**ACO**	**GBLADSR**	**GA**
30	0.51	2.90	7.87	10.46	14.03	17.41
60	5.29	8.07	12.05	13.44	19.80	30.14
90	5.68	10.46	14.63	21.19	28.15	36.90
120	5.49	10.86	17.02	23.38	39.88	52.01
150	6.08	9.26	19.01	29.74	49.82	60.36

Table 6 and Figure 7 show the cluster average lifetime (CALT) analysis of the EAGTOC-UIVN technique under various nodes. The experimental value implies that the EAGTOC-UIVN approach attained an improved performance under all nodes. For example, on 30 nodes, the EAGTOC-UIVN technique attained an improved CALT value of 67.61 s. On the other hand, the SIC, EALC, ACO, GBLADSR, and GA systems attained minimized CALT values of 66.21 s, 65.20 s, 63.17 s, 54.66 s, and 47.81 s, correspondingly. Meanwhile, on 150 nodes, the EAGTOC-UIVN approach gained an improved CALT value of 55.04 s. In contrast, the SIC, EALC, ACO, GBLADSR, and GA techniques attained reduced CALT values of 50.35 s, 46.03 s, 40.45 s, 37.02 s, and 30.30 s, correspondingly.

Table 7 and Figure 8 show the number of alive nodes (NOAN) investigation of the EAGTOC-UIVN technique under various rounds. The experimental value implies that the EAGTOC-UIVN approach gained improved performance under all rounds. For example, on 400 rounds, the EAGTOC-UIVN system reached an improved NOAN value of 100. In contrast, the SIC, EALC, ACO, GBLADSR, and GA approaches attained reduced NOAN values of 98, 95, 93, 92, and 83, correspondingly. Meanwhile, on 1800 rounds, the EAGTOC-UIVN model gained an improved NOAN value of 71. In contrast, the SIC, EALC, ACO, GBLADSR, and GA methods attained improved NOAN values of 44, 20, 11, 4, and 0, correspondingly.

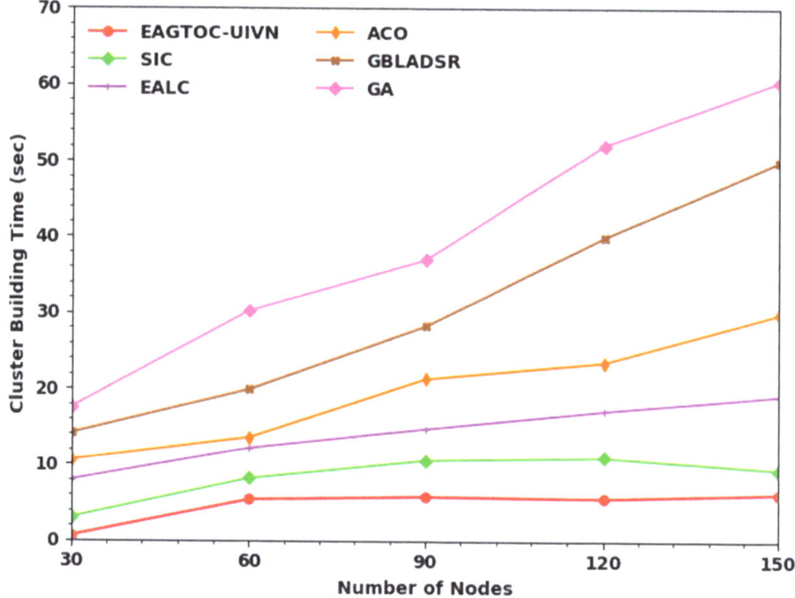

Figure 6. CBT analysis of EAGTOC-UIVN technique under varying nodes.

Table 6. CALT analysis of EAGTOC-UIVN method with other techniques under varying nodes.

Cluster Average Lifetime (s)						
Number of Nodes	EAGTOC-UIVN	SIC	EALC	ACO	GBLADSR	GA
30	67.61	66.21	65.20	63.17	54.66	47.81
60	67.10	64.56	61.14	57.96	48.57	41.59
90	61.77	54.28	53.52	51.24	44.00	34.61
120	58.22	51.24	48.95	46.03	36.01	31.06
150	55.04	50.35	46.03	40.45	37.02	30.30

Table 7. NOAN analysis of EAGTOC-UIVN technique with other systems under varying rounds.

No. of Alive Nodes						
No. of Rounds	EAGTOC-UIVN	SIC	EALC	ACO	GBLADSR	GA
0	100	100	100	100	100	100
200	100	100	99	96	95	92
400	100	98	95	93	92	83
600	99	98	90	85	76	70
800	97	92	88	77	62	61
1000	95	87	83	60	55	51
1200	94	75	68	52	43	30
1400	89	69	59	38	33	15
1600	80	55	45	24	15	3
1800	71	44	20	11	4	0
2000	58	28	9	0	0	0

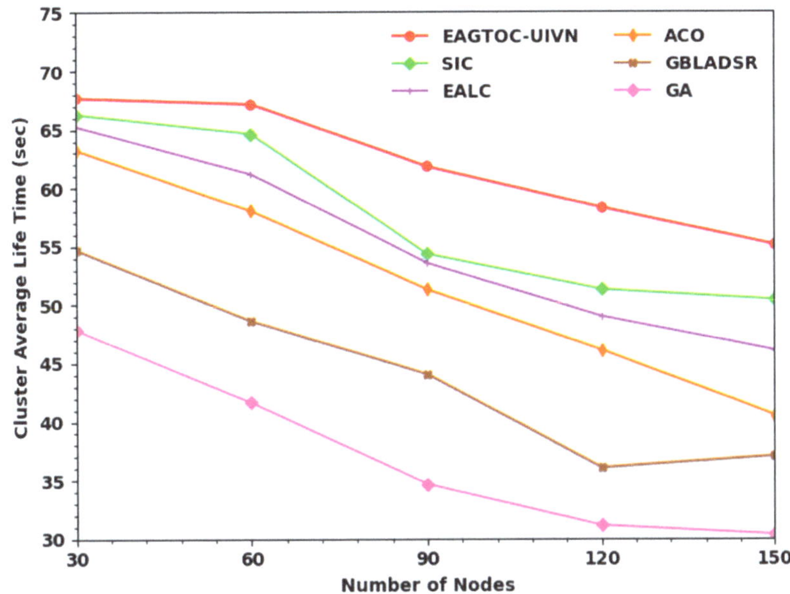

Figure 7. CALT analysis of EAGTOC-UIVN algorithm under varying nodes.

In Table 8 and Figure 9, a brief total energy consumption (TECON) assessment of the EAGTOC-UIVN technique with recent approaches is given. The outcomes imply that the GA technique has failed to describe effectual outcomes with maximal values of TECON.

Table 8. TECON analysis of EAGTOC-UIVN technique with other systems under varying rounds.

	Total Energy Consumption (J)					
No. of Rounds	**EAGTOC-UIVN**	**SIC**	**EALC**	**ACO**	**GBLADSR**	**GA**
0	0.00	0.00	0.00	0.00	0.00	0.00
200	38.61	70.97	80.92	100.84	123.24	247.68
400	98.35	165.55	200.39	225.28	250.17	466.71
600	182.97	265.11	337.29	392.04	441.82	568.76
800	270.09	369.64	476.67	553.83	563.78	643.43
1000	377.11	511.51	578.72	630.98	708.14	718.10
1200	494.09	578.72	660.85	787.79	812.68	790.28
1400	603.61	693.21	770.37	862.46	899.79	884.86
1600	673.30	795.26	867.44	912.24	952.06	937.13
1800	755.43	859.97	917.22	949.57	979.44	976.95
2000	770.37	869.93	947.08	954.55	984.42	996.86

Simultaneously, the EALC, ACO, and GBLADSR techniques obtained closer TECON values. Even though the SIC system tried to show a reasonable TECON value, the EAGTOC-UIVN approach attained effectual outcomes with minimal TECON values. Note that the EAGTOC-UIVN method attained a minimum TECON value of 38.61 J under 200 rounds. From these results, it is evident that the presented model improves the overall network efficacy.

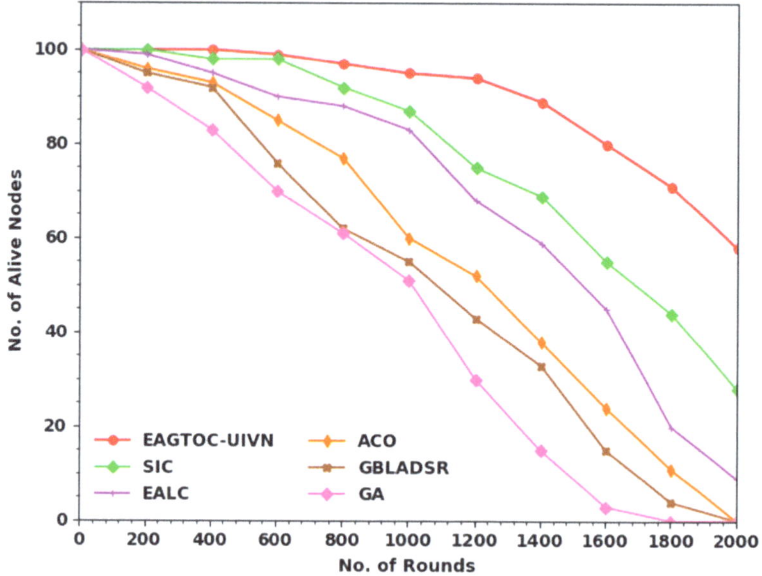

Figure 8. NOAN analysis of EAGTOC-UIVN algorithm under varying rounds.

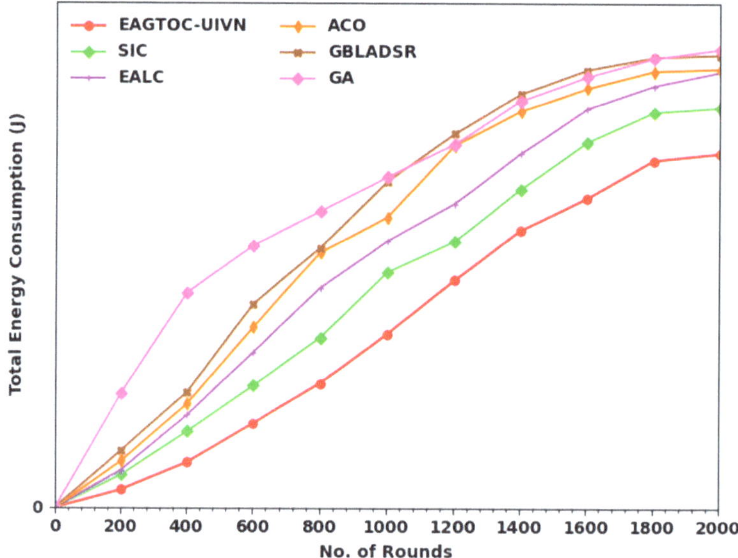

Figure 9. TECON analysis of EAGTOC-UIVN algorithm under varying rounds.

5. Conclusions

In this study, a new EAGTOC-UIVN technique was developed for clustering the UAV-assisted VN. The major aim of the EAGTOC-UIVN technique exists in the grouping of the nodes in UAV-based VN to achieve maximum lifetime and energy efficiency. In the presented EAGTOC-UIVN technique, the EAGTO algorithm is primarily designed by the use of the circle chaotic mapping technique. Moreover, the EAGTOC-UIVN technique computes a fitness function with the inclusion of multiple parameters. To depict the improved performance of the EAGTOC-UIVN technique, a widespread simulation analysis was performed. The comparison study reported the enhancements of the EAGTOC-UIVN technique over other recent approaches. In the future, data aggregation and localization techniques can be designed to increase the overall network effectiveness of the UAV-based VNs.

Author Contributions: Conceptualization, H.A.; methodology, J.S.A.; software, M.M.A.; validation, J.S.A., A.S.A.A. and A.M.H.; formal analysis, M.A. and M.M.A; investigation, M.M..; resources, H.A.; data curation, A.Q.; writing—original draft preparation, H.A., J.S.A., M.M., M.A., A.Q., A.S.A.A. and A.M.H.; writing—review and editing, M.M.A.; visualization, A.M.H.; supervision, M.A.; project administration, A.M.H.; funding acquisition, H.A. All authors have read and agreed to the published version of the manuscript.

Funding: The authors extend their appreciation to the Deanship of Scientific Research at King Khalid University for funding this work through Small Groups Project under grant number (241/43). Princess Nourah bint Abdulrahman University Researchers Supporting Project number (PNURSP2022R303), Princess Nourah bint Abdulrahman University, Riyadh, Saudi Arabia. The authors would like to thank the Deanship of Scientific Research at Umm Al-Qura University for supporting this work by Grant Code: 22UQU4340237DSR54.

Institutional Review Board Statement: Not applicable.

Informed Consent Statement: Not applicable.

Data Availability Statement: Data sharing is not applicable to this article, as no datasets were generated during the current study.

Conflicts of Interest: The authors declare that they have no conflict of interest. The manuscript was written through contributions of all authors. All authors have given approval to the final version of the manuscript.

References

1. Fatemidokht, H.; Rafsanjani, M.K.; Gupta, B.B.; Hsu, C.H. Efficient and secure routing protocol based on artificial intelligence algorithms with UAV-assisted for vehicular ad hoc networks in intelligent transportation systems. *IEEE Trans. Intell. Transp. Syst.* **2021**, *22*, 4757–4769. [CrossRef]
2. Amantayeva, A.; Yerzhanova, M.; Kizilirmak, R.C. UAV location optimization for UAV-to-vehicle multiple access channel with visible light Communication. In Proceedings of the 2019 Wireless Days (WD), Manchester, UK, 24–26 April 2019; pp. 1–4.
3. Huang, S.; Huang, C.; Wu, D.; Yin, Y.; Ashraf, M.; Fu, B. UAV-Assisted Sensor Data Dissemination in mmWave Vehicular Networks Based on Network Coding. *Wirel. Commun. Mob. Comput.* **2022**, *2022*, 2576182. [CrossRef]
4. Al-Hilo, A.; Samir, M.; Assi, C.; Sharafeddine, S.; Ebrahimi, D. A cooperative approach for content caching and delivery in UAV-assisted vehicular networks. *Veh. Commun.* **2021**, *32*, 100391. [CrossRef]
5. Barka, E.; Kerrache, C.A.; Lagraa, N.; Lakas, A. Behavior-aware UAV-assisted crowd sensing technique for urban vehicular environments. In Proceedings of the 2018 15th IEEE Annual Consumer Communications & Networking Conference (CCNC), Las Vegas, NV, USA, 12–15 January 2018; pp. 1–7.
6. Mokhtari, S.; Nouri, N.; Abouei, J.; Avokh, A.; Plataniotis, K.N. Relaying Data with Joint Optimization of Energy and Delay in Cluster-based UAV-assisted VANETs. *IEEE Int. Things J.* **2022**. [CrossRef]
7. Khabbaz, M.; Antoun, J.; Assi, C. Modeling and performance analysis of UAV-assisted vehicular networks. *IEEE Trans. Veh. Technol.* **2019**, *68*, 8384–8396. [CrossRef]
8. Hu, J.; Chen, C.; Cai, L.; Khosravi, M.R.; Pei, Q.; Wan, S. UAV-assisted vehicular edge computing for the 6G internet of vehicles: Architecture, intelligence, and challenges. *IEEE Commun. Stand. Mag.* **2021**, *5*, 12–18. [CrossRef]
9. Xing, N.; Zong, Q.; Dou, L.; Tian, B.; Wang, Q. A game theoretic approach for mobility prediction clustering in unmanned aerial vehicle networks. *IEEE Trans. Veh. Technol.* **2019**, *68*, 9963–9973. [CrossRef]
10. Fawaz, W. Effect of non-cooperative vehicles on path connectivity in vehicular networks: A theoretical analysis and UAV-based remedy. *Veh. Commun.* **2018**, *11*, 12–19. [CrossRef]
11. Nomikos, N.; Gkonis, P.K.; Bithas, P.S.; Trakadas, P. A Survey on UAV-Aided Maritime Communications: Deployment Considerations, Applications, and Future Challenges. *arXiv* **2022**, arXiv:2209.09605.
12. Wang, H.-M.; Zhang, X.; Jiang, J.-C. UAV-involved wireless physical-layer secure communications: Overview and research directions. *IEEE Wirel. Commun.* **2019**, *26*, 32–39. [CrossRef]
13. Oubbati, O.S.; Chaib, N.; Lakas, A.; Lorenz, P.; Rachedi, A. UAV-assisted supporting services connectivity in urban VANETs. *IEEE Trans. Veh. Technol.* **2019**, *68*, 3944–3951. [CrossRef]
14. Al-Hilo, A.; Samir, M.; Assi, C.; Sharafeddine, S.; Ebrahimi, D. UAV-assisted content delivery in intelligent transportation systems-joint trajectory planning and cache management. *IEEE Trans. Intell. Transp. Syst.* **2020**, *22*, 5155–5167. [CrossRef]

15. He, Y.; Nie, L.; Guo, T.; Kaur, K.; Hassan, M.M.; Yu, K. A NOMA-enabled framework for relay deployment and network optimization in double-layer airborne access VANETs. *IEEE Trans. Intell. Transp. Syst.* **2022**, *23*, 22452–22466. [CrossRef]

16. Khabbaz, M.; Assi, C.; Sharafeddine, S. Multihop V2U Path Availability Analysis in UAV-Assisted Vehicular Networks. *IEEE Int. Things J.* **2021**, *8*, 10745–10754. [CrossRef]

17. Zheng, K.; Sun, Y.; Lin, Z.; Tang, Y. UAV-assisted online video downloading in vehicular networks: A reinforcement learning approach. In Proceedings of the 2020 IEEE 91st Vehicular Technology Conference (VTC2020-Spring), Antwerp, Belgium, 25–28 May2020; pp. 1–5.

18. Raza, A.; Bukhari, S.H.R.; Aadil, F.; Iqbal, Z. An UAV-assisted VANET architecture for intelligent transportation system in smart cities. *Int. J. Distrib. Sens. Netw.* **2021**, *17*, 15501477211031750. [CrossRef]

19. Wu, H.; Lyu, F.; Zhou, C.; Chen, J.; Wang, L.; Shen, X. Optimal UAV caching and trajectory in aerial-assisted vehicular networks: A learning-based approach. *IEEE J. Sel. Areas Commun.* **2020**, *38*, 2783–2797. [CrossRef]

20. Ghazzai, H.; Khattab, A.; Massoud, Y. Mobility and energy aware data routing for UAV-assisted VANETs. In Proceedings of the 2019 IEEE International Conference on Vehicular Electronics and Safety (ICVES), Cairo, Egypt, 4–6 September 2019; pp. 1–6.

21. Huang, S.; Huang, C.; Yin, Y.; Wu, D.; Ashraf, M.W.A.; Fu, B. UAV-assisted data dissemination based on network coding in vehicular networks. *IET Intell. Transp. Syst.* **2022**, *16*, 421–433. [CrossRef]

22. Alioua, A.; Senouci, S.M.; Moussaoui, S.; Sedjelmaci, H.; Messous, M.A. Efficient data processing in software-defined UAV-assisted vehicular networks: A sequential game approach. *Wirel. Pers. Commun.* **2018**, *101*, 2255–2286. [CrossRef]

23. Arafat, M.Y.; Moh, S. Localization and clustering based on swarm intelligence in UAV networks for emergency communications. *IEEE Int. Things J.* **2019**, *6*, 8958–8976. [CrossRef]

24. Mohamed, T.A.; Mustafa, M.K. Adaptive trainer for multi-layer perceptron using artificial gorilla troops optimizer algorithm. *Int. J. Nonlinear Anal. Appl.* **2022**. [CrossRef]

25. Lakshmanna, K.; Subramani, N.; Alotaibi, Y.; Alghamdi, S.; Khalafand, O.I.; Nanda, A.K. Improved metaheuristic-driven energy-aware cluster-based routing scheme for IoT-assisted wireless sensor networks. *Sustainability* **2022**, *14*, 7712. [CrossRef]

Article

Physical-Layer Security for UAV-Assisted Air-to-Underwater Communication Systems with Fixed-Gain Amplify-and-Forward Relaying

Yi Lou [1], Ruofan Sun [2], Julian Cheng [3], Gang Qiao [2] and Jinlong Wang [1,*]

[1] College of Information Science and Engineering, Harbin Institute of Technology (Weihai), Weihai 264209, China; yilou@hit.edu.cn

[2] College of Underwater Acoustic Engineering, Harbin Engineering University, Harbin 150001, China; sunruofan@hrbeu.edu.cn (R.S.); qiaogang@hrbeu.edu.cn (G.Q.)

[3] School of Engineering, The University of British Columbia, Kelowna, BC V1Y 8L6, Canada; julian.cheng@ubc.ca

* Correspondence: jlw@hit.edu.cn

Citation: Lou, Y.; Sun, R.; Cheng, J.; Qiao, G.; Wang, J. Physical-Layer Security for UAV-Assisted Air-to-Underwater Communication Systems with Fixed-Gain Amplify-and-Forward Relaying. *Drones* **2022**, *6*, 341. https://doi.org/10.3390/drones6110341

Academic Editor: Emmanouel T. Michailidis

Received: 17 October 2022
Accepted: 28 October 2022
Published: 3 November 2022

Publisher's Note: MDPI stays neutral with regard to jurisdictional claims in published maps and institutional affiliations.

Abstract: We analyze a secure unmanned aerial vehicle-assisted two-hop mixed radio frequency (RF) and underwater wireless optical communication (UWOC) system using a fixed-gain amplify-and-forward (AF) relay. The UWOC channel was modeled using a mixture exponential-generalized Gamma distribution to consider the combined effects of air bubbles and temperature gradients on transmission characteristics. Both legitimate and eavesdropping RF channels were modeled using flexible α-μ distributions. Specifically, we first derived both the probability density function (PDF) and cumulative distribution function (CDF) of the received signal-to-noise ratio of the system. Based on the PDF and CDF expressions, we derived the closed-form expressions for the tight lower bound of the secrecy outage probability (SOP) and the probability of non-zero secrecy capacity (PNZ), which are both expressed in terms bivariate Fox's H-function. To utilize these analytical expressions, we derived asymptotic expressions of SOP and PNZ using only well-known functions. We also used asymptotic expressions to determine the suboptimal transmitting power to maximize energy efficiency. Furthermore, we investigated the effect of levels of air bubbles and temperature gradients in the UWOC channel, and studied the nonlinear characteristics of the transmission medium and the number of multipath clusters of the RF channel on the secrecy performance. Finally, all analyses were validated using a simulation.

Keywords: amplify-and-forward (AF); α-μ distribution; non-zero capacity (PNZ); performance analysis; underwater wireless optical communication (UWOC); secrecy outage probability (SOP)

1. Introduction

The rise of the underwater Internet of Things requires the support of a high-performance underwater communication network having high data rates, low latency, and long communication range. Underwater wireless optical communication (UWOC) is one of the essential technologies for this communication network. Unlike radio frequency (RF) [1–3] and acoustic technologies, UWOC technology can achieve ultra-high data rates of Gpbs over a moderate communication range when selecting blue or green light with wavelengths located in the transmission window [4]. Furthermore, a light-emitting diode or laser diode as a light source provides the versatility to select between communication range and coverage area within the constraints of the range-beamwidth tradeoff to meet the needs of a specific application scenario.

Using relay technology to construct a communication system in a multi-hop fashion is one of the primary techniques to extend the communication range. Based on the modality of processing and forwarding signals, relays can be divided into two main categories:

decode-and-forward relays (DF) and amplify-and-forward (AF) relays. In DF relaying systems, the relay down-converts the received signals to the baseband, decodes, re-encodes, and up-converts them to the RF band, and forwards the signal to the destination node. In AF relaying systems, the relay amplifies the received signals directly in the passband based on an amplification factor, then forwards them directly in the RF band. Since the AF scheme does not require time-consuming decoding and spectral shifting, it can significantly reduce complexity while still providing good performance [5]. Depending on the different channel state information (CSI) information required by the AF relay, AF relaying can be divided into the variable-gain AF (VG) and fixed-gain AF (FG). In a VG scheme, the relay requires instantaneous CSI of the source-to-relay link, whereas in an FG scheme, only statistical CSI of the SR link is required [6]. Therefore, from an engineering standpoint, the FG scheme is more attractive because of its low implementation complexity.

To maximize the utilization of the different transmission environments of each hop and to improve the overall performance of the multi-hop relaying system, mixed communication systems using different communication technologies have been proposed, and are widely used in unmanned aerial vehicles (UAV)-assisted vehicle communication systems [7–9]. For example, the mixed communication system using both RF and free-space optical (FSO) technologies has been proposed to take advantage of the robustness of the RF links and the high bandwidth characteristics of the FSO links. Further, RF sub-systems offer low-cost and non-line-of-sight communication capabilities, while FSO sub-systems offer low transmission latency and ultra-high transmission rates. Therefore, a mixed RF/FSO system is a cost-effective solution to the last-mile problem in wireless communication networks, where the high-bandwidth FSO sub-system of a mixed RF/FSO system is used to connect seamlessly the fiber backbone and RF sub-system access networks [10–13]. Achieving ultra-high-speed communication between underwater and airborne nodes across the sea surface medium is challenging due to the low data rate of underwater acoustic communications. To solve this problem, using an ocean buoy or a marine ship as a relay node, the mixed RF/UWOC system for UAV and autonomous underwater vehicle (AUV) communication is proposed, in which the high-speed UWOC is used instead of underwater acoustic communication, to achieve higher overall communication rates [14–18].

Accurate modeling of the UWOC channel, including absorption, scattering, and turbulence, is a prerequisite for proper performance analysis and algorithm development of the UWOC system [19,20]. Absorption and scattering have been extensively studied [21–23], where absorption limits the transmission distance of underwater light, while scattering diffuses the receiving radius of underwater light transmission and deflects the transmission path, thus reducing the received optical power. Due to changes in the random refractive index variation, turbulence can cause fluctuations in the received irradiance, i.e., scintillation, which can limit the performance and affect the stability of the UWOC system [4]. In early research, UWOC turbulence was modeled by borrowing models of atmospheric turbulence, e.g., weak turbulence is modeled by the Lognormal distribution [24–26], and moderate-to-strong turbulence is modeled by the Gamma-Gamma distribution [27–30].

However, the statistical distributions used to model atmospheric turbulence cannot accurately characterize UWOC systems due to the fundamental differences between aqueous and atmospheric mediums. Recently, based on experimental data, the mixed exponential-lognormal distribution has been proposed to model moderate to strong UWOC turbulence in the presence of air bubbles in both fresh water and salty water [31]. Later, the mixture exponential-generalized Gamma (EGG) distribution was proposed to model turbulence in the presence of air bubbles and temperature gradients in either fresh or salt water [32]. The EGG distribution not only can model turbulence of various intensities, but also has an analytically tractable mathematical form. Therefore, useful system performance metrics, such as ergodic capacity, outage probability, and bit-error rate (BER), can be easily obtained.

Due to the broadcast nature of RF signals, secrecy performance has always been one of the most important considerations for the mixed RF/FSO communication systems [11,33–37]. In [34], the expressions of the lower bound of the secrecy outage probability (SOP) and average secrecy capacity (ASC) for mixed RF/FSO systems using VG or FG relaying schemes, were both derived in closed-form, where the RF and FSO links are modeled by the Nakagami-m and GG distributions, respectively. The authors in [35] used Rayleigh and GG distributions to model RF and FSO links, respectively. Considering the impact of imperfect channel state information (CSI), both the exact and asymptotic expressions of the lower bound for SOP of a mixed RF/FSO system using VG or FG relay are derived. The same authors then extended the analysis to multiple-input and multiple-output configuration and analyzed the impact of different transmit antenna selection schemes on the secrecy performance of the mixed RF/FSO system using a DF relay, where RF and FSO links are modeled by the Nakagami-m and $\mathcal{M}$-distributions, respectively. Assuming the CSI of the FSO and RF links are imprecise and outdated, the authors derived the bound and asymptotic expressions of the effective secrecy throughput of the system. In [36], using more generalized η-μ and $\mathcal{M}$-distributions to model RF and FSO links, respectively, and assuming that the eavesdropper is only at the relay location, the authors derived the analytical results for the SOP and the average secrecy rate of the mixed RF/FSO system using the FG or VG relaying scheme. To quantify the impact of the energy harvesting operation on the system secrecy performance, the authors in [11] derived exact closed-form and asymptotic expressions for the SOP of the downlink simultaneous wireless information and power transfer system using DF relaying scheme, under the assumption that RF and FSO links are modeled using the Nakagami-m and GG distributions, respectively.

However, research on the secrecy performance of mixed RF/UWOC systems is still in its infancy despite the growing number of underwater communication applications. The authors in [16] investigated the secrecy performance of a two-hop mixed RF/UWOC system using a VG or FG multiple-antennas relay and maximal ratio combining scheme, where RF and UWOC links are modeled by Nakagami-m and the mixed exponential-Gamma (EG) distributions, respectively. Assuming that only the source-to-relay link is eavesdropped by unauthorized users, the authors in [16] derived the exact closed expressions of the ASC and SOP of the mixed RF/UWOC systems. Later, based on the same channel model as in [16], the same authors extended the analysis to the mixed RF/UWOC system using a multi-antennas DF relay with the selection combining scheme [15]. Both the exact closed-form and asymptotic expressions of the SOP were derived.

However, while the EG distribution is suitable for modeling turbulence of various intensities in both fresh water and salty water, this distribution fails to model the effects of air bubbles and temperature gradients on UWOC turbulence [32]. Further, the Nakagami-m distribution is only applicable to certain specific scenarios and cannot accurately characterize the effects of the properties of the transmission medium and multipath clusters on channel fading. It is shown that the impact of the medium on the signal propagation is mainly determined by the nonlinearity characteristics of the medium [38]. The α-μ distribution is a more general, flexible, and mathematically tractable model of channel fading whose parameters α and μ are correlated with the nonlinearity of the propagation medium and the number of clusters of multipath transmission, respectively. Further, by setting α and μ to specific values, the α-μ distribution can be reduced to several classical channel fading models, including Nakagami-m, Gamma, one-sided Gaussian, Rayleigh, and Weibull distributions. Recently, the secrecy performance of a two-hop mixed RF/UWOC system using DF relaying where RF and UWOC links are, respectively, modeled by flexible α-μ and water tank experimental data based EGG distributions has been analyzed in [39]; however, only the lower bound and asymptotic expressions of the SOP are derived. Furthermore, the overall end-to-end latency of the DF relaying based mixed RF/UWOC communication system is much higher than that of the FG relaying based one, due to the decoding and forwarding and spectral shifting operations required by DF relaying.

However, to the best of the authors' knowledge, this is the *first* comprehensive secrecy performance analysis of the mixed RF/UWOC communications system using a low-complexity FG relaying scheme. Unlike previous UWOC channel models that do not adequately characterize the underwater optical propagation and RF channel models that use various simplifying assumptions, we model the RF channels between UAV and relay and the UWOC channel between relay and AUV, using the more general and accurate α-μ and EGG distributions, respectively, to analyze the effects of a variety of realistic channel phenomena, such as different temperature gradients and levels of air bubbles of UWOC channels and different grades of medium nonlinearity, and the number of multipath clusters of the RF channels on the secrecy performance of the mixed RF/UWOC communication systems. We propose a novel analytical framework to derive the closed-form expressions of the SOP and the non-zero secrecy capacity (PNZ) metrics by the bivariate Fox's H-function. Moreover, our secrecy performance study provides a generalized framework for several fading models for both RF and UWOC channels, such as Rayleigh, Weibull for RF channels and EG and Generalized Gamma for UWOC channels. We first derive the probability density function (PDF) and cumulative distribution function (CDF) of the end-to-end signal-to-noise ratio (SNR) for the mixed RF/UWOC communication system in exact closed-form in terms of bivariate H-function. Depending on these expressions, we derive the exact closed-form expressions of the lower bound of the SOP and the PNZ. Furthermore, we also derive asymptotic expressions for both SOP and PNZ containing only simple functions at high SNRs. Addtionally, based on the asymptotic expressions for SOP and PNZ, we provide a straightforward approach to determine the suboptimal source transmission power to maximize energy efficiency for given performance goals of both SOP and PNZ. Finally, we use Monte Carlo simulation to validate all the derived analytical expressions and theoretical analyses.

The rest of this paper is organized as follows. In Section 2, the channel and system models are presented. In Section 3, the end-to-end statistics are studied. Both exact and asymptotic expressions for the SOP and PNZ are derived in Section 4. The numerical results and discussions are discussed in Section 5, which is followed by the conclusion in Section 6.

2. System and Channel Models

A mixed RF/UWOC system is considered in Figure 1 where a UAV acts as a source node (S) in the air transmits its private data to the legitimate destination node (D) acted by an AUV located underwater via a trusted relay node (R), which can be a buoy or a surface ship. The RF channel from S to R and underwater optical channel from the R to the D node is assumed to follow α-μ and EGG distributions, respectively. During transmission, one unauthorized receiver (E) attempts to eavesdrop on RF signals received by the R. In this paper, we consider a FG AF relay where the relay amplifies the received signal by a fixed factor and then forwards the amplified message to the destination node.

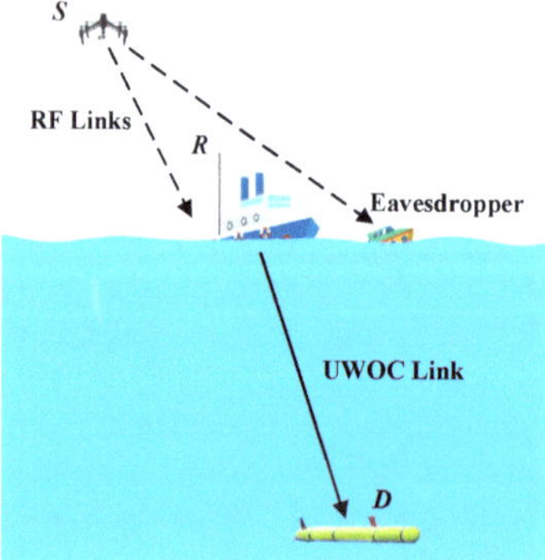

Figure 1. A two-hop mixed RF/UWOC system using FG relaying with one legitimate receiver in the presence of eavesdropping.

2.1. RF Channel Model

The RF SR link is modeled by α-μ flat fading models, where the PDF of the received SNR, denoted by γ_1, can be expressed as [38]

$$f_{\gamma_1}(\gamma_1) = \frac{\alpha}{2\Gamma(\mu)} \frac{\mu^\mu}{(\tilde{\gamma}_1)^{\frac{\alpha\mu}{2}}} \gamma_1^{\frac{\alpha\mu}{2}-1} \exp\left(-\mu\left(\frac{\gamma_1}{\tilde{\gamma}_1}\right)^{\frac{\alpha}{2}}\right) \tag{1}$$

where $\gamma_1 \geq 0$, $\mu \geq 0$, $\alpha \geq 0$, and $\Gamma(\cdot)$ denotes the gamma function. The fading model parameters α and μ are associated with the non-linearity and multi-path propagation of the channel. Furthermore, the PDF of the received SNR at the eavesdropping node E, denoted by $f_{\gamma_e}(\gamma_e)$, also follows α-μ with parameters α_e and μ_e.

Based on the definition of the Fox's H-function, the CDF of γ_1, which is defined as $F_{\gamma_1}(\gamma_1) = \int_0^{\gamma_1} f_{\gamma_1}(\gamma_1) d\gamma_1$, can be expressed as

$$\begin{aligned}
F_{\gamma_1}(\gamma_1) &\overset{(a)}{=} \kappa \int_0^\gamma H_{0,1}^{1,0}\left[\gamma\Lambda \,\middle|\, \left(-\tfrac{1}{\alpha}+\mu, \tfrac{1}{\alpha}\right)\right] d\gamma \\
&= -\frac{i\kappa}{2\pi} \int_{\mathcal{L}} \Lambda^{-s} \Gamma\left(\frac{s}{\alpha}+\mu-\frac{1}{\alpha}\right) \int_0^\gamma \gamma^{-s} d\gamma \, ds \\
&= \frac{\kappa}{\Lambda} H_{1,2}^{1,1}\left[\gamma\Lambda \,\middle|\, \begin{matrix}(1,1)\\ \left(\mu,\tfrac{1}{\alpha}\right),(0,1)\end{matrix}\right]
\end{aligned} \tag{2}$$

where we use ([40], Equation (1.60)) and ([40], Equation (1.125)) to express $f_{\gamma_1}(\gamma_1)$ in the right side of equity (a) into the form of H-function, where $H_{\cdot,\cdot}^{\cdot,\cdot}[\cdot|\cdot]$ is the H-Function ([40], Equation (1.2)), $\kappa = \frac{\beta}{\Gamma(\mu)\tilde{\gamma}_1}$, $\Lambda = \frac{\beta}{\tilde{\gamma}_1}$, and $\beta = \frac{\Gamma\left(\frac{1}{\alpha}+\mu\right)}{\Gamma(\mu)}$. Note that, the present form of $F_{\gamma_1}(\gamma_1)$ in (2) is more suitable for deriving secrecy performance of a two-hop mixed RF/UWOC than the form proposed in ([41], Equation (2)) for the point-to-point system over single-input multiple-output α-μ channels.

2.2. UWOC Channel Model

To characterize the combined effects of different levels of air bubbles and temperature gradients on the light intensity received at underwater node D, we model the UWOC channel from R to D using the EGG distribution [32], where the PDF of the received SNR, defined as $\gamma_2 = (\eta I)^r / N_{0_2}$, has been derived in closed-form in terms of Meijer-G functions ([42], Equation (3)), and N_{0_2} is the received noise power at D. Based on ([40], Equation (1.112)), we can re-write the PDF of γ_2 using H-functions as

$$\begin{aligned}
f_{\gamma_2}(\gamma_2) =\ & \frac{c(1-\omega)}{\gamma r \Gamma(a)} H_{0,1}^{1,0}\left[b^{-c}\left(\frac{\gamma_2}{\mu_r}\right)^{\frac{c}{r}} \,\middle|\, (a,1)\right] \\
& + \frac{\omega}{\gamma_2 r} H_{0,1}^{1,0}\left[\frac{1}{\lambda}\left(\frac{\gamma_2}{\mu_r}\right)^{\frac{1}{r}} \,\middle|\, (1,1)\right]
\end{aligned} \tag{3}$$

where the parameters ω, a, b and c can be estimated using the maximum-likelihood criterion with expectation maximization algorithm. The parameter ω is the mixed weight of the distribution; λ is the parameter related to the exponential distribution; parameters a, b, and c are related to the exponential distribution; r is a parameter dependent on the detection scheme, specifically, $r = 1$ for heterodyne detection and $r = 2$ for intensity modulation and direct detection ([43], Equation (31)).

The EGG distribution can provide the best fit with the measured data form laboratory water tank experiments in the presence of temperature gradients and air bubbles [42]. Therefore, by using the EGG distribution to model the UWOC link, we can gain more insight into the relationship between characteristics of the UWOC link and the secrecy performance of the mixed RF/UWOC communication system.

Using the definition of complementary cumulative distribution function (CCDF), i.e., $\bar{F}_{\gamma_2}(\gamma_2) = \int_0^{\gamma_2} f_{\gamma_2}(\gamma_2)d\gamma_2$, and an approach similar to that used to derive (2), we can derive the CCDF of γ_2 as

$$
\begin{aligned}
\bar{F}_{\gamma_2}(\gamma_2) =& -\frac{i(1-\omega)}{2\pi\Gamma(a)} \int_{\mathcal{L}} \Gamma\left(a+\frac{rs}{c}\right) b^{rs} \mu_r^s \int_\gamma^\infty \frac{1}{\gamma^{s+1}} d\gamma \, ds \\
& -\frac{i\omega}{2\pi} \int_{\mathcal{L}} \Gamma(rs+1) \lambda^{rs} \mu_r^s \int_\gamma^\infty \frac{1}{\gamma^{s+1}} d\gamma \, ds \\
=& \frac{(1-\omega)}{\Gamma(a)} H_{1,2}^{2,0}\left[\frac{b^{-r}\gamma}{\mu_r}\middle| \begin{matrix}(1,1)\\(0,1),(a,\frac{r}{c})\end{matrix}\right] \\
& +r\omega H_{0,1}^{1,0}\left[\frac{\gamma\lambda^{-r}}{\mu_r}\middle| (0,r)\right].
\end{aligned}
\tag{4}
$$

It is worth to mention that the expression in (4) is useful to derive the closed-form CDF expression of the end-to-end SNR of the mixed RF/UWOC communication system.

3. End-to-End SNR

In this section, we derive the exact closed-form expressions for PDF and CDF of the end-to-end SNR of mixed RF/UWOC communication system. We then use these expressions to derive closed-form and asymptotic expressions for the system secrecy metrics in the following section.

The end-to-end instantaneous SNR of the mixed RF/UWOC system using the FG relaying scheme is given as [6]

$$
\gamma_{eq} = \frac{\gamma_1 \gamma_2}{\gamma_2 + C}
\tag{5}
$$

where C denotes the FG amplifying constant and is inversely proportional to the square of the relay transmitting power, and this constant is defined as $C = 1/(G^2 N_{0_1})$, where N_{0_1} is the received noise power at R, and the FG amplifying factor G is defined as $G = \sqrt{1/(\mathbb{E}[N_0(\gamma_1+1)])}$.

Using the definition of H-function and (1), we can readily express G^2 in terms of the H-functions

$$
G^2 = \kappa H_{1,2}^{2,1}\left[\Lambda\middle| \begin{matrix}(0,1)\\(0,1),(-\frac{1}{\alpha}+\mu,\frac{1}{\alpha})\end{matrix}\right].
\tag{6}
$$

It is worth noting that the FG relaying requires only the statistical CSI of the RF channel from S to R, and is therefore more convenient than VG relaying, which requires the instantaneous CSI, from the perspective of practical system deployment.

Theorem 1. *The CDF of the end-to-end SNR of the mixed RF/UWOC communication system using the FG relaying scheme $F_{\gamma_{eq}}(\gamma_{eq})$, defined in (5), can be obtained in exact closed-form as*

$$
\begin{aligned}
F_{\gamma_{eq}}(\gamma_{eq}) =& 1 - \frac{\gamma\kappa(1-\omega)}{\Gamma(a)} \\
& \times H_{1,0:0,2;1,1}^{0,1:2,0;0,1}\left[\frac{b^{-r}C}{\mu_r}\middle| \begin{matrix}(2,1,1): & ;\left(1+\frac{1}{\alpha}-\mu,\frac{1}{\alpha}\right)\\ : (0,1),(a,\frac{r}{c}); & (1,1)\end{matrix}\right] \\
& - H_{1,0:0,2;1,1}^{0,1:2,0;0,1}\left[\frac{C\lambda^{-r}}{\mu_r}\middle| \begin{matrix}(2,1,1): & ;\left(1+\frac{1}{\alpha}-\mu,\frac{1}{\alpha}\right)\\ : (1,1),(0,r); & (1,1)\end{matrix}\right] \\
& \times \gamma\kappa r\omega
\end{aligned}
\tag{7}
$$

in terms of bivariate H-functions, where $H_{\cdots}^{\cdots}[\cdot|\cdot]$ is the bivariate H-Function defined as ([40], Equation (2.55)).

Proof. See Appendix A. □

Note that the current implementation of bivariate H-function for numerical computation is mature and efficient, including GPU-accelerated versions, and has been implemented using the most popular software, including MATLAB® [44], Mathematica® [45], and Python [46]. Addtionally, the exact-closed expression for the CDF in (7) is a key analytical tool to derive the SOP metric of the mixed RF/UWOC system.

Theorem 2. *The PDF of the end-to-end SNR, which is defined in (5), of the mixed RF/UWOC communication system using the FG relaying scheme, denoted by $f_{\gamma_{eq}}(\gamma_{eq})$, can be obtained in exact closed-form as*

$$
\begin{aligned}
f_{\gamma_{eq}}(\gamma_{eq}) = {}& \frac{\kappa(1-\omega)}{\Gamma(a)} \\
&\times H^{0,1:0,1;2,0}_{1,0:1,1;0,2}\left[\begin{matrix} \frac{1}{\gamma\Lambda} \\ \frac{b^{-r}C}{\mu_r} \end{matrix} \,\middle|\, \begin{matrix} (2,1,1)\colon \left(1+\frac{1}{\alpha}-\mu,\frac{1}{\alpha}\right); \\ \colon \qquad (2,1) \qquad ;(0,1),(a,\frac{r}{c}) \end{matrix}\right] \\
&+ H^{0,1:0,1;2,0}_{1,0:1,1;0,2}\left[\begin{matrix} \frac{1}{\gamma\Lambda} \\ \frac{C\lambda^{-r}}{\mu_r} \end{matrix} \,\middle|\, \begin{matrix} (2,1,1)\colon \left(1+\frac{1}{\alpha}-\mu,\frac{1}{\alpha}\right); \\ \colon \qquad (2,1) \qquad ;(1,1),(0,r) \end{matrix}\right] \\
&\times \kappa r\omega.
\end{aligned}
\tag{8}
$$

Proof. See Appendix B. □

It is worth noting that the PDF expression in (8) is the most critical step required to evaluate the PNZ performance metric, as will be shown in the next section.

4. Performance Metrics

his section presents analytical results for the critical secrecy performance metrics of a mixed RF/UWOC communication system, including both SOP and PNZ, in the presence of air bubbles and temperature gradients in the UWOC channel and medium nonlinearity in the RF channel.

4.1. SOP

SOP is defined as the probability that the secrecy capacity C_s falls below a target rate of confidential information R_s and it can be expressed as

$$
\begin{aligned}
\mathcal{P}_{out}(R_s) &= \Pr\left\{\log_2\left(\frac{1+\gamma_{eq}}{1+\gamma_e}\right) < R_s\right\} \\
&= \int_0^\infty F_{eq}(\Theta\gamma_e + \Theta - 1)f_e(\gamma_e)d\gamma_e
\end{aligned}
\tag{9}
$$

where $\Theta = e^{R_s}$.

4.1.1. Lower Bound

Referring to [47,48], a tight lower bound for the SOP can be given as

$$
\mathcal{P}_{out,L} = \int_0^\infty F_{\gamma_{eq}}(\Theta\gamma)f_{\gamma_e}(\gamma)d\gamma.
\tag{10}
$$

Theorem 3. *The lower bound for the SOP of the mixed RF/UWOC communication system using the FG relaying scheme defined in (10) can be obtained in exact closed-form as*

$$
\mathcal{P}_{out,L} = 1 - \frac{\Theta\kappa}{\Lambda_e\Gamma(\mu_e)}\left(\frac{(1-\omega)}{\Gamma(a)}\right.
$$

$$
\times H^{0,1:1,1;2,0}_{1,0:1,2;0,2}\left[\begin{array}{c}\frac{\Lambda_e}{\Theta\Lambda}\\ \frac{b^{-r}C}{\mu_r}\end{array}\middle|\begin{array}{c}(2,1,1):\quad\left(1+\frac{1}{\alpha}-\mu,\frac{1}{\alpha}\right)\quad;\\ :\left(\frac{1}{\alpha_e}+\mu_e,\frac{1}{\alpha_e}\right),(1,1)\,;(0,1),(a,\frac{r}{c})\end{array}\right]
$$

$$
- H^{0,1:1,1;2,0}_{1,0:1,2;0,2}\left[\begin{array}{c}\frac{\Lambda_e}{\Theta\Lambda}\\ \frac{C\lambda^{-r}}{\mu_r}\end{array}\middle|\begin{array}{c}(2,1,1):\quad\left(1+\frac{1}{\alpha}-\mu,\frac{1}{\alpha}\right)\quad:\\ :\left(\frac{1}{\alpha_e}+\mu_e,\frac{1}{\alpha_e}\right),(1,1):(1,1),(0,r)\end{array}\right]
$$

$$
\left.\times r\omega\right). \tag{11}
$$

Proof. See Appendix C. $\square$

Special case. When the RF channel follows Rayleigh fading (i.e., $\alpha = \alpha_e = 1$, $\mu = \mu_e = 1$) and the thermally uniform UWOC channel (i.e., $c = 1$) use heterodyne detection (i.e., $r = 1$), using definition of bivariate H-functions and ([49], Equation (07.34.03.0397.01)), Equation (5) can be simplified into the following form

$$
\mathcal{P}_{out,L} = \frac{\kappa\kappa_e}{\Lambda(\Lambda_e + \Theta\Lambda)}\left(a(\omega - 1)\exp\left(\frac{\mathcal{J}}{b}\right)E_{a+1}\left(\frac{\mathcal{J}}{b}\right)\right.
$$

$$
\left.- \omega G^{1,2}_{2,1}\left[\frac{\lambda}{\mathcal{J}}\middle|\begin{array}{c}0,1\\ 1\end{array}\right]\right) + 1 \tag{12}
$$

where $\mathcal{J} = \frac{\mathcal{C}\Theta\Lambda}{\mu_r(\Lambda_e+\Theta\Lambda)}$, $E_i(x)$ and $E_n(x)$ both denote the exponential integral ([50], Equation (8.211.1)). We emphasize that the distribution in (12) contains only elementary functions and leads to straightforward secrecy performance evaluation of two-hop mixed RF/UWOC systems.

4.1.2. Asymptotic Results

To gain more insight into the SOP performance and the dependency between the link quality of both RF and UWOC channels, we now derive asymptotic expressions for SOP. We consider two scenarios, namely $\gamma_1 \to \infty$ and $\gamma_e \to \infty$.

Corollary 1. *For scenarios $\gamma_1 \to \infty$ and $\gamma_e \to \infty$, the asymptotic expressions of SOP of a mixed RF/UWOC communication system using FG relaying scheme can be given as*

$$
\mathcal{P}_{out,a,1} = 1 - \frac{1}{\Gamma(\mu)\Gamma(\mu_e)}\left(\frac{(1-\omega)}{\Gamma(a)}\right.
$$

$$
\times H^{1,3}_{3,2}\left[\frac{b^r\Lambda_e\mu_r}{\mathcal{C}\Theta\Lambda}\middle|\begin{array}{c}(1,1),(1-a,\frac{r}{c}),(1-\mu,\frac{1}{\alpha})\\ \left(\mu_e,\frac{1}{\alpha_e}\right),(0,1)\end{array}\right]
$$

$$
\left.- r\omega H^{1,2}_{2,1}\left[\frac{\lambda^r\Lambda_e\mu_r}{\mathcal{C}\Theta\Lambda}\middle|\begin{array}{c}(1,r),(1-\mu,\frac{1}{\alpha})\\ \left(\mu_e,\frac{1}{\alpha_e}\right)\end{array}\right]\right) \tag{13}
$$

and

$$\mathcal{P}_{out,a,e}=1-\frac{\alpha_e\Gamma\left(\mu+\frac{\alpha_e\mu_e}{\alpha}\right)}{\Gamma(\mu)\Gamma(\mu_e)\Gamma(\alpha_e\mu_e+1)}\left(\frac{\Lambda_e}{\Theta\Lambda}\right)^{\alpha_e\mu_e}$$
$$\times\left(\frac{(1-\omega)}{\Gamma(a)}H_{2,1}^{1,2}\left[\frac{b^r\mu_r}{C}\middle|\begin{array}{c}(1,1),(1-a,\frac{r}{c})\\(\alpha_e\mu_e,1)\end{array}\right]\right.$$
$$\left.-r\omega H_{2,1}^{1,2}\left[\frac{\lambda^r\mu_r}{C}\middle|\begin{array}{c}(0,1),(1,r)\\(\alpha_e\mu_e,1)\end{array}\right]\right)\tag{14}$$

in terms of H-functions, respectively.

Proof. See Appendix D. □

Note that in contrast to the closed expression of the lower bound of the SOP in (11) in terms of bivariate *H*-functions, which requires numerical evaluation of double line integrals, the asymptotic expressions in (13) and (14) only require the numerical calculation of single line integrals, thus reducing the complexity of the calculations. Furthermore, as shown in Section 5, for a target SOP performance, the asymptotic expressions in (13) and (14) can be used to determine rapidly the suboptimal transmitting power to maximize energy efficiency.

Special case. A two-hop mixed RF/UWOC communication system over Rayleigh RF links and a thermally uniform UWOC channel, we can further simplify the asymptotic expressions in (13) and (14) by setting $c=1$, $\alpha=\alpha_e=1$, $\mu=\mu_e=1$. For example, Equation (13) can be simplified into

$$\mathcal{P}_{out,a,1}=\frac{1}{\lambda\Lambda_e\mu_r\Gamma(\mu)\Gamma(\mu_e)}\left(a\lambda(\omega-1)\Lambda_e\mu_r\exp\left(\frac{C\Theta\Lambda}{b\Lambda_e\mu_r}\right)\right.$$
$$\times E_{a+1}\left(\frac{C\Theta\Lambda}{b\Lambda_e\mu_r}\right)-C\Theta\Lambda\omega\exp\left(\frac{C\Theta\Lambda}{\lambda\Lambda_e\mu_r}\right)$$
$$\left.\times E_i\left(-\frac{C\Theta\Lambda}{\lambda\Lambda_e\mu_r}\right)\right)+\frac{\Lambda\Lambda_e-\kappa\omega\kappa_e}{\Lambda\Lambda_e}\tag{15}$$

4.2. PNZ

PNZ is another critical metric for to evaluate the secrecy performance of a communication system, which is defined as $\Pr(C_s>0)$, where C_s is the secrecy capacity. PNZ is generally related to channel conditions of all the channels in the mixed RF/UWOC systems. In this section, we derive the exact closed-form and asymptotic expressions for PNZ and analyze the relationship between channel parameters and PNZ performance.

4.2.1. Exact Results

According to [48], PNZ can be reformed as

$$\mathcal{P}_{nz}=\Pr(\gamma_{eq}>\gamma_e)=\int_0^\infty f_{eq}(\gamma_{eq})F_e(\gamma_{eq})d\gamma_{eq}.\tag{16}$$

Theorem 4. *The exact PNZ of the mixed RF/UWOC communication system using the FG relaying scheme defined in (16) can be obtained in exact closed-form as*

$$\mathcal{P}_{nz} = \frac{\kappa}{\Gamma(a)\Lambda_e\Gamma(\mu_e)}\Bigg((1-\omega)$$

$$\times H^{0,1:2,0;1,1}_{1,0:0,2;1,2}\left[\begin{matrix}\frac{b^{-r}c}{\mu_r}\\[2pt]\frac{\Lambda_e}{\Lambda}\end{matrix}\left|\begin{matrix}(2,1,1):\qquad\qquad;\quad\left(1+\frac{1}{\alpha}-\mu,\frac{1}{\alpha}\right)\\[4pt]:(0,1),(a,\frac{r}{c});\left(\frac{1}{\alpha_e}+\mu_e,\frac{1}{\alpha_e}\right),(1,1)\end{matrix}\right.\right]$$

$$+H^{0,1:2,0;1,1}_{1,0:0,2;1,2}\left[\begin{matrix}\frac{c\lambda^{-r}}{\mu_r}\\[2pt]\frac{\Lambda_e}{\Lambda}\end{matrix}\left|\begin{matrix}(2,1,1)\qquad\qquad\left(1+\frac{1}{\alpha}-\mu,\frac{1}{\alpha}\right)\\[4pt](1,1),(0,r)\left(\frac{1}{\alpha_e}+\mu_e,\frac{1}{\alpha_e}\right),(1,1)\end{matrix}\right.\right]$$

$$\times r\omega\Gamma(a)\Bigg) \tag{17}$$

Proof. See Appendix E. $\square$

    Special case. For a RF/UWOC system with Rayleigh and uniform temperature EGG distributions, using a similar approach to the derivation of (12), we can simplify (17) into

$$\mathcal{P}_{nz}=-\frac{\kappa\kappa_e}{\lambda\Lambda(\Lambda_e+\Lambda)}\Bigg(\omega\mathcal{H}\exp(\mathcal{H}/\lambda)E_1\left(\frac{\mathcal{H}}{\lambda}\right)$$

$$+\lambda\left(a(\omega-1)\exp\left(\frac{\mathcal{H}}{b}\right)E_{a+1}\left(\frac{\mathcal{H}}{b}\right)-\omega\right)\Bigg) \tag{18}$$

where $\mathcal{H}=\frac{c\Lambda}{(\Lambda_e+\Lambda)\mu_r}$.

4.2.2. Asymptotic Results

    To gain more insight into the PNZ performance and the dependency between the link quality of both RF and UWOC channels, we now derive asymptotic expressions for PNZ. We consider two scenarios, namely $\gamma_1 \to \infty$ and $\gamma_e \to \infty$.

Corollary 2. *For scenarios $\gamma_1 \to \infty$ and $\gamma_e \to \infty$, the asymptotic expressions of PNZ of a mixed RF/UWOC communication system using the FG relaying scheme are given as*

$$\mathcal{P}_{nz,1}=\frac{1}{\Gamma(a)\Gamma(\mu)\Gamma(\mu_e)}\Bigg((1-\omega)$$

$$\times H^{3,1}_{2,3}\left[\frac{b^{-r}C\Lambda}{\Lambda_e\mu_r}\left|\begin{matrix}\left(1-\mu_e,\frac{1}{\alpha_e}\right),(1,1)\\[2pt](0,1),(a,\frac{r}{c}),(\mu,\frac{1}{\alpha})\end{matrix}\right.\right]$$

$$+r\omega\Gamma(a)H^{2,1}_{1,2}\left[\frac{C\lambda^{-r}\Lambda}{\Lambda_e\mu_r}\left|\begin{matrix}\left(1-\mu_e,\frac{1}{\alpha_e}\right)\\[2pt](0,r),(\mu,\frac{1}{\alpha})\end{matrix}\right.\right]\Bigg) \tag{19}$$

and

$$\mathcal{P}_{nz,e}=\frac{\alpha_e\left(\frac{\Lambda_e}{\Lambda}\right)^{\alpha_e\mu_e}\Gamma\left(\mu+\frac{\alpha_e\mu_e}{\alpha}\right)}{\Gamma(a)\Gamma(\mu)\Gamma(\mu_e)\Gamma(\alpha_e\mu_e+1)}\Bigg((1-\omega)$$

$$\times H^{1,2}_{2,1}\left[\frac{b^r\mu_r}{C}\left|\begin{matrix}(1,1),\left(1-a,\frac{r}{c}\right)\\[2pt](\alpha_e\mu_e,1)\end{matrix}\right.\right]$$

$$+r\omega\Gamma(a)H^{1,2}_{2,1}\left[\frac{\lambda^r\mu_r}{C}\left|\begin{matrix}(0,1),(1,r)\\[2pt](\alpha_e\mu_e,1)\end{matrix}\right.\right]\Bigg) \tag{20}$$

in terms of H-functions, respectively.

Proof. Observing that the expressions for the lower bound of the SOP in (11) and exact PNZ in (15) have a similar structure; therefore, Equations (19) and (20) can be easily obtained using the same techniques as those used for deriving (13) and (14), and the proof is complete. □

Note that, similar to the asymptotic expressions of the SOP in (13) and (14), for a target PNZ performance, the asymptotic expressions of PNZ in (19) and (20) are also suitable for fast numerical calculations and are useful to determine the suboptimal transmitting power to maximize energy efficiency.

5. Numerical Results and Discussion

In this section, we provide some numerical results to verify the analytic and asymptotic expressions of SOP and PNZ derived in Section 4, and thoroughly investigate the combined effect of the channel quality of both RF and UWOC channels on the secrecy performance of the two-hop mixed RF/UWOC communication system. All practical environmental physical factors that can affect channel quality, including levels of air bubbles, temperature gradients, and salinity of the UWOC channel [32], as well as the medium nonlinearity and multipath cluster characteristics of the RF channel [48], are taken into account. For brevity, we use $[\cdot, \cdot]$ to denote the value set of [air bubbles level, temperature gradient] in this section.

In Figures 2–6, we investigate the combined effect of the channel quality of both RF and UWOC channels on the SOP metric of the two-hop mixed RF/UWOC communication system.

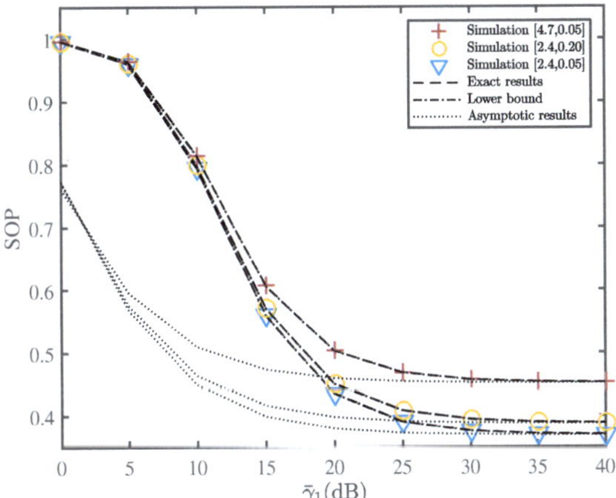

Figure 2. SOP versus $\bar{\gamma}_1$ with various fading parameters when $\alpha = \alpha_e = 1.6$, $\mu = \mu_e = 1.5$, $R_s = 0.01$, and $\bar{\gamma}_e = \bar{\gamma}_2 = 10$ dB.

Figure 2 shows the lower bound and the asymptotic SOP with average SNR of the SR link γ_1 for a mixed two-hop RF/UWOC system under different quality scenarios of UWOC channel. Both RF SR and SE links follow the α-μ distribution and have the same parameters, where $\alpha = \alpha_e = 1.6$, $\mu = \mu_e = 1.5$. The average SNR of the SE and RD links are both set as $\bar{\gamma}_e = \bar{\gamma}_2 = 10$ dB [48]. As shown in Figure 2, the exact theoretical results are almost identical to the simulation results, and both closely agree with the derived lower bound. Asymptotic results are tight when the average SNR is greater than 30 dB. Further, when the average SNR increases from 0 to 30 dB, SOP rapidly decreases. Additionally, SOP tends to saturate when the average SNR is between 30 and 40 dB. Given the cost of the relay battery replacement and engineering difficulties, the communication system should guarantee the SOP while cutting down on energy consumption. In practice, one should therefore select the suboptimal transmission power corresponding to the saturation starting point.

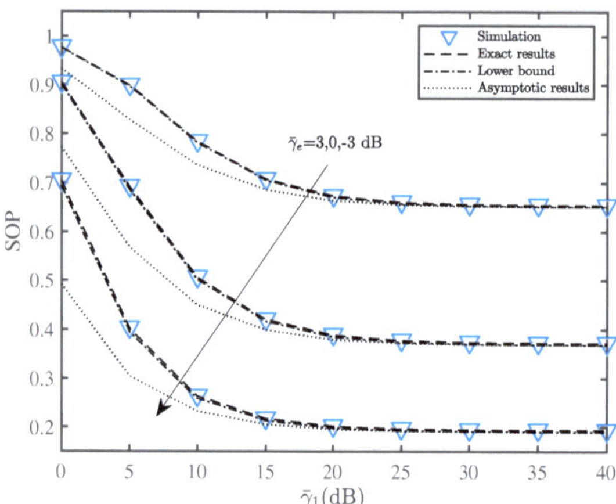

Figure 3. SOP versus $\bar{\gamma}_1$ with various fading parameters when $\alpha = \alpha_e = 1.6$, $\mu = \mu_e = 1.5$, $R_s = 0.01$, and $\bar{\gamma}_2 = 0$ dB.

Figure 3 depicts the SOP variation versus the SR average SNR γ_1 for the mixed two-hop RF/UWOC system under three different eavesdropper interference levels, i.e., $\bar{\gamma}_e = 3, 0, -3$ dB. Parameters in Figure 3 are set as follows: $\alpha = \alpha_e = 1.6$, $\mu = \mu_e = 1.5$, UWOC channel parameter is [2.4, 0.05], and $\bar{\gamma}_2 = 0$ dB. It can be observed that the lower bounds closely match the exact results in the whole SNR region. The asymptotic result curve gradually coincides with the exact result curve when $\bar{\gamma}_1$ takes higher values starting from 20 dB. We can also observe that the SOP is monotonically decreasing with $\bar{\gamma}_1$, assuming that the SNR of the SE link is a fixed value. Comparing the SOP curves for three different eavesdropping interference levels, one can conclude that as the quality of the SE channel improves, the secrecy performance of the system deteriorates.

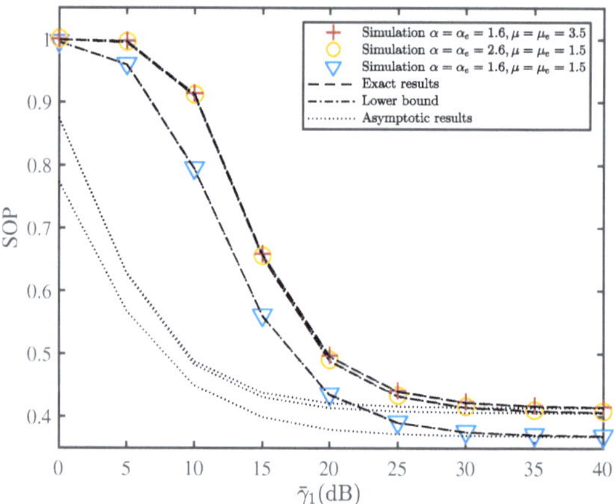

Figure 4. SOP versus $\bar{\gamma}_1$ with various fading parameters $R_s = 0.01$, $\bar{\gamma}_2 = \bar{\gamma}_e = 10$ dB, and UWOC channel parameter is [2.4, 0.05].

Figure 4 indicates the effect of the variation in average SNR of the SR link on the SOP metric of a two-hop mixed RF/UWOC, with three different RF channel qualities. Evidently, SOP monotonically decreases with the increase of $\bar{\gamma}_1$, and SOP tends to saturate when $\bar{\gamma}_1 \geq 30$ dB. Moreover, Figure 4 depicts that as the α-μ value increases, the two-hop mixed RF/UWOC system secrecy performance worsens, and vice versa. This is because of the phenomena of severe nonlinearity and sparse clustering when the signals are propagating in a high α-μ value RF channel, and poor RF channel quality makes it easier for eavesdroppers to intercept signals. As shown in Figure 5, as the $\bar{\gamma}_e$ progressively increases, the SOP value increases, the information intercepted by the eavesdropper increases, and the system secrecy performance gradually decreases. Moreover, the asymptotic result is more accurate at $\bar{\gamma}_e$ greater than 15 dB.

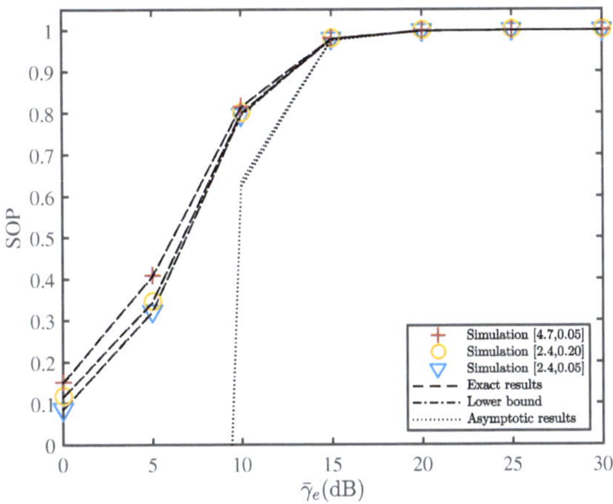

Figure 5. SOP versus $\bar{\gamma}_e$ with various fading parameters when $\alpha = \alpha_e = 1.6$, $\mu = \mu_e = 1.5$, $R_s = 0.01$, and $\bar{\gamma}_1 = \bar{\gamma}_2 = 10$ dB.

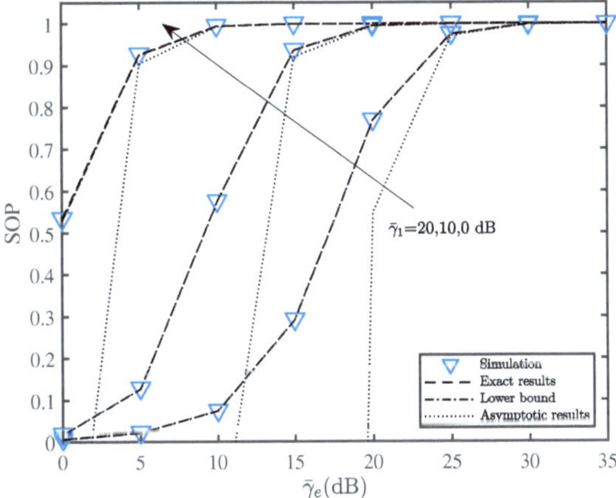

Figure 6. SOP versus $\bar{\gamma}_e$ with various fading parameters when $\alpha = \alpha_e = 1.6$, $\mu = \mu_e = 1.5$, $R_s = 0.01$, and $\bar{\gamma}_2 = 20$ dB.

In Figure 6, we set the same channel parameters as in Figure 3, except for setting the UWOC average SNR, i.e., $\bar{\gamma}_2 = 20$ dB. Figure 6 shows that SOP increases with $\bar{\gamma}_e$ when the other parameters remain unchanged. The same interpretation of Figure 5 can also be applied to Figure 6. Additionally, the rate at which the asymptotic results approach exact results varies for different SR average SNR. For $\bar{\gamma}_1 = 0$ dB, the asymptotic results begin to match the exact result starting at $\bar{\gamma}_e = 5$ dB. Moreover, the close match of the lower bound and the exact results demonstrate the robustness and accuracy of (11).

In Figures 7–10, We investigate the combined effect of the channel quality of both RF and UWOC channels on the PNZ metric of the two-hop mixed RF/UWOC communication system.

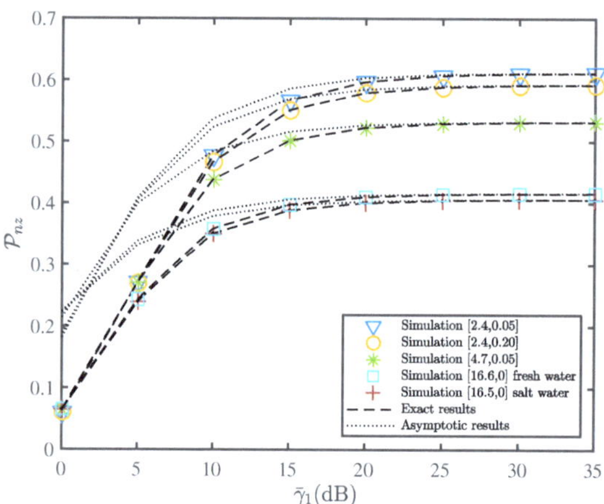

Figure 7. $\mathcal{P}_{nz}$ versus $\bar{\gamma}_1$ with various fading parameters when $\alpha = \alpha_e = 2.1$, $\mu = \mu_e = 1.4$ and $\bar{\gamma}_e = \bar{\gamma}_2 = 0$ dB.

Figure 7 shows the effect of the SR link average SNR $\bar{\gamma}_1$ on the PNZ of the mixed RF/UWOC for different UWOC channel parameters. PNZ increases incrementally as $\bar{\gamma}_1$ increases, which indicates an increase in secrecy performance. It can be observed that PNZ decreases as the degree of turbulence increases, i.e., the higher the level of air bubbles and the larger the temperature gradient, the worse the secrecy performance in the system. Additionally, we depict the effects of salinity on UWOC performance in Figure 7. The salinity affects the system secrecy performance to a much lesser extent than the level of the air bubble and temperature gradient. This is because the generation and break-up of the air bubbles in the UWOC channels causes dramatic and random fluctuations of the underwater optical signals, which can significantly deteriorate the secrecy performance of the system. Figure 7 shows that eavesdroppers may benefit from a low UWOC channel quality. On the contrary, in a high-quality UWOC channel, the likelihood of an eavesdropper successfully eavesdropping is greatly reduced. Therefore, in practical applications, increasing the channel quality can increase the system transmission capacity and thus improve the system secrecy performance. Figure 7 also shows that asymptotic results can quickly approach the exact result for poorer channels. For example, for a UWOC channel with channel parameters of [16.5, 0], the asymptotic result can achieve a match with the exact value at $\bar{\gamma}_1 \geq 20$ dB. When the channel parameter set is [2.4, 0.05], the asymptotic result can only be accurate at $\bar{\gamma}_1 > 25$ dB. The remaining parameters are set as follows, $\bar{\gamma}_e = \bar{\gamma}_2 = 0$ dB, $\alpha = \alpha_e = 2.1$, $\mu = \mu_e = 1.4$.

In Figure 8, the RF channel parameters are $\alpha = \alpha_e = 1.5$, $\mu = \mu_e = 0.8$, and the UWOC channel parameters are [2.4, 0.05]. We can explain the curves in Figure 8 using a principle similar to Figure 7. In particular, Figure 8 demonstrates the PNZ curves for three different SE link channel qualities. Obviously, as $\bar{\gamma}_e$ decreases, the secrecy performance of the system improves.

In addition to Figure 8, we analyzed the effect of the average SNR $\bar{\gamma}_1$ on the PNZ, as shown by Figures 9 and 10. The difference is that in Figure 9 $\alpha = \alpha_e = 2.1$, $\mu = \mu_e = 1.4$ and $\bar{\gamma}_1 = \bar{\gamma}_2 = 20$ dB. whereas the RF channel parameters in Figure 10 are $\alpha = \alpha_e = 1.5$ and $\mu = \mu_e = 0.8$. It can be inferred from Figures 9 and 10 that the asymptotic result only matches the exact value when $\bar{\gamma}_e$ is large, and the PNZ gradually decreases until it reaches zero.

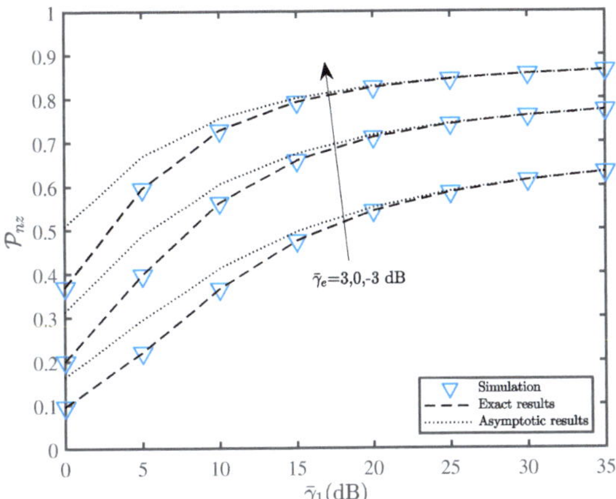

Figure 8. $\mathcal{P}_{nz}$ versus $\bar{\gamma}_1$ with various fading parameters when $\alpha = \alpha_e = 1.5$, $\mu = \mu_e = 0.8$ and $\bar{\gamma}_2 = 0$ dB.

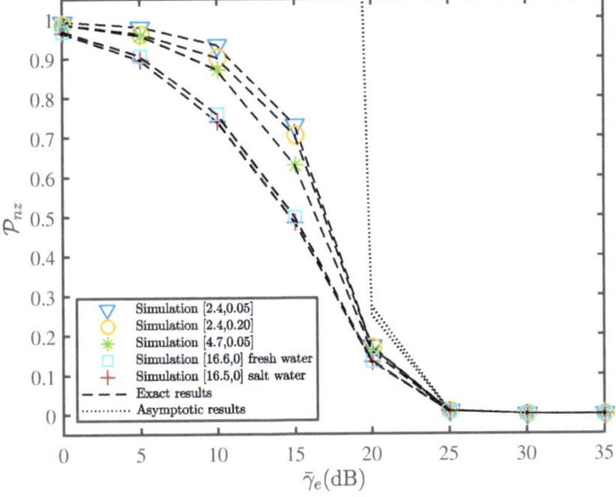

Figure 9. $\mathcal{P}_{nz}$ versus $\bar{\gamma}_e$ with various fading parameters when $\alpha = \alpha_e = 2.1$, $\mu = \mu_e = 1.4$ and $\bar{\gamma}_1 = \bar{\gamma}_2 = 20$ dB.

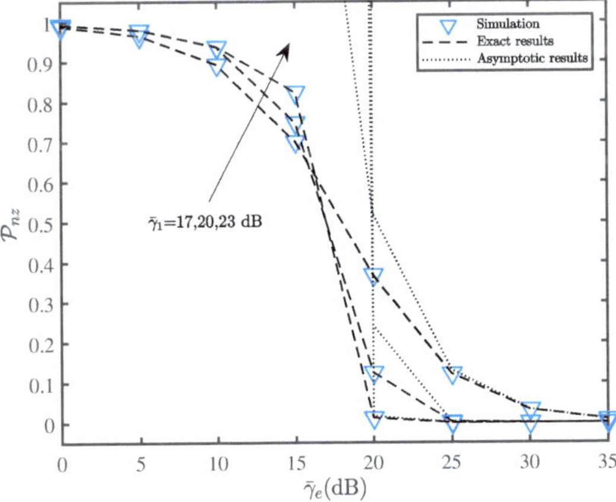

Figure 10. $\mathcal{P}_{nz}$ versus $\bar{\gamma}_e$ with various fading parameters when $\alpha = \alpha_e = 1.5$, $\mu = \mu_e = 0.8$ and $\bar{\gamma}_2 = 20$ dB.

6. Conclusions

We investigated the secrecy performance of a UAV-assisted two-hop mixed RF/UWOC communication system using fixed-gain AF relaying. To allow the results to be more generic and applicable to more realistic physical scenarios, we modeled RF channels using the α-μ distribution, which considers both the nonlinear of the transmission medium and multipath cluster characteristics, and modeled UWOC channels using the laboratory EGG distribution, which can account for different levels of air bubbles, temperature gradients, and salinity. Closed-form expressions for the PDF and the CDF of the two-hop end-to-end SNR were both derived in terms of the bivariate H-function. Based on these results, we obtained a tight closed-form expression of the lower bound of the SOP and the exact closed-form expression of the PNZ. Furthermore, we also derived asymptotic expressions in simple functions for both SOP and PNZ to allow rapid numerical evaluation. Moreover, based on the asymptotic results, we presented an approach to determine the suboptimal transmitting power to maximize the energy efficiency, for given target performance of both SOP and PNZ. We fully investigated the effects of various existing phenomena of both RF and UWOC channels on the secrecy performance of the mixed RF/UWOC communication system. Additionally, our generalized theoretical framework is also applicable to various classical RF and underwater optical channel models including Rayleigh and Nakagami for RF channels and EG and Generalized Gamma for UWOC channels. Our results can be used in a practical mixed security RF/UWOC communication systems design. The interesting topics for future work include: (i) to investigate the secrecy performance of a mixed RF/UWOC communication system using an energy-harvesting enabled relay to improve the system lifetime; (ii) to investigate the secrecy performance of a mixed RF/UWOC communication system using multiple relays with appropriate relaying selection algorithms.

Author Contributions: Conceptualization, Y.L. and R.S.; methodology, Y.L.; software, J.W.; validation, Y.L., R.S. and J.C.; formal analysis, Y.L.; investigation, R.S.; resources, G.Q.; data curation, R.S.; writing—original draft preparation, Y.L.; writing—review and editing, J.C.; visualization, J.W.; supervision, G.Q.; project administration, G.Q.; funding acquisition, Y.L. All authors have read and agreed to the published version of the manuscript.

Funding: This research was funded by the National Natural Science Foundation of China (62101152).

Data Availability Statement: Not applicable.

Conflicts of Interest: The authors declare no conflict of interest.

Appendix A. Proof of Theorem 1

Using (5), we write the CDF of the end-to-end SNR in the following form

$$F_{\gamma_{eq}}(\gamma_{eq}) = \int_0^\infty \Pr\left[\frac{\gamma_1 \gamma_2}{\gamma_2 + C} \leq \gamma \mid \gamma_2\right] f_{\gamma_2}(\gamma_2)\, d\gamma_2$$

$$= 1 - \int_\gamma^\infty \bar{F}_{\gamma_2}\left(\frac{C\gamma}{x - \gamma}\right) f_{\gamma_1}(x)\, dx. \tag{A1}$$

Substituting (1) and (4) into (A1) and replacing the integral variable x with $z = x + \gamma$, after some simplifications, we can express (A1) as

$$F_{\gamma_{eq}}(\gamma_{eq}) = 1 + I_1 + I_2 \tag{A2}$$

where

$$I_1 = -\frac{\kappa(1-\omega)}{\Gamma(a)} \int_0^\infty H_{0,1}^{1,0}\left[(z+\gamma)\Lambda \,\middle|\, \left(-\tfrac{1}{\alpha}+\mu, \tfrac{1}{\alpha}\right)\right]$$

$$\times H_{1,2}^{2,0}\left[\frac{b^{-r}C\gamma}{z\mu_r} \,\middle|\, \begin{matrix}(1,1)\\(0,1),(a,\tfrac{r}{c})\end{matrix}\right] dz \tag{A3}$$

and

$$I_2 = -\kappa r \omega \int_0^\infty H_{0,1}^{1,0}\left[\frac{C\gamma\lambda^{-r}}{z\mu_r}\,\middle|\,(0,r)\right]$$
$$\times H_{0,1}^{1,0}\left[(z+\gamma)\Lambda\,\middle|\,\left(-\tfrac{1}{\alpha}+\mu,\tfrac{1}{\alpha}\right)\right]dz. \tag{A4}$$

To solve (A3), we convert all the H-functions in (A3) into a line integral, and place the integral with respect to x in the innermost part by rearranging the order of multiple integrals. Then, we have

$$I_1 = \frac{\kappa(1-\omega)}{4\pi^2\Gamma(a)}\int_{\mathcal{L}}^t \frac{\Gamma(t)}{\Gamma(t+1)}\Gamma\left(a+\frac{rt}{c}\right)\left(\frac{b^r\mu_r}{C\gamma}\right)^t$$
$$\times \int_{\mathcal{L}}^s \Lambda^{-s}\Gamma\left(\frac{s}{\alpha}+\mu-\frac{1}{\alpha}\right)\int_0^\infty z^t(z+\gamma)^{-s}dz\,ds\,dt. \tag{A5}$$

By utilizing ([50], Equation (3.197/1)) to solve the integration of z, after some simplifications and using the definition of the bivariate H-function ([40], Equation (2.57)), we can finally express I_1 in (A3) in the following form

$$I_1 = -\frac{\gamma\kappa(1-\omega)}{\Gamma(a)}$$
$$\times H_{1,0:0,2;1,1}^{0,1:2,0;0,1}\left[\begin{matrix}\frac{b^{-r}C}{\mu_r}\\[4pt]\frac{1}{\gamma\Lambda}\end{matrix}\,\middle|\,\begin{matrix}(2,1,1): & & :\left(1+\tfrac{1}{\alpha}-\mu,\tfrac{1}{\alpha}\right)\\ & :(0,1),(a,\tfrac{r}{c}): & (1,1)\end{matrix}\right]. \tag{A6}$$

We can solve (A4) in a similar way as we have solved (A3). All H-functions are converted to the form of the line integrals and by rearranging the multiple integrals, the integral regarding z is placed in the innermost part of the expression. Then, we have

$$I_2 = \frac{\kappa r \omega}{4\pi^2}\int_{\mathcal{L}}^s \Gamma(rs)\left(\frac{\lambda^r\mu_r}{C\gamma}\right)^s \int_{\mathcal{L}}^t \Lambda^{-t}\Gamma\left(\frac{t}{\alpha}+\mu-\frac{1}{\alpha}\right)$$
$$\times \int_0^\infty z^s(z+\gamma)^{-t}dz\,dt\,ds. \tag{A7}$$

Again, we use ([50], Equation (3.197/1)) to solve the integration regarding z. Then use ([40], Equation (2.57)) and some simplification, we obtain the following expression

$$I_2 = -\gamma\kappa r \omega$$
$$\times H_{1,0:0,2;1,1}^{0,1:2,0;0,1}\left[\begin{matrix}\frac{C\lambda^{-r}}{\mu_r}\\[4pt]\frac{1}{\gamma\Lambda}\end{matrix}\,\middle|\,\begin{matrix}(2,1,1): & & ;\left(1+\tfrac{1}{\alpha}-\mu,\tfrac{1}{\alpha}\right)\\ & :(1,1),(0,r); & (1,1)\end{matrix}\right]. \tag{A8}$$

Substituting (A6) and (A8) into (A2), we obtain the exact closed-form expression for the CDF as shown by (7).

Appendix B. Proof of Theorem 2

The PDF of the end-to-end SNR can be obtained by using

$$f(\gamma_{eq}) = \frac{dF(\gamma_{eq})}{d\gamma_{eq}}. \tag{A9}$$

Substituting (7) into (A9), after some simplifications, we have

$$f_{\gamma_{eq}}(\gamma_{eq}) = \frac{dJ_1}{d\gamma_{eq}} + \frac{dJ_2}{d\gamma_{eq}} \tag{A10}$$

where

$$J_1 = \frac{\gamma_{eq}\kappa(1-\omega)}{4\pi^2\Gamma(a)} \int_{\mathcal{L}}^{t} \int_{\mathcal{L}}^{s} \frac{1}{\Gamma(t)} \left(\frac{1}{\gamma_{eq}\Lambda}\right)^{t} \Gamma(-s)\Gamma\left(a - \frac{rs}{c}\right)$$
$$\times \Gamma(s+t-1)\Gamma\left(\frac{t}{\alpha} + \mu - \frac{1}{\alpha}\right)\left(\frac{b^{-r}C}{\mu_r}\right)^{s} dsdt \tag{A11}$$

and

$$J_2 = \frac{\gamma_{eq}\kappa r\omega}{4\pi^2} \int_{\mathcal{L}}^{t} \int_{\mathcal{L}}^{s} \frac{1}{\Gamma(t)} \left(\frac{1}{\gamma_{eq}\Lambda}\right)^{t} \Gamma(1-s)\Gamma(-rs)\Gamma(s+t-1)$$
$$\times \Gamma\left(\frac{t}{\alpha} + \mu - \frac{1}{\alpha}\right)\left(\frac{C\lambda^{-r}}{\mu_r}\right)^{s} dsdt. \tag{A12}$$

By enabling the differential operation in (A10), after some rearrangements, we can represent the first and the second terms on the right of the Equation (A10) as

$$\frac{dJ_1}{d\gamma_{eq}} = \frac{\kappa(1-\omega)}{4\pi^2\Gamma(a)} \int_{\mathcal{L}}^{t} \int_{\mathcal{L}}^{s} \frac{(1-t)}{\Gamma(t)} C^{s}\Gamma(-s)(\gamma_{eq}\Lambda)^{-t}b^{-rs}\mu_r^{-s}$$
$$\times \Gamma(s+t-1)\Gamma\left(\frac{t}{\alpha} + \mu - \frac{1}{\alpha}\right)\Gamma\left(a - \frac{rs}{c}\right)dsdt \tag{A13}$$

and

$$\frac{dJ_2}{d\gamma_{eq}} = \frac{\kappa r\omega}{4\pi^2} \int_{\mathcal{L}}^{t} \int_{\mathcal{L}}^{s} \frac{1}{\Gamma(t)} (1-t)C^{s}\Gamma(1-s)(\gamma_{eq}\Lambda)^{-t}\lambda^{-rs}$$
$$\times \mu_r^{-s}\Gamma(-rs)\Gamma(s+t-1)\Gamma\left(\frac{t}{\alpha} + \mu - \frac{1}{\alpha}\right)dsdt, \tag{A14}$$

respectively.

After substitute (A13) and (A14) to (A10) and use the definition of bivariate H-function, we can derive the exact closed-form expression of PDF as shown in (8).

Appendix C. Proof of Theorem 3

Substituting (1) and (7) into (10), after some rearrangements, we have

$$\mathcal{P}_{out,L} = 1 + Q_1 + Q_2 \tag{A15}$$

where

$$Q_1 = -\frac{\Theta\kappa(1-\omega)\kappa_e}{\Gamma(a)} \int_0^{\infty} \gamma H_{0,1}^{1,0}\left[\gamma\Lambda_e \left| \left(-\frac{1}{\alpha_e} + \mu_e, \frac{1}{\alpha_e}\right)\right. \right]$$
$$\times H_{1,0:0,2;1,1}^{0,1:2,0;0,1}\left[\left.\begin{array}{c} \frac{b^{-r}C}{\mu_r} \\ \frac{1}{\gamma\Theta\Lambda} \end{array}\right| \begin{array}{ccc} (2,1,1): & ; & \left(1 + \frac{1}{\alpha} - \mu, \frac{1}{\alpha}\right) \\ : (0,1),(a,\frac{r}{c}); & & (1,1) \end{array}\right]d\gamma \tag{A16}$$

and

$$Q_2 = -r\gamma\Theta\kappa\omega\kappa_e \int_0^{\infty} \gamma H_{0,1}^{1,0}\left[\gamma\Lambda_e \left| \left(-\frac{1}{\alpha_e} + \mu_e, \frac{1}{\alpha_e}\right)\right. \right]$$
$$\times H_{1,0:0,2;1,1}^{0,1:2,0;0,1}\left[\left.\begin{array}{c} \frac{C\lambda^{-r}}{\mu_r} \\ \frac{1}{\gamma\Theta\Lambda} \end{array}\right| \begin{array}{ccc} (2,1,1): & ; & \left(1 + \frac{1}{\alpha} - \mu, \frac{1}{\alpha}\right) \\ : (1,1),(0,r); & & (1,1) \end{array}\right]d\gamma. \tag{A17}$$

To simplify (A16) further, we first express the bivariate H-functions in (A16) into the form of a double line integral, and then place the curve integral regarding γ to the innermost level by rearranging (A16), we have

$$
\begin{aligned}
Q_1 = {} & \frac{\Theta\kappa(1-\omega)\kappa_e}{4\pi^2\Gamma(a)} \int_{\mathcal{L}}^t \frac{(\Theta\Lambda)^{-t}}{\Gamma(t)} \Gamma\left(\frac{t}{\alpha}+\mu-\frac{1}{\alpha}\right) \\
& \times \int_{\mathcal{L}}^s \Gamma(-s)\Gamma\left(a-\frac{rs}{c}\right)\Gamma(s+t-1)\left(\frac{b^{-r}C}{\mu_r}\right)^s \\
& \times \int_0^\infty \gamma^{1-t} H_{0,1}^{1,0}\left[\gamma\Lambda_e \,\middle|\, \left(-\frac{1}{\alpha_e}+\mu_e,\frac{1}{\alpha_e}\right)\right] d\gamma\, ds\, dt.
\end{aligned}
\tag{A18}
$$

Then, using ([51], Equation (2.25.2/1)), we can transform (A18) into

$$
\begin{aligned}
Q_1 = {} & \frac{\Theta\kappa(1-\omega)\kappa_e}{4\pi^2\Gamma(a)} \int_{\mathcal{L}}^t \frac{(\Theta\Lambda)^{-t}}{\Gamma(t)} \Gamma\left(\frac{t}{\alpha}+\mu-\frac{1}{\alpha}\right) \\
& \times \int_{\mathcal{L}}^s \Gamma(-s)\Gamma\left(a-\frac{rs}{c}\right)\Gamma(s+t-1) \\
& \times \Gamma\left(\frac{2-t}{\alpha_e}+\mu_e-\frac{1}{\alpha_e}\right)\Lambda_e^{t-2}\left(\frac{b^{-r}C}{\mu_r}\right)^s ds\, dt.
\end{aligned}
\tag{A19}
$$

Finally, converting the double curve integral into bivariate H-function using ([40], Equation (2.57)), after some simplifications, we obtain from (A19) in an exact closed-form as

$$
\begin{aligned}
Q_1 = {} & -\frac{\Theta\kappa(1-\omega)\kappa_e}{\Gamma(a)\Lambda_e^2} \\
& \times H_{1,0:1,2;0,2}^{0,1:1,1;2,0}\left[\begin{array}{c}\frac{\Lambda_e}{\Theta\Lambda}\\[2pt]\frac{b^{-r}C}{\mu_r}\end{array}\middle|\begin{array}{c}(2,1,1): \left(1+\frac{1}{\alpha}-\mu,\frac{1}{\alpha}\right) ;\\[4pt] : \left(\frac{1+\alpha_e\mu_e}{\alpha_e},\frac{1}{\alpha_e}\right),(1,1);(0,1),\left(a,\frac{r}{c}\right)\end{array}\right].
\end{aligned}
\tag{A20}
$$

To process (A17) further, we first convert the bivariate H-function in (A17) into the form of one double curve integral using ([40], Equation (2.55)). After placing the line integral of γ into the innermost layer, we can transform (A17) into

$$
\begin{aligned}
Q_2 = {} & \frac{\Theta\kappa r\omega\kappa_e}{4\pi^2} \int_{\mathcal{L}}^t \frac{(\Theta\Lambda)^{-t}}{\Gamma(t)} \Gamma\left(\frac{t}{\alpha}+\mu-\frac{1}{\alpha}\right) \\
& \times \int_{\mathcal{L}}^s \Gamma(1-s)\Gamma(-rs)\Gamma(s+t-1)\left(\frac{C\lambda^{-r}}{\mu_r}\right)^s \\
& \times \int_0^\infty \gamma^{1-t} H_{0,1}^{1,0}\left[\gamma\Lambda_e \,\middle|\, \left(-\frac{1}{\alpha_e}+\mu_e,\frac{1}{\alpha_e}\right)\right] d\gamma\, ds\, dt.
\end{aligned}
\tag{A21}
$$

Subsequently, using ([51], Equation (2.25.2/1)), we express the innermost curve integral in (A21) in the form of the product of Gamma functions. Then, we can write (A21) as

$$
\begin{aligned}
Q_2 = {} & \frac{\Theta\kappa r\omega\kappa_e}{4\pi^2} \int_{\mathcal{L}}^t \frac{(\Theta\Lambda)^{-t}}{\Gamma(t)} \Gamma\left(\frac{t}{\alpha}+\mu-\frac{1}{\alpha}\right) \\
& \times \int_{\mathcal{L}}^s \Gamma(1-s)\Gamma(-rs)\Gamma(s+t-1) \\
& \times \Gamma\left(\frac{2-t}{\alpha_e}+\mu_e-\frac{1}{\alpha_e}\right)\Lambda_e^{t-2}\left(\frac{C\lambda^{-r}}{\mu_r}\right)^s ds\, dt.
\end{aligned}
\tag{A22}
$$

Subsequently, based on the same steps as for the derivation of (A20), Equation (A22) can be expressed in exact closed-form as

$$Q_2 = -\frac{\Theta \kappa r \omega \kappa_e}{\Lambda_e^2}$$

$$\times H^{0,1:1,1;2,0}_{1,0:1,2;0,2}\left[\begin{array}{c}\frac{\Lambda_e}{\Theta\Lambda}\\[4pt]\frac{C\lambda^{-r}}{\mu_r}\end{array}\middle|\begin{array}{l}(2,1): \left(1+\tfrac{1}{\alpha}-\mu,\tfrac{1}{\alpha}\right) \ ; \\[6pt] :\left(\tfrac{1+\alpha_e\mu_e}{\alpha_e},\tfrac{1}{\alpha_e}\right),(1,1)\,;(1,1),(0,r)\end{array}\right]. \tag{A23}$$

After substituting (A20) and (A23) into (A15), we can finally obtain the closed-form expression of $\mathcal{P}_{out,L}$ in (11).

Appendix D. Proof of Corollary 1

To derive the asymptotic expression of SOP, we need to derive the asymptotic expressions of the first and the second bivariate H-function on the right-hand side of (11), which are denoted by O_1 and O_2, respectively. We consider two cases: (a) $\gamma_1 \to \infty$ and (b) $\gamma_e \to \infty$.

Appendix D.1. Case $\gamma_1 \to \infty$

For the case $\gamma_1 \to \infty$, we first focus on deriving asymptotic expression for O_1. Observe that as γ_1 tends to infinity, $\frac{\theta\Lambda}{\Lambda e}$ tends to zero. Thus, we first express the bivariate H-function in the form of one double curve integral, and express the curve integral containing $\frac{\theta\Lambda}{\Lambda e}$ in the form of an H- function. Then, we have

$$O_1 = \frac{i\Theta\kappa(1-\omega)\kappa_e}{2\pi\Gamma(a)\Lambda_e^2}\int_{\mathcal{L}}^{t}\Gamma(-t)\Gamma\left(a-\frac{rt}{c}\right)\left(\frac{b^{-r}C}{\mu_r}\right)^t$$

$$\times H^{2,1}_{2,2}\left[\frac{\Theta\Lambda}{\Lambda_e}\middle|\begin{array}{l}\left(1-\tfrac{1}{\alpha_e}-\mu_e,\tfrac{1}{\alpha_e}\right),(0,1)\\[4pt](-1+t,1),\left(-\tfrac{1}{\alpha}+\mu,\tfrac{1}{\alpha}\right)\end{array}\right]dt. \tag{A24}$$

It is easy to observe that the H-function in (A24) contains two poles: $(1-t)$ and $(1-\alpha\mu)$. According to [46], when the argument tends to zero, the asymptotic value of the H-function can be expressed as the residue of the closest pole to the left of the integration path l. Therefore, by utilizing ([52], Equation (1.8.4)), we can express (A24) as

$$O_1 = \frac{i\kappa(1-\omega)\kappa_e}{2\pi\Lambda\Gamma(a)\Lambda_e}\int_{\mathcal{L}}^{t}\frac{\Gamma(-t)}{\Gamma(1-t)}\Gamma\left(a-\frac{rt}{c}\right)\Gamma\left(\mu-\frac{t}{\alpha}\right)$$

$$\times\Gamma\left(\frac{t}{\alpha_e}+\mu_e\right)\left(\frac{b^{-r}C\Theta\Lambda}{\Lambda_e\mu_r}\right)^t dt. \tag{A25}$$

Following some simplifications, and using the definition of the H-function, we can transform (A25) into the following form

$$O_1 = -\frac{\kappa(1-\omega)\kappa_e}{\Lambda\Gamma(a)\Lambda_e}H^{1,3}_{3,2}\left[\frac{b^r\Lambda_e\mu_r}{C\Theta\Lambda}\middle|\begin{array}{l}(1,1),\left(1-a,\tfrac{r}{c}\right),\left(1-\mu,\tfrac{1}{\alpha}\right)\\[4pt]\left(\mu_e,\tfrac{1}{\alpha_e}\right),(0,1)\end{array}\right]. \tag{A26}$$

Next, we derive the asymptotic expression for O_2. Observing that O_2 and O_1 have a similar structure, we can readily transform O_2 into the following form

$$O_2 = -\frac{i\Theta\kappa r\omega\kappa_e}{2\pi\Lambda_e^2}\int_{\mathcal{L}}^{t}\Gamma(1-t)\Gamma(-rt)\left(\frac{C\lambda^{-r}}{\mu_r}\right)^t$$

$$\times H^{2,1}_{2,2}\left[\frac{\Theta\Lambda}{\Lambda_e}\middle|\begin{array}{l}\left(1-\tfrac{1}{\alpha_e}-\mu_e,\tfrac{1}{\alpha_e}\right),(0,1)\\[4pt](-1+t,1),\left(-\tfrac{1}{\alpha}+\mu,\tfrac{1}{\alpha}\right)\end{array}\right]dt. \tag{A27}$$

Similarly, we again use the residue of the pole $(1-t)$ to represent the asymptotic value of the H-function in (A27) as the argument tends to zero. Then, we have

$$O_2 = \frac{i\kappa r \omega \kappa_e}{2\pi\Lambda\Lambda_e} \int_{\mathcal{L}} \Gamma(-rt)\Gamma\left(\mu - \frac{t}{\alpha}\right)\Gamma\left(\frac{t}{\alpha_e} + \mu_e\right) \times \left(\frac{C\Theta\lambda^{-r}\Lambda}{\Lambda_e\mu_r}\right)^t dt. \tag{A28}$$

By using the definition of the H-function, we can transform (A28) into the following form

$$O_2 = -\frac{\kappa r \omega \kappa_e}{\Lambda\Lambda_e} H_{2,1}^{1,2}\left[\frac{\lambda^r\Lambda_e\mu_r}{C\Theta\Lambda}\ \middle|\ \begin{matrix}(1,r),(1-\mu,\frac{1}{\alpha})\\(\mu_e,\frac{1}{\alpha_e})\end{matrix}\right]. \tag{A29}$$

Substituting (A26) and (A29) into (11), we obtain the asymptotic expression for SOP for the case $\gamma_1 \to \infty$ as shown in (13).

Appendix D.2. Case $\gamma_e \to \infty$

Now, we focus on the case $\gamma_e \to \infty$. Obviously, as γ_e tends to infinity, $\frac{\theta\Lambda}{\Lambda_e}$ tends to infinity. Thus, using ([52], Equation (1.5.9)) and a similar approach to that used in case $\gamma_1 \to \infty$, we can easily obtain closed-form expressions for O_1 and O_2 for case $\gamma_e \to \infty$, as

$$O_1 = \frac{(\omega-1)\alpha_e}{\Gamma(a)\Gamma(\mu)\Gamma(\mu_e)\Gamma(\alpha_e\mu_e+1)}\Gamma\left(\mu + \frac{\alpha_e\mu_e}{\alpha}\right)\left(\frac{\Lambda_e}{\Theta\Lambda}\right)^{\alpha_e\mu_e}$$
$$\times H_{2,1}^{1,2}\left[\frac{b^r\mu_r}{C}\ \middle|\ \begin{matrix}(1,1),(1-a,\frac{r}{c})\\(\alpha_e\mu_e,1)\end{matrix}\right] \tag{A30}$$

and

$$O_2 = \frac{-r\omega\alpha_e}{\Gamma(\mu)\Gamma(\mu_e)\Gamma(\alpha_e\mu_e+1)}\Gamma\left(\mu + \frac{\alpha_e\mu_e}{\alpha}\right)\left(\frac{\Lambda_e}{\Theta\Lambda}\right)^{\alpha_e\mu_e}$$
$$\times H_{2,1}^{1,2}\left[\frac{\lambda^r\mu_r}{C}\ \middle|\ \begin{matrix}(0,1),(1,r)\\(\alpha_e\mu_e,1)\end{matrix}\right] \tag{A31}$$

respectively.

Substituting (A30) and (A31) into (11), we obtain the asymptotic expression for SOP for the case $\gamma_e \to \infty$ as shown in (14).

Appendix E. Proof of Theorem 4

Substituting (2) and (8) into (16), after some simplifications, we can transform the PNZ expression in (16) to

$$\mathcal{P}_{nz} = T_1 + T_2 \tag{A32}$$

where

$$T_1 = \int_0^\infty \frac{\kappa(1-\omega)\kappa_e}{\Gamma(a)\Lambda_e} H_{1,2}^{1,1}\left[\gamma\Lambda_e\ \middle|\ \begin{matrix}(1,1)\\(\mu_e,\frac{1}{\alpha_e}),(0,1)\end{matrix}\right]$$
$$\times H_{1,0:1,1;0,2}^{0,1:0,1;2,0}\left[\begin{matrix}\frac{1}{\gamma\Lambda}\\\frac{b^{-r}C}{\mu_r}\end{matrix}\ \middle|\ \begin{matrix}(2,1,1):\left(1+\frac{1}{\alpha}-\mu,\frac{1}{\alpha}\right);\\(2,1)\quad\quad;(0,1),(a,\frac{r}{c})\end{matrix}\right]d\gamma \tag{A33}$$

and

$$T_2 = \int_0^\infty \frac{r\kappa\omega\kappa_e}{\Lambda_e} H_{1,2}^{1,1}\left[\gamma\Lambda_e \,\middle|\, \begin{matrix} (1,1) \\ \left(\mu_e, \frac{1}{\alpha_e}\right),(0,1) \end{matrix}\right]$$

$$\times H_{1,0:1,1;0,2}^{0,1:0,1;2,0}\left[\begin{matrix} \frac{1}{\gamma\Lambda} \\ \frac{C\lambda^{-r}}{\mu_r} \end{matrix}\,\middle|\, \begin{matrix} (2,1,1): \left(1+\frac{1}{\alpha}-\mu,\frac{1}{\alpha}\right); \\ : \quad (2,1) \quad ;(1,1),(0,r) \end{matrix}\right] d\gamma. \tag{A34}$$

Representing the bivariate H-function into the form of one double line integral and moving the line integral regarding γ to the innermost level, we can re-write (A33) as

$$T_1 = -\frac{\kappa(1-\omega)\kappa_e}{4\pi^2\Gamma(a)\Lambda_e}\int_{\mathcal{L}}^t \Gamma(-t)\Gamma\left(a-\frac{rt}{c}\right)\left(\frac{b^{-r}C}{\mu_r}\right)^t$$

$$\times \int_{\mathcal{L}}^s \frac{\Lambda^{-s}}{\Gamma(s-1)}\Gamma(s+t-1)\Gamma\left(\frac{s}{\alpha}+\mu-\frac{1}{\alpha}\right)$$

$$\times \int_0^\infty \gamma^{-s} H_{1,2}^{1,1}\left[\gamma\Lambda_e \,\middle|\, \begin{matrix} (1,1) \\ \left(\mu_e, \frac{1}{\alpha_e}\right),(0,1) \end{matrix}\right] d\gamma\, ds\, dt. \tag{A35}$$

Afterwards, using the same technique as that used for deducing (A20) and (A23), we can express (A35) as

$$T_1 = \frac{\kappa(1-\omega)\kappa_e}{\Gamma(a)\Lambda_e^2}$$

$$\times H_{1,0:0,2;1,2}^{0,1:2,0;1,1}\left[\begin{matrix} \frac{b^{-r}C}{\mu_r} \\ \frac{\Lambda_e}{\Lambda} \end{matrix}\,\middle|\, \begin{matrix} (2,1,1): \quad ; \quad \left(1+\frac{1}{\alpha}-\mu,\frac{1}{\alpha}\right) \\ : (0,1),(a,\frac{r}{c}); \left(\frac{1+\alpha_e\mu_e}{\alpha_e},\frac{1}{\alpha_e}\right),(1,1) \end{matrix}\right]. \tag{A36}$$

Similarly, T_2 in (A34) can be expressed as

$$T_2 = \frac{\kappa r\omega\kappa_e}{\Lambda_e^2}$$

$$\times H_{1,0:0,2;1,2}^{0,1:2,0;1,1}\left[\begin{matrix} \frac{C\lambda^{-r}}{\mu_r} \\ \frac{\Lambda_e}{\Lambda} \end{matrix}\,\middle|\, \begin{matrix} (2,1,1): \quad ; \quad \left(1+\frac{1}{\alpha}-\mu,\frac{1}{\alpha}\right) \\ : (1,1),(0,r); \left(\frac{1+\alpha_e\mu_e}{\alpha_e},\frac{1}{\alpha_e}\right),(1,1) \end{matrix}\right] \tag{A37}$$

using ([51], Equation (2.25.2/1)).

References

1. Wang, D.; He, Y.; Yu, K.; Srivastava, G.; Nie, L.; Zhang, R. Delay-Sensitive Secure NOMA Transmission for Hierarchical HAP–LAP Medical-Care IoT Networks. *IEEE Trans. Ind. Inf.* **2021**, *18*, 5561–5572. [CrossRef]
2. Wang, D.; Zhou, F.; Lin, W.; Ding, Z.; Al-Dhahir, N. Cooperative Hybrid Non-Orthogonal Multiple Access Based Mobile-Edge Computing in Cognitive Radio Networks. *IEEE Trans. Cogn. Commun. Netw.* **2022**, *8*, 1104–1117. [CrossRef]
3. Wang, D.; He, T.; Zhou, F.; Cheng, J.; Zhang, R.; Wu, Q. Outage-driven link selection for secure buffer-aided networks. *Sci. China Inf. Sci* **2022**, *65*, 182303 . [CrossRef]
4. Zeng, Z.; Fu, S.; Zhang, H.; Dong, Y.; Cheng, J. A Survey of Underwater Optical Wireless Communications. *IEEE Commun. Surv. Tutor.* **2017**, *19*, 204–238. [CrossRef]
5. Dohler, M.; Li, Y. *Cooperative Communications: Hardware, Channel and PHY*, 1st ed.; Wiley: New York, NY, USA, 2010.
6. Shin, H.D.; Song, J.B. MRC analysis of cooperative diversity with fixed-gain relays in Nakagami-*M* Fading Channels. *IEEE Trans. Wirel. Commun.* **2008**, *7*, 2069–2074. [CrossRef]
7. Wang, G.; Xiang, W.; Yuan, J. Outage performance for compute-and-forward in generalized multi-way relay channels. *IEEE Commun. Lett.* **2012**, *16*, 2099–2102. [CrossRef]
8. Zhang, L.; Xiang, W.; Tang, X. An efficient bit-detecting protocol for continuous tag recognition in mobile RFID systems. *IEEE Trans. Mob. Comput.* **2017**, *17*, 503–516. [CrossRef]
9. Long, H.; Xiang, W.; Wang, J.; Zhang, Y.; Wang, W. Cooperative jamming and power allocation with untrusty two-way relay nodes. *IET Commun.* **2014**, *8*, 2290–2297. [CrossRef]
10. Upadhya, A.; Dwivedi, V.K.; Karagiannidis, G.K. On the Effect of Interference and Misalignment Error in Mixed RF/FSO Systems over Generalized Fading Channels. *IEEE Trans. Commun.* **2020**, *68*, 3681–3695. [CrossRef]

11. Lei, H.; Dai, Z.; Park, K.; Lei, W.; Pan, G.; Alouini, M. Secrecy outage analysis of mixed RF-FSO downlink SWIPT systems. *IEEE Trans. Commun.* **2018**, *66*, 6384–6395. [CrossRef]
12. Zedini, E.; Soury, H.; Alouini, M.S. On the Performance Analysis of Dual-Hop Mixed FSO/RF Systems. *IEEE Trans. Wirel. Commun.* **2016**, *15*, 3679–3689. [CrossRef]
13. Djordjevic, G.T.; Petkovic, M.I.; Cvetkovic, A.M.; Karagiannidis, G.K. Mixed RF/FSO Relaying with Outdated Channel State Information. *IEEE J. Sel. AREAS Commun.* **2015**, *33*, 1935–1948. [CrossRef]
14. Lei, H.J.; Zhang, Y.Y.; Park, K.H.; Ansari, I.S.; Pan, G.F.; Alouini, M.S. Performance Analysis of Dual-Hop RF-UWOC Systems. *IEEE Photonics J.* **2020**, *12*, 7901915. [CrossRef]
15. Illi, E.; El Bouanani, F.; Benevides da Costa, D.; Sofotasios, P.C.; Ayoub, F.; Mezher, K.; Muhaidat, S. Physical Layer Security of a Dual-Hop Regenerative Mixed RF/UOW System. *IEEE Trans. Sustain. Comput.* **2019**, *6*, 90–104. [CrossRef]
16. Illi, E.; El Bouanani, F.; Da Costa, D.B.; Ayoub, F.; Dias, U.S. Dual-Hop Mixed RF-UOW Communication System: A PHY Security Analysis. *IEEE Access* **2018**, *6*, 55345–55360. [CrossRef]
17. Christopoulou, C.; Sandalidis, H.G.; Ansari, I.S. Outage Probability of a Multisensor Mixed UOWC–FSO Setup. *IEEE Sens. Lett.* **2019**, *3*, 7501104. [CrossRef]
18. Xing, F.Y.; Yin, H.X.; Ji, X.Y.; Leung, V.C.M. An Adaptive and Energy-Efficient Algorithm for Surface Gateway Deployment in Underwater Optical/Acoustic Hybrid Sensor Networks. *IEEE Commun. Lett.* **2018**, *22*, 1810–1813. [CrossRef]
19. Johnson, L.; Green, R.; Leeson, M. A survey of channel models for underwater optical wireless communication. In Proceedings of the 2013 2nd International Workshop on Optical Wireless Communications (IWOW), Newcastle upon Tyne, UK, 21 October 2013; pp. 1–5.
20. Zedini, E.; Oubei, H.M.; Kammoun, A.; Hamdi, M.; Ooi, B.S.; Alouini, M.S. A New Simple Model for Underwater Wireless Optical Channels in the Presence of Air Bubbles. In Proceedings of the GLOBECOM 2017—2017 IEEE Global Communications Conference, Singapore, 4–8 December 2017; pp. 1–6.
21. Jaruwatanadilok, S. Underwater Wireless Optical Communication Channel Modeling and Performance Evaluation Using Vector Radiative Transfer Theory. *IEEE J. Sel. AREAS Commun.* **2008**, *26*, 1620–1627. [CrossRef]
22. Cochenour, B.; Mullen, L.; Muth, J. Temporal Response of the Underwater Optical Channel for High-Bandwidth Wireless Laser Communications. *IEEE J. Ocean. Eng.* **2013**, *38*, 730–742. [CrossRef]
23. Nabavi, P.; Haq, A.; Yuksel, M. Empirical Modeling and Analysis of Water-to-Air Optical Wireless Communication Channels. In Proceedings of the 2019 IEEE International Conference on Communications Workshops (ICC Workshops), Shanghai, China, 20–24 May 2019; pp.1–6.
24. Jamali, M.V.; Chizari, A.; Salehi, J.A. Performance Analysis of Multi-Hop Underwater Wireless Optical Communication Systems. *IEEE Photonics Technol. Lett.* **2017**, *29*, 462–465. [CrossRef]
25. Jamali, M.V.; Salehi, J.A.; Akhoundi, F. Performance Studies of Underwater Wireless Optical Communication Systems with Spatial Diversity: MIMO Scheme. *IEEE Trans. Commun.* **2017**, *65*, 1176–1192. [CrossRef]
26. Nezamalhosseini, S.A.; Chen, L.R. Optimal Power Allocation for MIMO Underwater Wireless Optical Communication Systems Using Channel State Information at the Transmitter. *IEEE J. Ocean. Eng.* **2020**, *46*, 319–325. [CrossRef]
27. Luan, X.; Yue, P.; Yi, X. Scintillation index of an optical wave propagating through moderate-to-strong oceanic turbulence. *JOSA A* **2019**, *36*, 2048–2059. [CrossRef] [PubMed]
28. Boucouvalas, A.C.; Peppas, K.P.; Yiannopoulos, K.; Ghassemlooy, Z. Underwater Optical Wireless Communications with Optical Amplification and Spatial Diversity. *IEEE Photonics Technol. Lett.* **2016**, *28*, 2613–2616. [CrossRef]
29. Shin, M.; Park, K.H.; Alouini, M.S. Statistical Modeling of the Impact of Underwater Bubbles on an Optical Wireless Channel. *IEEE Open J. Commun. Soc.* **2020**, *1*, 808–818. [CrossRef]
30. Elamassie, M.; Uysal, M. Vertical Underwater Visible Light Communication Links: Channel Modeling and Performance Analysis. *IEEE Trans. Wirel. Commun.* **2020**, *19*, 6948–6959. [CrossRef]
31. Jamali, M.V.; Mirani, A.; Parsay, A.; Abolhassani, B.; Nabavi, P.; Chizari, A.; Khorramshahi, P.; Abdollahramezani, S.; Salehi, J.A. Statistical Studies of Fading in Underwater Wireless Optical Channels in the Presence of Air Bubble, Temperature, and Salinity Random Variations. *IEEE Trans. Commun.* **2018**, *66*, 4706–4723. [CrossRef]
32. Zedini, E.; Oubei, H.M.; Kammoun, A.; Hamdi, M.; Ooi, B.S.; Alouini, M.S. Unified Statistical Channel Model for Turbulence-Induced Fading in Underwater Wireless Optical Communication Systems. *IEEE Trans. Commun.* **2019**, *67*, 2893–2907. [CrossRef]
33. Hamamreh, J.M.; Furqan, H.M.; Arslan, H. Classifications and Applications of Physical Layer Security Techniques for Confidentiality: A Comprehensive Survey. *IEEE Commun. Surv. Tutor.* **2019**, *21*, 1773–1828. [CrossRef]
34. Lei, H.J.; Dai, Z.J.; Ansari, I.S.; Park, K.H.; Pan, G.F.; Alouini, M.S. On Secrecy Performance of Mixed RF-FSO Systems. *IEEE Photonics J.* **2017**, *9*, 7904814. [CrossRef]
35. Lei, H.J.; Luo, H.L.; Park, K.H.; Ren, Z.; Pan, G.F.; Alouini, M.S. Secrecy Outage Analysis of Mixed RF-FSO Systems with Channel Imperfection. *IEEE Photonics J.* **2018**, *10*, 7904113. [CrossRef]
36. Yang, L.; Liu, T.; Chen, J.C.; Alouini, M.S. Physical-Layer Security for Mixed $\eta - \mu$ and M-Distrib. Dual-Hop RF/FSO Syst. *IEEE Trans. Veh. Technol.* **2018**, *67*, 12427–12431. [CrossRef]
37. Lei, H.; Luo, H.; Park, K.H.; Ansari, I.S.; Lei, W.; Pan, G.; Alouini, M.S. On Secure Mixed RF-FSO Systems With TAS and Imperfect CSI. *IEEE Trans. Commun.* **2020**, *68*, 4461–4475. [CrossRef]

38. Yacoub, M.D. The α-μ Distribution: A Physical Fading Model for the Stacy Distribution. *IEEE Trans. Veh. Technol.* **2007**, *56*, 27–34. [CrossRef]
39. Lou, Y.; Sun, R.; Cheng, J.; Nie, D.; Qiao, G. Secrecy Outage Analysis of Two-Hop Decode-and-Forward Mixed RF/UWOC Systems. *arXiv* **2020**, arXiv:2009.00328.
40. Mathai, A.M.; Saxena, R.K.; Haubold, H.J. *The H-Function Theory and Applications*; Springer: New York, NY, USA, 2010.
41. Kong, L.; Kaddoum, G.; Rezki, Z. Highly Accurate and Asymptotic Analysis on the SOP Over SIMO $\alpha - \mu$ Fading Channels. *IEEE Commun. Lett.* **2018**, *22*, 2088–2091. [CrossRef]
42. Zedini, E.; Kammoun, A.; Soury, H.; Hamdi, M.; Alouini, M.S. Performance Analysis of Dual-Hop Underwater Wireless Optical Communication Systems over Mixture Exponential-Generalized Gamma Turbulence Channels. *IEEE Trans. Commun.* **2020**, *68*, 5718–5731. [CrossRef]
43. Lapidoth, A.; Moser, S.M.; Wigger, M.A. On the Capacity of Free-Space Optical Intensity Channels. *IEEE Trans. Inf. Theory* **2009**, *55*, 4449–4461. [CrossRef]
44. Chergui, H.; Benjillali, M.; Alouini, M.S. Rician K-Factor Anal. XLOS Serv. Probab. 5G Outdoor Ultra-Dense Networks. *IEEE Wirel. Commun. Lett.* **2019**, *8*, 428–431.
45. Almeida Garcia, F.D.; Flores Rodriguez, A.C.; Fraidenraich, G.; Santos Filho, J.C.S. CA-CFAR Detection Performance in Homogeneous Weibull Clutter. *IEEE Geosci. Remote Sens. Lett.* **2019**, *16*, 887–891. [CrossRef]
46. Alhennawi, H.R.; El Ayadi, M.M.H.; Ismail, M.H.; Mourad, H.A.M. Closed-Form Exact and Asymptotic Expressions for the Symbol Error Rate and Capacity of the H-Function Fading Channel. *IEEE Trans. Veh. Technol.* **2016**, *65*, 1957–1974. [CrossRef]
47. Lei, H.J.; Ansari, I.S.; Pan, G.F.; Alomair, B.; Alouini, M.S. Secrecy Capacity Analysis Over $\alpha - \mu$ Fading Channels. *IEEE Commun. Lett.* **2017**, *21*, 1445–1448. [CrossRef]
48. Kong, L.; Kaddoum, G.; Chergui, H. On Physical Layer Security Over Fox's H-Function Wiretap Fading Channels. *IEEE Trans. Veh. Technol.* **2019**, *68*, 6608–6621. [CrossRef]
49. Research., W. The Wolfram Functions Site. Available online: http://functions.wolfram.com (accessed on 29 September 2021).
50. Gradshteyn, I.S.; Ryzhik, I.M. *Table of Integrals, Series, and Products*, 7th ed.; Academic Press: San Diego, CA, USA, 2007.
51. Verma, R.U. On some integrals involving Meijer's G-fucntion of two variables. *Proc. Nat. Inst. Sci. India* **1966**, *39*, 509–515.
52. Kilbas, A.A.; Saigo, M. *H-transforms: Theory and Applications (Analytical Method and Special Function)*, 1st ed.; CRC Press: Boca Raton, FL, USA, 2004.

 *drones*

Article

Joint Placement and Power Optimization of UAV-Relay in NOMA Enabled Maritime IoT System

Woping Xu [1,2,*], Junhui Tian [1], Li Gu [1] and Shaohua Tao [3,4,5]

1 College of Information Engineering, Shanghai Maritime University, Shanghai 201306, China
2 Department of Computer Science, University of Victoria, Victoria, BC V8P 5C2, Canada
3 College of Information Sciences and Technology, Donghua University, Shanghai 200051, China
4 School of Information Engineering, Xu Chang University, Xuchang 461002, China
5 Engineering Research Center of Digitized Textile & Apparel Technology, Ministry of Education, Donghua University, Shanghai 200051, China
* Correspondence: wpxu@shmtu.edu.cn

Abstract: In this paper, an unmanned aerial vehicle is utilized as an aerial relay to connect onshore base station with offshore users in a maritime IoT system with uplink non-orthogonal multiple access enabled. A coordinated direct and relay transmission scheme is adopted in the proposed system, where close shore maritime users directly communicate with onshore BS and offshore maritime users need assistance of an aerial relay to communicate with onshore BS. We aim to minimize the total transmit energy of the aerial relay by jointly optimizing the UAV hovering position and transmit power allocation. The minimum rate requirements of maritime users and transmitters' power budgets are considered. The formulated optimization problem is non-convex due to its non-convex constraints. Therefore, we introduce successive convex optimization and block coordinate descent to decompose the original problem into two subproblems, which are alternately solved to optimize the UAV energy consumption with satisfying the proposed constraints. Numerical results indicate that the proposed algorithm outperformed the benchmark algorithm, and shed light on the potential of exploiting the energy-limited aerial relay in IoT systems.

Keywords: maritime communication system; optimization; Uplink NOMA; UAV relay network

Citation: Xu, W.; Tian, J.; Gu, L.; Tao, S. Joint Placement and Power Optimization of UAV-Relay in NOMA Enabled Maritime IoT System. *Drones* **2022**, *6*, 304. https://doi.org/10.3390/drones6100304

Academic Editors: Dawei Wang and Ruonan Zhang

Received: 22 September 2022
Accepted: 14 October 2022
Published: 18 October 2022

Publisher's Note: MDPI stays neutral with regard to jurisdictional claims in published maps and institutional affiliations.

1. Introduction

One significant challenge for maritime communication networks is the rapid increasing demand of broadband wireless services, especially for offshore maritime users [1]. For near shore maritime users, it is possible to enjoy broadband wireless service in locations where either mobile operator coverage is available or Wi-Fi-based long distance links can be deployed [2]. However, recently, it is still a challenge to provide seamless mobile broadband coverage for offshore maritime users located over tens of kilometers from shore since the communication infrastructures are difficult to deploy in the ocean [3]. For years, MF/HF/WHF-based communication dominated offshore user wireless communications. However, this communication technology cannot afford a high-rate transmission service on accounts of higher propagation delay and insufficient bandwidth [4]. As a conventional solution for offshore high rate transmission service, satellite communication offers a better quality-of-service as well as higher system maintenance cost and the problem of flexibility [5]. Different from satellites, unmanned aerial vehicles (UAVs) have been considered as an economic on-demand data service solution for offshore maritime users in diverse maritime activities owning to its advantage of highly maneuverable and flexible deployment, especially for various mission-critical applications such as emergency deployment and maritime search and rescue [6,7].

Although UAV-assisted communications have been widely studied for terrestrial communication scenarios [8–11], UAV-integrated maritime communication is still an open research field. To provide an effective aerial relay service, the distribution of users has great importance in resource allocation and trajectory optimization. Unlike in the terrestrial scenario, it is difficult to acquire either accurate location information or real-time channel state information (CSI) of maritime users since the distribution is scattered within a vast area [12], which increases the complexity of fly trajectory design. In addition, the energy-efficient issue is another vital problem for UAV relay communication due to the limited life time, especially for maritime communication scenarios. After departure, a UAV relay cannot land on the sea until the assigned mission is accomplished.

1.1. Recent Works

Recently, UAV-assisted communication techniques have attracted lots of attention in maritime communication systems for their deployment flexibility and line of sight (LoS) transmission ability. There are some works on UAV deployment and resource allocation problems in UAV-assisted maritime communication systems [13–17]. In [13], the authors study the optimal UAV placement problem to achieve the maximum system rate in a maritime downlink caching UAV-assisted decode-and-forward (DF) relay communication system with both air-to-ground and air-to-sea models considered. In [14], a UAV-assisted communication system is used to extend the coverage of the onshore BSs in the downlink communication scenario. The non-orthogonal multiple access (NOMA) protocol is adopted to enable the aerial BSs and simultaneously sever multiple ships. The authors have proposed a joint UAV transmit power and transmission duration optimization scheme to maximize the sum rate of ships. In [15], to facilitate spectrum sharing and efficient backhaul, UAV-added coverage enhancement is studied for maritime communication in a hybrid space–air–ground integrated network. The UAV trajectory design and transmitted power allocation have been jointly optimized by considering the constraints on UAV kinematics, tolerable interference, backhaul, and UAV transmit power budget. In [16], UAV-assisted ocean monitoring network architecture has been constructed for a remote oceanic data collection. In [17], a fermat-point theory-based fast trajectory planning scheme is proposed to improve received data throughput of UAV. Although the power optimization problem of UAV relay is studied in [13–15], these works focused on the rate maximization problem in a downlink maritime communication scenario and neglect resource allocation issues in uplink scenarios. References [16,17] focused on trajectory design issues subject to the constraint on the UAV flying energy budget in UAV-assisted maritime data collection systems and ignored the resource allocation problem of UAV relay. Moreover, all of the aforementioned works presented a power allocation problem of UAV relay by formulating a throughput maximization problem with constraints of the UAV power budget. However, the power minimization problem of UAV relay should be discussed to shed light on the potential of utilizing an energy-limited aerial platform in a maritime communication scenario.

1.2. Motivation and Contributions

Motivated by [18,19], we study the joint UAV hovering position and power allocation in UAV-assisted maritime communication systems. Particularly, we focus on a UAV transmit power consumption minimization issue subject to maritime user's minimum rate requirements and transmit power constraints. NOMA-enabled relay is introduced to the maritime IoT system to improve the spectrum efficiency [20,21]. It worth noting that the uplink NOMA scheme is introduced to boost the system spectrum reuse in our work, which is different from the downlink relay system discussed in [13,15,18]. Furthermore, the NOMA scheme was not employed in [19]. The main contributions of this paper are outlined as follows:

1. In this paper, we study the power minimization problem subject to user's minimum rate requirement and UAV transmit power budget in a maritime IoT system with A2A and A2S link model considered. A coordinated direct and relay transmission scheme employing uplink NOMA scheme is proposed and investigated, where maritime close-shore users (MCU) directly communication with onshore BS, whereas maritime remote users (MRU) communicate with the onshore BS by a half-duplex DF UAV relay.
2. In the proposed maritime IoT system, an interference cancellation parameter is introduced to summarized UAV's received data expression in transmission phase 1, which facilitates solving the proposed UAV power transmission minimization problem.
3. The successive convex approximation method is applied to deal with non-convex inequality constraints of the formulated optimization problem. The block coordinate descent method (BCD) is used to decouple the original problem into two subproblems, namely power allocation and optimal UAV placement. After that, an iterative algorithm is proposed to optimize power allocation coefficients and optimal UAV coordinates alternately.

1.3. Paper Organization

The rest of this work is organized as follows. In Section 2, the interested UAV-assisted maritime IoT system model and formulated optimization problem are proposed. Section 3 presents joint power allocation and an optimal UAV placement solution to the optimization problem. Numerical results are presented in Section 4. Finally, Section 5 concludes our paper.

2. System Model and Problem Formulation

2.1. System Model

As shown in Figure 1, a UAV-assisted maritime IoT system model, including one onshore base station (BS), one mobile UAV relay, and multiple maritime users deployed on a certain area for data collection, is considered. These maritime users are divided into two groups, including K maritime close-shore users (MCU) and K maritime remote users (MRU), according to communication service type. $\mathcal{K}_c = \{k'|k' \in \mathcal{K}_c, |\mathcal{K}_c| = K\}$ and $\mathcal{K}_r = \{k|k \in \mathcal{K}_r, |\mathcal{K}_r| = K\}$ denote the MRUs set and MCUs set, respectively. It is assumed that all transmission nodes in this model are equipped with a single antenna, $\mathcal{K}_r \cap \mathcal{K}_c = \varnothing$. MCUs are deployed closely along the coastline and can be served directly by the onshore BS. MRUs are deployed far away from the coastline and must rely on a UAV relay for data ferrying. A coordinated direct and relay transmission is introduced, where an MCU directly communicates with an onshore BS, whereas an MRU communicates with an onshore BS by a half-duplex DF UAV relay. A two-user uplink NOMA scheme is considered by an MRU coexisting with an MCU in a spectrum resource block. In this paper, we focus on total transmit power minimum optimization of UAV relay by jointly optimizing power allocation and UAV hovering position. It is assumed that $|\mathcal{K}_r| = |\mathcal{K}_c| = K$. Thus, there are K NOMA pairs in the proposed networks.

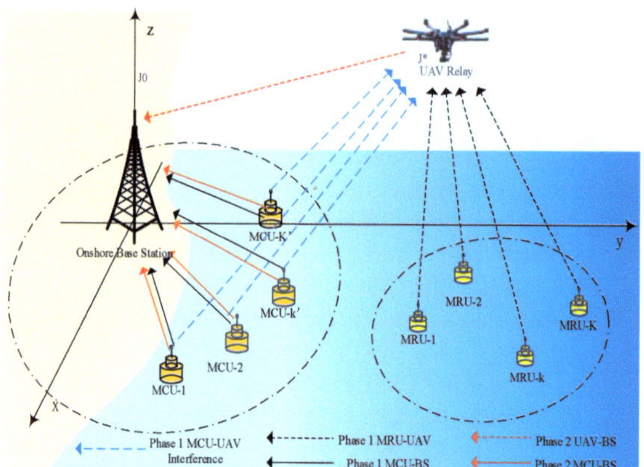

Figure 1. System model of maritime IoT system with a UAV relay.

The three-dimensional Cartesian coordinate is considered, where the coastline is approximated as the x-axis, the y-axis extends into the ocean, and the z-axis represents the altitude. The coordinate of the onshore BS is $c_B = (0, 0, h)$. Although locations of maritime users may change along with time according to sea surface waves, their coordinates can still be regarded as approximately fixed in a certain period, which can be denoted as $c_k = (x_k, y_k, 0), \forall k \in \mathcal{K}_r$ and $c_{k'} = (x_{k'}, y_{k'}, 0), \forall k' \in \mathcal{K}_c$, respectively. The UAV departs from the starting point J_0, then flys to the optimal hovering position J^* for the data relaying mission, and finally returns to the original location. According to the pre-planned deployed position of maritime users, the optimal hovering position of the UAV can be predetermined before its mission executes. Thus, the coordinates of the UAV can be expressed as (x_u, y_u, h_u) during the transmission mission's executed duration. Denote the distances between BS and the UAV, between the UAV and the k-th MRU, and between the k'-th MCU, respectively, as:

$$d_{UB} = \|c_u - c_B\| = \sqrt{(x_u - x_B)^2 + (y_u - y_B)^2 + h_u^2} \tag{1}$$

$$d_{Bk'} = \|c_B - c_{k'}\| = \sqrt{(x_B - x_{k'})^2 + (y_B - y_{k'})^2 + h^2} \tag{2}$$

$$d_{Uk} = \|c_u - c_k\| = \sqrt{(x_u - x_k)^2 + (y_u - y_k)^2 + h_u^2}, \forall k \in K_r \tag{3}$$

$$d_{Uk'} = \|c_u - c_{k'}\| = \sqrt{(x_u - x_{k'})^2 + (y_u - y_{k'})^2 + h_u^2}, \forall k' \in K_c \tag{4}$$

We assume that the UAV flies high enough to enable an LoS transmission. An air-to-sea (A2S) channel model is composed by large-scale and small-scale fading [15]. Thus, the channel between UAV and k-th MRU can be represented as:

$$h_{Uk} = \frac{1}{(L_{Uk})^{1/2}} \widetilde{h}_{Uk}, \forall k \tag{5}$$

where L_{Uk} denotes the path loss component and $\widetilde{h}_{Uk}$ denotes the Rician fading component. Then, the path loss model can be expressed as:

$$L_{Uk}(\text{dB}) = A_U + 10\varsigma_U \log 10\left(\frac{d_{Uk}}{d_0}\right) + X_U \tag{6}$$

where d_0 refers to the reference distance, A_U denotes the path loss at d_0, ς_U denotes the path loss exponent, and X_U is a zero-mean Gaussian random variable with standard deviation σ_{X_U} [22,23]. Rician fading can be represented as:

$$\tilde{h}_{Uk} = \sqrt{\frac{S_U}{1+S_U}} + \sqrt{\frac{1}{1+S_U}}g_{Uk} \tag{7}$$

where $g_{Uk} \sim \mathcal{CN}(0,1)$ and S_U indicates the Rician factor that corresponds to the ratio between the LoS power and the scattering power [24]. In the proposed system, the maritime users are deployed on pre-planned positions, then their location data can be pre-measured. Although locations of MRUs and MCUs may change along with time according to sea surface waves, their coordinates can still be regarded as approximately fixed in a certain period. Thus, their corresponding large-scaled channel state information (CSI) can be obtained. Similarly, the A2S channel between onshore BS and k'-th MCU is denoted as:

$$h_{Bk'} = \left(\frac{d_0}{d_{Bk'}}\right)^{\frac{\varsigma_U}{2}} 10^{-\frac{A_U+X_U}{20}}\left(\sqrt{\frac{S_U}{1+S_U}} + \sqrt{\frac{1}{1+S_U}}g_{Bk'}\right) \tag{8}$$

where $g_{Bk'} \sim \mathcal{CN}(0,1)$. The interference A2S link between UAV and k'-th MCU is denoted as:

$$h_{Uk'} = \left(\frac{d_0}{d_{Uk'}}\right)^{\frac{\varsigma_U}{2}} 10^{-\frac{A_U+X_U}{20}}\left(\sqrt{\frac{S_U}{1+S_U}} + \sqrt{\frac{1}{1+S_U}}g_{Uk'}\right) \tag{9}$$

where $g_{Uk'} \sim \mathcal{CN}(0,1)$. On the other hand, The air-to-air (A2A) channel between onshore BS and UAV has a high LoS probability, which can be represented as:

$$h_{UB} = \left(\frac{d_0}{d_{UB}(t)}\right)^{\frac{\varsigma_{UB}}{2}} 10^{-\frac{A_{UB}+X_{UB}}{20}}\left(\sqrt{\frac{S_{UB}}{1+S_{UB}}} + \sqrt{\frac{1}{1+S_{UB}}}g_{UB}\right) \tag{10}$$

where A_{UB} denotes the path loss at d_0, ς_{UB} denotes the path loss exponent, X_{UB} is a zero-mean Gaussian random variable with standard deviation, S_{UB} is the Rician factor, and $\sigma_{X_{UB}}$ and $g_{UB} \sim \mathcal{CN}(0,1)$.

For each transmission duration, the UAV receives data from an MRU in Phase 1 and then forwards it to BS in phase 2; meanwhile, an MCU transmits data to BS in the same subchannel with the MRU in both phases by uplink NOMA. With loss of generality, we consider $|h_{Bk'}|^2 \geq |h_{UB}|^2 \geq |h_{Bk}|^2$.

1. Phase-1 (t_1)

   In an uplink NOMA transmission scenario, an MCU and an MRU transmit symbols x_1 and y_1 simultaneously with $\alpha_{k'}P_t$ and $\alpha_k P_t$, where P_t denotes the total transmit power in phase 1. $\alpha_{k'}$ and α_k are the power allocation coefficient in phase 1. To guarantee an efficient SIC decoding at the NOMA receiver, it is assumed that $\alpha_{k'} + \alpha_k = 1$. Thus, data received at onshore BS and the UAV in Phase 1 can be given, respectively, by:

$$y_{Bk'}^{t_1} = |h_{Bk'}|\sqrt{\alpha_{k'}P_t}x_1^{t_1} + w \tag{11}$$

$$y_{Uk}^{t_1} = |h_{Uk}|\left(\sqrt{\alpha_{k'}P_t}x_1^{t_1} + \sqrt{\alpha_k P_t}y_1^{t_1}\right) + w \tag{12}$$

where $w \sim \mathcal{CN}(0,\sigma^2)$ denotes the background noise. Due to the half-duplex relay scheme, the achievable rate of the k'-th MCU at BS in phase 1 can be represented as:

$$R_{Bk'}^{t_1} = \frac{1}{2}\log_2\left(1 + \frac{\alpha_{k'}P_t|h_{Bk'}|^2}{\sigma^2}\right) = \frac{1}{2}\log_2\left(1 + \alpha_{k'}\rho_t|h_{Bk'}|^2\right) \tag{13}$$

where $\frac{P_t}{\rho^2} = \rho_t$. Because of the simultaneous transmission of MRU and MCU, UAV is able to receive the signal from both of them. We assume perfect time synchronization between MRU and MCU. According to the uplink NOMA principle, the UAV relay obtains the decoded symbol y_1 by considering the following two conditions.

- $|h_{Uk}|^2 \geq |h_{Uk'}|^2$

$$R_{Uk}^{t_1} = \frac{1}{2}\log_2\left(1 + \frac{\alpha_k P_t |h_{Uk}|^2}{\alpha_{k'} P_t |h_{Uk'}|^2 + \sigma^2}\right) = \frac{1}{2}\log_2\left(1 + \frac{\alpha_k \rho_t |h_{Uk}|^2}{\alpha_{k'} \rho_t |h_{Uk'}|^2 + 1}\right) \tag{14}$$

- $|h_{Uk}|^2 \leq |h_{Uk'}|^2$

$$R_{Uk}^{t_1} = \frac{1}{2}\log_2\left(1 + \frac{\alpha_k P_t |h_{Uk}|^2}{\sigma^2}\right) = \frac{1}{2}\log_2\left(1 + \alpha_k \rho_t |h_{Uk}|^2\right) \tag{15}$$

By introducing $\kappa \in \{0, 1\}$ as the interference cancellation parameter, the received data rate at UAV in Phase 1 can be summarized as:

$$
\begin{aligned}
R_{Uk}^{t_1} &= \log_2\left(1 + \frac{\alpha_k \rho_t |h_{Uk}|^2}{\kappa \alpha_{k'} \rho_t |h_{Uk'}|^2 + 1}\right) \\
&= \begin{cases}
\log_2\left(1 + \alpha_k \rho_t |h_{Uk}|^2\right) & \kappa = 0, |h_{Uk}|^2 \leq |h_{Uk'}|^2 \\
\log_2\left(1 + \dfrac{\alpha_k \rho_t |h_{Uk}|^2}{\alpha_{k'} \rho_t |h_{Uk'}|^2 + 1}\right) & \kappa = 1, |h_{Uk}|^2 \leq |h_{Uk'}|^2
\end{cases}
\end{aligned} \tag{16}
$$

2. Phase-2 (t_2)

In phase 2, both MCU and UAV transmit symbols x_2 and y_1 simultaneously to onshore BS with powers $\beta_{k'} P_t$ and $\beta_k P_t$, where $\beta_{k'}$, β_k are the power allocation coefficient in phase 2 and $\beta_{k'} + \beta_k = 1$. Thus, received data at onshore BS can be represented as:

$$y_{BUk'}^{t_2} = |h_{UB}|\sqrt{P_k^2} y_1^{t_2} + |h_{Bk'}|\sqrt{P_{k'}^2} x_2^{t_2} + w \tag{17}$$

Since $|h_{UB}|^2 \leq |h_{Bk'}|^2$, the achievable data rates of the UAV relay and MCU are presented, respectively, by:

$$R_{UB}^{t_2} = \frac{1}{2}\log_2\left(1 + \frac{\beta_k P_t |h_{UB}|^2}{\sigma^2}\right) = \frac{1}{2}\log_2\left(1 + \beta_k \rho_t |h_{UB}|^2\right) \tag{18}$$

$$R_{Bk'}^{t_2} = \frac{1}{2}\log_2\left(1 + \frac{\beta_k' P_t |h_{Bk'}|^2}{\beta_k P_t |h_{UB}|^2 + \sigma^2}\right) = \frac{1}{2}\log_2\left(1 + \frac{\beta_{k'} \rho_t |h_{Bk'}|^2}{\beta_k \rho_t |h_{UB}|^2 + 1}\right) \tag{19}$$

3. Sum Capacity

Using Equations (13) and (19), the sum capacity of the k'-th MCU is given as:

$$R_{k'} = R_{Bk'}^{t_1} + R_{Bk'}^{t_2} \tag{20}$$

On the other hand, the end-to-end capacity of a two-hop cooperative link is the minimum one of the two hops. Thus, according to Equations (16) and (18), the capacity of MRU is obtained as:

$$R_k = \min\left(R_{Uk}^{t_1}, R_{UB}^{t_2}\right) \tag{21}$$

2.2. Problem Formulation

It is assumed that the UAV fights with a fixed altitude, which means h_u is constant. Since the transmission duration and fighting trajectory for each relay mission is predetermined, the propulsion energy and hovering energy required for the UAV is also considered before. The UAV transmit power minimization problem has formulated constraints on maritime users' minimum rate requirements, maritime maximum users' transmit power thresholds and UAV power budget, which can be summarized as:

$$\text{(P1)} \quad \min_{\alpha,\beta,(x_u,y_u)} \quad P_{UAV}$$

$$\text{s.t. } C1 : P_{UAV} \leq \frac{\bar{P}_{Ut}}{\delta^2}, \forall k$$

$$C2 : \alpha_k \geq 0, \alpha_{k'} \geq 0, \beta_k \geq 0, \beta_{k'} \geq 0, \forall k, k'$$

$$C3 : \alpha_{k'} + \alpha_k = 1, \beta_{k'} + \beta_k = 1, \forall k, k'$$

$$C4 : \min\left(R_{Uk}^{t_1}, R_{UB}^{t_2}\right) \geq \bar{R}_k, \forall k$$

$$C5 : R_{Bk'}^{t_1} \geq \bar{R}_{k'}, \forall k'$$

$$C6 : R_{Bk'}^{t_2} \geq \bar{R}_{k'}, \forall k' \tag{22}$$

where $P_{UAV} = \sum_{k \in \mathcal{K}_r} \beta_k \rho_t$ denotes the total transmitted energy consumption of UAV during mission execution. $\alpha = [\alpha_1, \ldots, \alpha_k, \cdots, \alpha_K]$ and $\beta = [\beta_1, \ldots, \beta_k, \cdots, \beta_K]$ denote power coefficient vectors of K NOMA pairs. $\bar{P}_{Ut}$ denotes the transmit power budget of UAV. C1 guarantees the power supplied by UAV is not exceeding its transmit power budget. C2 and C3 are the power allocation coefficient constraints. C4–C6 can guarantee the minimum rate requirements of two types of maritime users. To solve (P1), we first introduce K slack variables $\eta_k = \min\left(R_{Uk}^{t_1}, R_{UB}^{t_2}\right), \forall k$ into the objective function such that it is reformulated as:

$$\text{(P2)} \quad \min_{\alpha,\beta,(x_u,y_u),\eta} \quad P_{UAV}$$

$$\text{s.t. } C7 : R_{Uk}^{t_1} \geq \eta_k, \forall k$$

$$C8 : R_{UB}^{t_2} \geq \eta_k, \forall k \tag{23}$$

$$C1 - C6$$

where $\eta = [\eta_1, \cdots, \eta_k, \cdots, \eta_K]$ denotes the slack variables vector of R_k. Thus, (P2) can be transformed as:

$$\text{(P3)} \quad \min_{\alpha,\beta,(x_u,y_u),\eta} \quad P_{UAV}$$

$$\text{s.t. } C4' : \alpha_k \geq \frac{\lambda_k|g_{Bk'}|^2 + \frac{\lambda_k}{\rho_t}}{|h_{Uk}|^2 + \lambda_k|g_{Bk'}|^2}, \beta_k \geq \frac{\lambda_k}{\rho_t|h_{UB}|^2}$$

$$C5' : \alpha_k \leq \frac{|h_{Bk'}|^2 - \frac{\lambda_{k'}}{\rho_t}}{|h_{Bk'}|^2} \tag{24}$$

$$C6' : \beta_k \leq \frac{|h_{Bk'}|^2 - \frac{\lambda_{k'}}{\rho_t}}{\lambda_{k'}|h_{UB}|^2 + |h_{Bk'}|^2}$$

$$C1 - 3, C7 - 8$$

where $\lambda_k = \left(2^{2\bar{R}_k} - 1\right)$ and $\lambda_{k'} = \left(2^{2\bar{R}_{k'}} - 1\right)$. (P3) is still challenging to solve since C7 and C8 are non-convex to α and β. For the UAV placement optimization problem, there are two sorts of optimization approaches, namely the deterministic optimization method and stochastic optimization method [25]. The BCD approach is a computationally-efficient deterministic approach that can be used to solve joint UAV placement and resource allocation problem by iteratively optimizing two block variables in turn [26]. Therefore, we introduce the BCD optimization method to decouple the original problem into a power allocation problem and optimal UAV placement and optimize two block variables alternately.

3. Proposed Optimization Solution

3.1. Power Minimization

With the given optimum UAV hovering placement, the power minimization problem can be transformed as:

$$(P4) \quad \min_{\alpha,\beta,\eta} P_{UAV}$$

$$\text{s.t. } C7 : R_{Uk}^{t_1} = \frac{1}{2}\log_2\left(1 + \frac{\alpha_k \rho_t |h_{Uk}|^2}{\kappa(1-\alpha_k)|h_{Uk'}|^2 + 1}\right) \geq \eta_k, \forall k \tag{25}$$

$$C8 : R_{UB}^{t_2} = \log_2\left(1 + \beta_k \rho_t |h_{UB}|^2\right) \geq \eta_k, \forall k$$

$$C1 - C3, C4', C5', C6'$$

where C8 is non-convex inequality constraint since $\log_2\left(1 + \beta_k \rho_t |h_{UB}|^2\right)$ is concave. C7 can be transformed as:

$$R_{Uk}^{t_1} = \log_2\left(\alpha_k \rho_t \left(|h_{Uk}|^2 - \kappa |h_{Uk'}|^2\right) + \rho_t |g_{Uk'}|^2 + 1\right)$$
$$-\log_2\left(\kappa(1-\alpha_k)\rho_t |h_{Uk'}|^2 + 1\right) \tag{26}$$
$$\geq 2\eta_k, \forall k$$

which is a non-convex inequality constraint since $\log_2\left(\alpha_k \rho_t \left(|h_{Uk}|^2 - \kappa |h_{Uk'}|^2\right) + \rho_t |g_{Uk'}|^2 + 1\right)$ is concave and $-\log_2\left(\kappa(1-\alpha_k)\rho_t |h_{Uk'}|^2 + 1\right)$ is convex. To tackle the non-convexity of C7 and C8, we introduce the successive convex approximation (SCA) method. By giving any local point $\bar{\alpha}_k$, the upper bound of $R_{Uk}^{t_1}$ can be obtained as:

$$R_{Uk}^{t_1(\text{upper})} = \log_2\left(\alpha_k \rho_t \left(|h_{Uk}|^2 - \kappa |h_{Uk'}|^2\right) + \kappa \rho_t |h_{Uk'}|^2 + 1\right)$$
$$-\log_2\left(\kappa(1-\alpha_k)\rho_t |h_{Uk'}|^2 + 1\right)$$
$$+\frac{\rho_t\left(|h_{Uk}|^2 - \kappa |h_{Uk'}|^2\right)(\alpha_k - \bar{\alpha}_k)}{\ln 2\left(\alpha_k \rho_t \left(|h_{Uk}|^2 - \kappa |h_{Uk'}|^2\right) + \kappa \rho_t |h_{Uk'}|^2 + 1\right)} \tag{27}$$

Similarly, the left-side of C8 is concave with respect to β_k. Given any local point $\bar{\beta}_k$, the upper bound of $R_{Uk}^{t_2}$ is obtained as:

$$R_{UB}^{t_2(\text{upper})} = \log_2\left(1 + \beta_k \rho_t |h_{UB}|^2\right) + \frac{\rho_t |h_{UB}|^2}{\ln 2\left(1 + \beta_k \rho_t |h_{UB}|^2\right)}(\beta_k - \bar{\beta}_k) \tag{28}$$

Then, (P4) can be approximated by:

$$(P5) \min_{\alpha,\beta,\eta} P_{UAV}$$

$$\text{s.t. } C7' : \eta_k \leq R_{Uk}^{t_1(\text{upper})} \tag{29}$$

$$C8' : \eta_k \leq R_{UB}^{t_2(\text{upper})}$$

$$C1 - C3, C4' - C6'$$

(P5) is convex with respect to α, β and η, which can be solved by the interior point method.

3.2. UAV Placement Optimization

With the given power allocation coefficients, (P1) can be rewritten as:

$$(\text{P6}) \min_{(x_u, y_u)} \sum_{k \in \mathcal{K}_r} \beta_k \rho_t$$

$$\text{s.t. } C6'' : \log_2\left(1 + \frac{\beta_{k'}\rho_t |h_{Bk'}|^2 (d_{UB})^{\varsigma u}}{\beta_k \rho_t \varphi_{UB}^2 + (d_{UB})^{\varsigma u}}\right) \geq 2\bar{R}_{k'}, \forall k$$

$$C7'' : \log_2\left(1 + \frac{\alpha_k \rho_t \varphi_{Uk}^2 (d_{Uk'})^{\varsigma u}}{\kappa \alpha_{k'} \rho_t \varphi_{Uk'}^2 (d_{Uk})^{\varsigma u} + (d_{Uk})^{\varsigma u}(d_{Uk'})^{\varsigma u}}\right) \geq 2\eta_k, \forall k \tag{30}$$

$$C8'' : \log_2\left(1 + \frac{\beta_k \rho_t \varphi_{UB}^2}{(d_{UB})^{\varsigma u}}\right) \geq 2\eta_k, \forall k$$

where,

$$\varphi_{UB} = 10^{-\frac{A_{UB}+X_{UB}}{20}} \cdot (d_0)^{\frac{\varsigma_{UB}}{2}} \left(\sqrt{\frac{S_{UB}}{1 + S_{UB}}} + \sqrt{\frac{1}{1 + S_{UB}}} g_{UB}\right) \tag{31}$$

$$\varphi_{Uk} = 10^{-\frac{A_U+X_U}{20}} \cdot (d_0)^{\frac{\varsigma_U}{2}} \left(\sqrt{\frac{S_U}{1 + S_U}} + \sqrt{\frac{1}{1 + S_U}} g_{Uk}\right), \tag{32}$$

$$\varphi_{Uk'} = 10^{-\frac{A_U+X_U}{20}} \cdot (d_0)^{\frac{\varsigma_U}{2}} \left(\sqrt{\frac{S_U}{1 + S_U}} + \sqrt{\frac{1}{1 + S_U}} g_{Uk'}\right). \tag{33}$$

(P6) is non-convex for its three non-convex constraints. By the SCA method, C6'' is transformed into:

$$\log_2\left(\beta_k \rho_t \varphi_{UB}^2 + (d_{UB})^{\varsigma u} + \beta_{k'}\rho_t |h_{Bk'}|^2 (d_{UB})^{\varsigma u}\right) - \log_2\left(\beta_k \rho_t \varphi_{UB}^2 + (d_{UB})^{\varsigma u}\right) \geq 2\bar{R}_{k'} \tag{34}$$

By introducing a slack variable Z_1, C6'' are equivalent to the following two constraints as:

$$\log(Z_1) - \log_2\left(\beta_k \rho_t \varphi_{UB}^2 + (d_{UB})^{\varsigma u}\right) \geq 2\bar{R}_{k'}$$
$$\log_2\left(\beta_k \rho_t \varphi_{UB}^2 + (d_{UB})^{\varsigma u} + \beta_{k'}\rho_t |h_{Bk'}|^2 (d_{UB})^{\varsigma u}\right) \geq Z_1 \tag{35}$$

Then, we approximate the above equations by their lower bounds. The item $-\log_2\left(\beta_k \rho_t \varphi_{UB}^2 + (d_{UB})^{\varsigma''}\right)$ in Equation (35) is convex with respect to $(d_{UB})^{\varsigma u}$. $(d_{UB})^{\varsigma u}$ is convex with respect to $c_u = (x_u, y_u, h_u)$, where h_u is constant. Thus, given any $\bar{c}_u = (\bar{x}_u, \bar{y}_u, h_u)$, Equation (35) can be approximated by their lower bound as.

$$\log(Z_1) - \log_2\left(\beta_k \rho_t \varphi_{UB}^2 + (d_{\bar{c}_u,c_B})^{\varsigma_{UB}}\right) - \frac{\varsigma_{UB}(d_{\bar{c}_u,c_B})^{(\varsigma_{UB}-1)}(\bar{c}_u - c_B)^{\mathrm{T}}(c_u - \bar{c}_u)}{\ln 2 (\beta_k \rho_t \varphi_{UB}^2 + (d_{\bar{c}_u,c_B})^{\varsigma_{UB}})} \geq 2\bar{R}_{k'} \tag{36}$$

$$\beta_k \rho_t \varphi_{UB}^2 + \left(1 + \beta_{k'}\rho_t |h_{Bk'}|^2\right)\left((d_{\bar{c}_u,c_B})^{\varsigma_{UB}} + d_{\bar{c}_u,c_B}^{(\varsigma_{UB}-1)}(\bar{c}_u - c_B)^{\mathrm{T}}(c_u - \bar{c}_u)\right) \geq Z_1 \tag{37}$$

where $d_{\bar{c}_u,c_B} = \|\bar{c}_u - c_B\|$. Apparently, the left-side of Equations (36) and (37) are concave with respect to $c_u = (x_u, y_u)$, η_k and Z_1. Similarly, the non-convex constraints C7'' and C8'' are approximated by their lower bounds:

$$\log(Z_2) - \log_2\left(\kappa \alpha_{k'}\rho_t \varphi_{Uk'}^2 (d_{Uk})^{\varsigma u} + (d_{Uk}d_{Uk'})^{\varsigma u}\right)$$
$$- \frac{\varsigma u \left(\kappa \alpha_{k'}\rho_t \varphi_{Uk'}^2 (d_{\bar{c}_u,c_k})^{-1} + (d_{\bar{c}_u,c_{k'}})^{\varsigma u-1}\right)(\bar{c}_u - c_k)^{\mathrm{T}}(c_u - \bar{c}_u)}{\ln 2 \left(\kappa \alpha_{k'}\rho_t \varphi_{Uk'}^2 + (d_{\bar{c}_u,c_{k'}})^{\varsigma u}\right)} \geq 2\bar{\eta}_k \tag{38}$$

$$\frac{\varsigma u \left(d_{\bar{c}_u,c_{k'}}\right)^{(\varsigma u-1)}(\bar{c}_u - c_k)^T(c_u - \bar{c}_u)}{\left(d_{\bar{c}_u,c_k}\right)^{(2-\varsigma u)}} + \kappa\alpha_{k'}\rho_t\varphi_{Uk'}^2\left(d_{\bar{c}_u,c_k}\right)^{\varsigma u}\left(1 + \frac{(\bar{c}_u - c_k)^T(c_u - \bar{c}_u)}{d_{\bar{c}_u,c_k}}\right)$$
$$+ \alpha_k\rho_t\varphi_{Uk}^2\left(d_{\bar{c}_u,c_{k'}}\right)^{\varsigma u}\left(1 + \frac{(\bar{c}_u - c_{k'})^T(c_u - \bar{c}_u)}{d_{\bar{c}_u,c_{k'}}}\right) \geq Z_2 \tag{39}$$

where $d_{\bar{c}_u,c_k} = \|\bar{c}_u - c_k\|$, $d_{\bar{c}_u,c_{k'}} = \|\bar{c}_u - c_{k'}\|$.

$$\log(Z_3) - \log_2\left((d_{\bar{c}_u,c_B})^{\varsigma UB}\right) - \frac{\varsigma UB(d_{\bar{c}_u,c_B})^{(\varsigma UB-1)}(\bar{c}_u - c_B)^T(c_u - \bar{c}_u)}{\ln 2(d_{\bar{c}_u,c_B})^{\varsigma UB}} \geq 2\eta_k \tag{40}$$

$$\beta_k\rho_t\varphi_{UB}^2 + (d_{\bar{c}_u,c_B})^{\varsigma UB} + (d_{\bar{c}_u,c_B})^{(\varsigma UB-1)}(\bar{c}_u - c_B)^T(c_u - \bar{c}_u) \geq Z_3 \tag{41}$$

where Z_2 and Z_3 are the slack variable introduced to C7″ and C8″, respectively, and $\bar{c}_u$ is the local point of the UAV coordinate obtained in the last iteration. Thus, the optimum solution of (P6) is always lower bounded by:

$$(\text{P7})\min_{c_u,\eta} \sum_{k\in\mathcal{K}_r} \beta_k\rho_t \tag{42}$$
$$\text{s.t. } (36), (37), (38), (39), (40), (41), \forall k, k'$$

3.3. Iterative Algorithm

Based on the solutions to the two sub-problems, we propose an iterative algorithm for (P2) by using the BCD method, which is guaranteed to cover a sub-optimum [27] (In [27]; its BCD-based algorithm contains an additional feasibility checking to guarantee the feasibility of its proposed optimization problem), as summarized in Algorithm 1. Different from [27], the optimal results are acquired in feasible sets obtained in Equation (29) and Equation (42) by Algorithm 1 without additional feasibility checking. q and ϵ are denoted as the number of iteration times and the algorithm convergence factor.

Algorithm 1 BCD Method for Joint Placement and Power Optimization

---

Initialize $\alpha^{(0)}$, $\beta^{(0)}$, $\eta^{(0)}$ and c_u let $q = 0$, $\epsilon = 10^{-5}$
repeat
    Solve (P5) for given $\left\{c_u^{(q)}\right\}$, and denote the optimal solution as $\left\{\alpha^{(q+1)}, \beta^{(q+1)}, \eta^{(q+1)}\right\}$
    Solve (P7) for given $\left\{\alpha^{(q+1)}, \beta^{(q+1)}\eta^{(q+1)}\right\}$, and denote the optimal solution
as $\left\{c_u^{(q+1)}\right\}$
    Update $q = q + 1$
until $\sum_{k\in\mathcal{K}_r} \beta_k^{(q+1)} - \sum_{k\in\mathcal{K}_r} \beta_k^{(q)} \geq \epsilon$

The flowchart of Algorithm 1 is present in Figure 2, which shows the iterative procedure of the proposed optimization solution of joint UAV placement and power optimization. In this paper, we introduce the BCD method to alternately solve the power allocation problem and UAV optimal placement problem by convex optimization. Firstly, initial values $\left\{\alpha^{(0)}, \beta^{(0)}, \eta^{(0)}\right\}$ and $\left\{c_u^{(0)}\right\}$ are given and the BCD method is applied to optimize the variables block with the other variables block fixed. Then, we repeat the iteration until the UAV transmitted power minimization is obtained due to its convergence. Finally, the optimal power allocation $\left\{\alpha^{(*)}, \beta^{(*)}, \eta^{(*)}\right\}$ and UAV position coordinates $\left\{c_u^{(*)}\right\}$ can be derived. In addition, Algorithm 1 yields a suboptimal solution, due to the optimal solution obtained by two suboptimal problems.

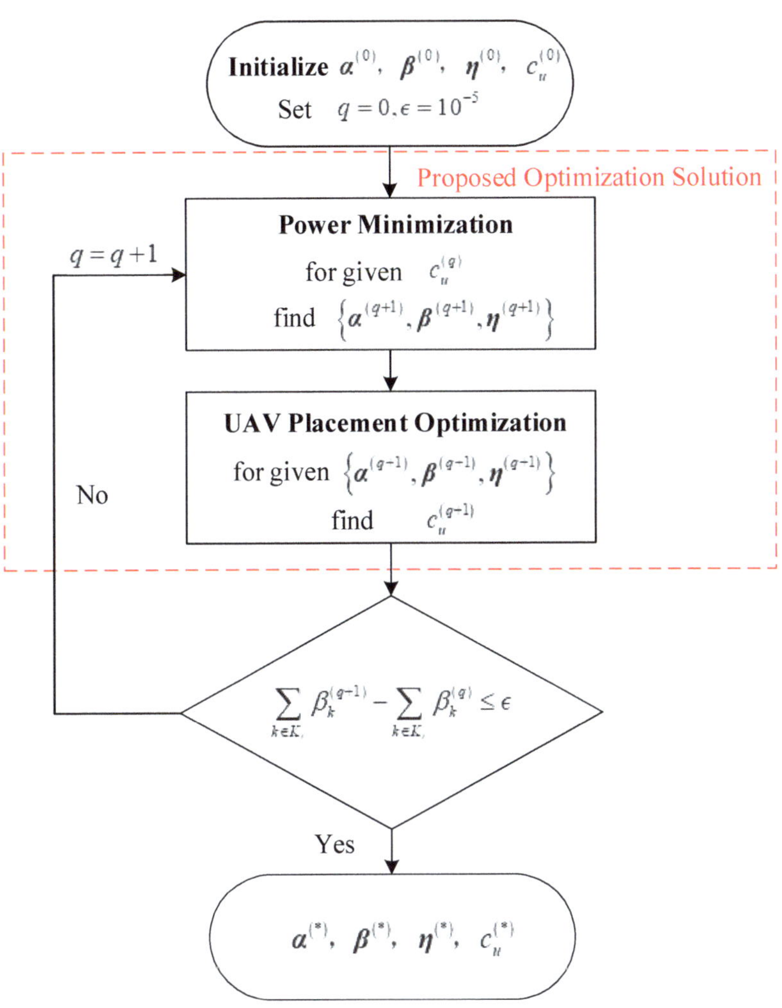

Figure 2. The flowchart of Algorithm 1.

4. Numerical Results and Discussion

In this section, we simulate a maritime IoT UAV-relaying network with K pairs of maritime users. The coordinates of onshore BS are $(0, 0, 150)$. The onshore BS coverage is a circle with 200 m as radius. K MCUs are randomly located along the coastline within the onshore BS service area. K MRUs are randomly located in a square area of 200×200 m^2 with coordinates $(0, 400, 0)$. The constant altitude of UAV is set as $h_u = 150$ m. Set UAV power budget $\bar{P}_{Ut} = 4$ W, $\delta^2 = -108$ dBm, the carrier frequency $f_c = 5$ MHz, $d_0 = 1$ m, light speed $c = 3 \times 10^8$ m/s. The parameters of the channel models in maritime propagation environment can be obtain by [22]. The simulation parameters are given in Table 1.

Table 1. Simulation parameters.

Parameter	Description	Value
c_B	Coordination of onshore BS	$(0, 150, 0)$
h_u	Flight altitude of UAV	150 m
d_0	Reference distance	1 m
f_c	Carrier frequency	5 MHz
σ^2	Background noise	-108 dBm
c	Light speed	3×10^8 m/s
$\bar{P}_{Ut}$	UAV transmit power budget	4 W
A_U	A2S link path loss at d_0	116.7
ς_U	A2S link path loss exponent	20
δ_{X_U}	standard deviation of X_U	0.1
S_U	A2S link Rician factor	30
A_{UB}	A2A linkpath loss at d_0	46.4
ς_{UB}	A2A link path loss exponent	15
$\delta_{X_{UB}}$	standard deviation of X_{UB}	0.1
S_{UB}	A2A link Rician factor	10

Figure 3 shows the optimized horizontal locations of UAV with different transmit power budget. Both the $K = 4$ NOMA pair scenario and $K = 6$ NOMA pair scenario are discussed in this part. We set the minimum rate requirement $\bar{R}_k = \bar{R}_{k'} = 0.5$ bps/Hz. The optimized UAV locations with a NOMA pair total transmit power $P_t = 1$ W, 1.5 W and 2 W are marked with colored star markers. It shows that there exists an optimal hovering position, which achieves the tradeoff between UAV transmit power minimization and MRU's minimum rate requirements. It can be seen in Figure 3 that with higher maritime user pair total transmit power, the UAV should hover closer to the onshore BS to enjoyed better A2A channels. Then, the UAV relay consumes less energy. Moreover, with the number of MRUs increasing, the UAV should hover closer to the onshore BS to serve more MRUs within its power budget. Figure 4 illustrates the power allocation results of UAV for each MRU data ferrying with various NOMA pair's total transmitted power, according to the optimal position in Figure 3. It is indicated that UAV consumes less power for hovering closer to onshore BS and enjoying a better A2A link. Furthermore, at the same hovering position, UAV allocates more power for the MRU that is closer to relay data since UAV receives a higher MRU rate with better A2S channel condition.

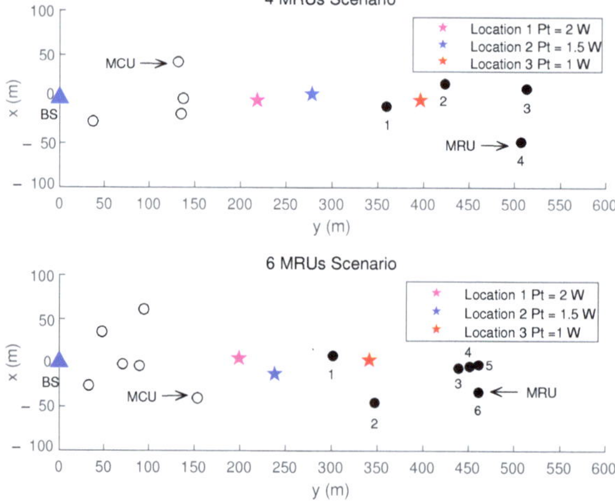

Figure 3. Horizontal locations of UAV with different NOMA total power pairs P_t.

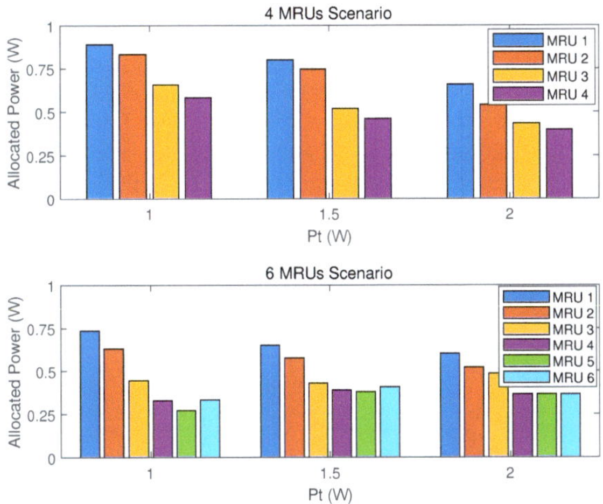

Figure 4. Power allocation of UAV to BS for each MRU with different NOMA total power pairs P_t.

Figure 5 demonstrates that the required UAV transmits power with various NOMA total transmitted power pairs P_t in the $K = 4$ NOMA pairs scenario. We set the same dMCU rate requirement as the MRUs. Specifically, randomly searched UAV placement scheme (In the randomly searched UAV placement scheme, the UAV coordinates are obtained after several times randomly searching in the feasible region. The number of searching times is predetermined. This scheme is also used as a benchmark in [8,18].) is employed as a benchmark with 10 times random search. The proposed optimized power allocation is applied as a power allocation scheme. It shows in Figure 5 that UAV requires more total transmitted power by increasing MRU's rate requirement. With the increasing of P_t, the required UAV's transmitted power decreases. It is shown that, compared with a randomly searched location scheme, the UAV transmit power decreases over 7% with $\bar{R}_k = 0.45$ bps/Hz, over 3% with $\bar{R}_k = 0.5$ bps/Hz, and over 3% with $\bar{R}_k = 0.55$ bps/Hz, respectively. Furthermore, it is illustrated that our proposed optimal algorithm outperforms the benchmark scheme.

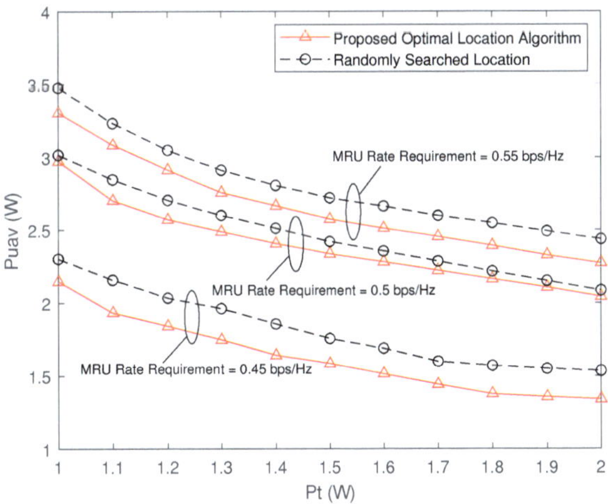

Figure 5. UAV transmit power with different NOMA pair total power P_t.

The impact of the MRU rate requirements to the required UAV transmitted power is evaluated in Figure 6 with a NOMA pair total transmitted power $P_t = 1$ W. The 1.5 W and 2 W $K = 4$ NOMA pair scenario is discussed in this part. We set the same MCU rate requirement as the MRUs. It shows that more UAV transmitted power is required by the increasing MRU rate requirement with more transmitted power to guarantee higher data rate requirements. It is shown that compared with randomly searched location scheme, the UAV transmit power decrease over 2% with $P_t = 1$ W, over 4% with $P_t = 1.5$ W and 7% with $P_t = 2$ W, respectively. It is worth noting that the curve of UAV transmitted power grows smoothly with higher rate requirements compared to lower rate requirements. The reason is that with the increasing user rate requirement, more transmitted power is allocated to MCU to guarantee its minimum rate requirement, which has no extra channel gain due to its fixed position, whereas the MRU can obtain a better data rate with better A2A channel condition.

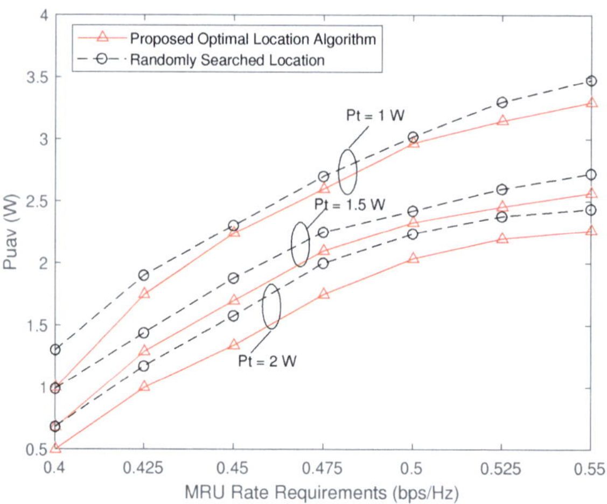

Figure 6. UAV transmitted power with different MRU rate requirements $\bar{R}_k$.

5. Conclusions

In this paper, a UAV was utilized as a relay to help offshore maritime users connect with onshore BS. The UP-NOMA scheme was proposed to increase system spectrum efficiency. The optimal UAV hovering placement and transmitted power allocation were jointly optimized to minimize the total transmitted power consumption of the UAV relay energy with constraints of the minimum maritime user rate requirements and transmitters' power budgets. The SCA method was applied to deal with the non-convexity of the proposed optimization problem and the BCD method was employed to generate an iterative algorithm for successively optimizing two sub-problems. Numerical results indicate that the proposed algorithm outperformed the benchmark algorithm and shed light on the potential for exploiting the energy-limited aerial relay in IoT systems.

Author Contributions: Conceptualization, W.X.; methodology, W.X.; software, W.X.; validation, W.X.; formal analysis, W.X.; investigation, W.X., J.T. and L.G.; resources, W.X.; data curation, W.X.; writing—original draft preparation, W.X.; writing—review and editing, W.X., J.T., L.G. and S.T.; visualization, W.X.; supervision, W.X.; project administration, W.X.; funding acquisition, W.X. All authors have read and agreed to the published version of the manuscript.

Funding: This research was funded by the Shanghai Sailing Program, grant number 20YF1416700, the Innovation Program of Shanghai Municipal Education Commission of China under Grant 2021-01-07-00-10-E00121, the National Natural Science Foundation of China (62271303).

Data Availability Statement: Not applicable.

Conflicts of Interest: The authors declare no conflict of interest.

References

1. Li, X.; Feng, W.; Wang, J.; Chen, Y.; Ge, N.; Wang, C.X. Enabling 5G on the ocean: A hybrid satellite-UAV-terrestrial network solution. *IEEE Wirel. Commun.* **2020**, *27*, 116–121. [CrossRef]
2. Lopes, M.J.; Teixeira, F.; Mamede, J.B.; Campos, R. Wi-Fi broadband maritime communications using 5.8 GHz band. In Proceedings of the Underwater Communications Networking, Levante, Italy, 3–5 September 2014.
3. Teixeira, F.B.; Campos, R.; Ricardo, M. Height optimization in aerial networks for enhanced broadband communications at sea. *IEEE Access* **2020**, *8*, 28311–28323. [CrossRef]
4. Ji, X.; Wang, J.; Li, Y.; Sun, Q.; Xu, C. Modulation recognition in maritime multipath channels: A blind equalization-aided deep learning approach. *China Commun.* **2020**,*17*, 12–25. [CrossRef]
5. Wei, T.; Feng, W.; Chen, Y.; Wang, C.X.; Ge, N.; Lu, J. Hybrid satellite-terrestrial communication networks for the maritime internet of things: Key technologies, opportunities, and challenges. *IEEE Internet Things J.* **2021**, *8*, 8910–8934. [CrossRef]
6. Wang, N.; Li, F.; Chen, D.; Liu, L.; Bao, Z. NOMA-based energy-efficiency optimization for UAV enabled space-air-ground intergarted relay networks. *IEEE Trans. Veh. Technol.* **2022**, *71*, 4129–4141. [CrossRef]
7. Liu, C.; Feng, W.; Chen, Y.; Wang, C.X.; Ge, N. Cell-Free Satellite-UAV Networks for 6G Wide-Area Internet of Things. *IEEE J. Sel. Areas Commun.* **2021**, *39*, 1116–1131. [CrossRef]
8. Liu, X.; Wang, J.; Zhao, N.; Chen, Y.; Zhang, S.; Ding, Z.; Yu, F.R. Placement and power allocation for NOMA-UAV networks. *IEEE Wirel. Commun. Lett.* **2019**, *8*, 965–968. [CrossRef]
9. Sharma, P.K.; Kim, D.I. UAV-enabled downlink wireless system with non-orthogonal multiple access. In Proceedings of the 2017 IEEE Globecom Workshops (GC Wkshps), Singapore, 4–8 December 2017.
10. Sohail, M.F.; Leow, C.Y.; Won, S. Non-orthogonal multiple access for unmanned aerial vehicle assisted communication. *IEEE Access* **2018**, *6*, 22716–22727. [CrossRef]
11. Nasir, A.A.; Tuan, H.D.; Duong, T.Q.; Poor, H.V. UAV-enabled communication using NOMA. *IEEE Trans. Commun.* **2019**, *67*, 5126–5138. [CrossRef]
12. Wang, J.; Zhou, H.; Li, Y.; Sun, Q.; Wu, Y.; Jin, S.; Quek, T.Q.S.; Xu, C. Wireless channel models for maritime communications. *IEEE Access* **2018**, *6*, 68070–68088. [CrossRef]
13. Zhang, J.; Liang, F.; Li, B.; Yang, Z.; Wu, Y.; Zhu, H. Placement optimization of caching mobile relay maritime communication. *China Commun.* **2020**, *17*, 209–219. [CrossRef]
14. Tang, R.; Feng, W.; Chen, Y.; Ge, N. NOMA-based UAV communications for maritime coverage enhancement. *China Commun.* **2021**, *18*, 230–243. [CrossRef]
15. Li, X.; Feng, W.; Chen, Y.; Wang, C.X.; Ge, N. Maritime coverage enhancement using UAVs coordinated with hybrid satellite-terrestrial networks. *IEEE Trans. Commun.* **2020**, *68*, 2355–2369. [CrossRef]
16. Ma, R.; Wang, R.; Liu, G.; Chen, H.H.; Qin, Z. UAV-assisted data collection for ocean monitoring networks. *IEEE Netw.* **2020**, *34*, 250–258. [CrossRef]
17. Lyu, L.; Chu, Z.; Lin, B.; Dai, Y.; Cheng, N. Fast trajectory planning for UAV-enabled maritime IoT systems: A Fermat-Point based approach. *IEEE Wirel. Commun. Lett.* **2022**, *11*, 328–332. [CrossRef]
18. Jiang, X.; Wu, Z.; Yin, Z.; Yang, Z.; Zhao, N. Power comsumption minimization of UAV relay in NOMA networks. *IEEE Wirel. Commun. Lett.* **2020**, *9*, 666–670. [CrossRef]
19. Guo, Y.; Yin, S.; Hao, J. Joint placement and resource optimization for multi-user UAV-relaying system with underlaid cellular networks. *IEEE Trans. Veh.* **2020**, *10*, 12374–12377. [CrossRef]
20. Wang, D.; Zhou, F.; Lin, W.; Ding, Z.; Al-Dhahir, N. Cooperative Hybrid Non-Orthogonal Multiple Access Based Mobile-Edge Computing in Cognitive Radio Networks. *IEEE Trans. Cogn. Commun. Netw.* **2022**, *8*, 1104–1117. [CrossRef]
21. Wang, D.; He, T.; Zhou, F.; Cheng, J.; Zhang, R.; Wu, Q. Outage-driven link selection for secure buffer-aided networks. *Sci. China Inf. Sci.* **2022**, *65*, 182303. [CrossRef]
22. Matolak, D.W.; Sun, R. Air-ground channel characterization for unmanned aircraft systems—Part I: Methods, measurements, and models for over-water settingse. *IEEE Trans. Veh. Technol.* **2017**, *66*, 26–44. [CrossRef]
23. Wu, S.; Wang, C.X.; Alwakeel, M.M.; You, X. A general 3-D non-stationary 5G wireless channel model. *IEEE Trans. Commun.* **2018**, *66*, 3065–3078. [CrossRef]
24. Zeng, L.; Cheng, X.; Wang, C.X.; Yin, X. Second order statistics of non-isotropic UAV Ricean fading channels. In Proceedings of the IEEE 86th Vehicle Technology Conference (VTC-Fall), Toronto, ON, Canada, 24–27 September 2017.
25. Kang, Z.; You, C.; Zhang, R. 3D Placement for Multi-UAV Relaying: An Iterative Gibbs-Sampling and Block Coordinate Descent Optimization Approach. *IEEE Trans. Commun.* **2021**, *69*, 2047–2062. [CrossRef]
26. Wang, D.; Wu, M.; He, Y.; Pang, L.; Xu, Q.; Zhang, R. An HAP and UAVs Collaboration Framework for Uplink Secure Rate Maximization in NOMA-Enabled IoT Networks. *Remote. Sens.* **2022**, *14*, 4501. [CrossRef]
27. Huang, W.; Yang, Z.; Pan, C.; Pei, L.; Chen, M.; Shikh-Bahaei, M.; Elkashlan, M.; Nallanathan, A. Joint power, altitude, location and bandwidth optimization for UAV with underlaid D2D communications. *IEEE Wirel. Commun. Lett.* **2019**, *8*, 524–527. [CrossRef]

MDPI AG

Grosspeteranlage 5

4052 Basel

Switzerland

Tel.: +41 61 683 77 34

Drones Editorial Office

E-mail: drones@mdpi.com

www.mdpi.com/journal/drones

www.ingramcontent.com/pod-product-compliance
Lightning Source LLC
LaVergne TN
LVHW071935160726
843515LV00011B/2619